Physics and Psychics

This is the first systematic exploration of the intriguing connections between Victorian physical sciences and the study of the controversial phenomena broadly classified as psychic, occult and paranormal. These phenomena included animal magnetism, spirit-rapping, telekinesis and telepathy. Richard Noakes shows that psychic phenomena interested far more Victorian scientists than we have previously assumed, challenging the view of these scientists as individuals clinging rigidly to a materialistic worldview. Physicists, chemists and other physical scientists studied psychic phenomena for a host of scientific, philosophical, religious and emotional reasons, and many saw such investigations as exciting new extensions to their theoretical and experimental researches. While these attempted extensions were largely unsuccessful, they laid the foundations of modern-day explorations of the connections between physics and psychic phenomena. This revelatory study challenges our view of the history of physics and deepens our understanding of the relationships between science and the occult, and science and religion.

Richard Noakes is a leading historian of nineteenth- and twentieth-century sciences and technology at the University of Exeter. He is the co-editor of *From Newton to Hawking: A History of Cambridge University's Lucasian Professors of Mathematics* (2003) and the co-author of *Science in the Nineteenth Century Periodical: Reading the Magazine of Nature* (2004).

SCIENCE IN HISTORY

Series Editors
Simon J. Schaffer, University of Cambridge
James A. Secord, University of Cambridge

Science in History is a major series of ambitious books on the history of the sciences from the mid-eighteenth century through the mid-twentieth century, highlighting work that interprets the sciences from perspectives drawn from across the discipline of history. The focus on the major epoch of global economic, industrial and social transformations is intended to encourage the use of sophisticated historical models to make sense of the ways in which the sciences have developed and changed. The series encourages the exploration of a wide range of scientific traditions and the interrelations between them. It particularly welcomes work that takes seriously the material practices of the sciences and is broad in geographical scope.

Physics and Psychics

The Occult and the Sciences in Modern Britain

Richard Noakes

University of Exeter

CAMBRIDGE
UNIVERSITY PRESS

CAMBRIDGE
UNIVERSITY PRESS

University Printing House, Cambridge CB2 8BS, United Kingdom

One Liberty Plaza, 20th Floor, New York, NY 10006, USA

477 Williamstown Road, Port Melbourne, VIC 3207, Australia

314-321, 3rd Floor, Plot 3, Splendor Forum, Jasola District Centre, New Delhi - 110025, India

79 Anson Road, #06-04/06, Singapore 079906

Cambridge University Press is part of the University of Cambridge.

It furthers the University's mission by disseminating knowledge in the pursuit of education, learning and research at the highest international levels of excellence.

www.cambridge.org
Information on this title: www.cambridge.org/9781316638569
DOI: 10.1017/9781316882436

© Richard Noakes 2019

First published 2019
First paperback edition 2020

A catalogue record for this publication is available from the British Library

ISBN 978-1-107-18854-9 Hardback
ISBN 978-1-316-63856-9 Paperback

For Abigail, Emily, Jill and Valerie

Contents

Figures and Tables

Figures

Tables

Acknowledgements

A book that has taken as long to write as this one necessarily incurs many debts. The oldest one is to Simon Schaffer, who supervised the doctoral dissertation that is now a very distant ancestor of the present work. Without his encouragement, encyclopaedic knowledge and often terrifying ability to suggest more sophisticated historical interpretations, *Physics and Psychics* would have been a far poorer work. There were times when Simon's predictions of the likely contents of obscure historical sources were so accurate that I half suspected he possessed some of the psychic abilities on which this book is focussed!

Over the past two decades, I have been enormously privileged to be able to discuss the ideas of this book with many scholars whose advice, encouragement and works have helped me in more ways than they know: Will Ashworth, Bill Brock, Graeme Gooday, Linda Henderson, the late Jeff Hughes, Bruce Hunt, Rob Iliffe, Myles Jackson, Frank James, Bernard Lightman, Jim Moore, Iwan Rhys Morus, Jaume Navarro, Graham Richards, Jim Secord, Crosbie Smith, Andreas Sommer, Matt Stanley, Andy Warwick, the late Alison Winter and David Wilson. Graeme Gooday and Jim Moore kindly read portions of the book manuscript, while Jim Secord read the whole work. I am so very grateful to all three of them, as well as the anonymous reader for the final manuscript, for their criticisms and suggestions.

As a postgraduate student and postdoctoral research fellow in the Department of History and Philosophy of Science, Cambridge (in 1992–8 and 2002–7 respectively), my historical knowledge and understanding benefited from exchanges with so many individuals, including Jim Bennett, Bob Brain, the late John Forrester, Arne Hessenbruch, Emese Lafferton, Otto Sibum, Richard Staley and Jennifer Tucker. Trevor Pinch and Sven Widmalm were visiting fellows during my time at Cambridge and their comments on my work were invaluable and reassuring. One former Cambridge colleague – Kevin Knox – needs to be singled out because our collaboration on a history of Cambridge

University's Lucasian professors of mathematics taught me as much about good prose style as incisive historical analysis.

From 1999 to 2002 I was a Leverhulme Postdoctoral Research Fellow on the 'Science in the Nineteenth Century Periodical' project at the Universities of Leeds and Sheffield. I am so grateful to project members Sam Alberti, Geoffrey Cantor, Gowan Dawson, Graeme Gooday, Louise Henson, Sally Shuttleworth and Jon Topham for stimulating my thoughts on questions of popularisation, publication and readership that underpins much of what follows. The periodicals project also brought me into contact with Ruth Barton, John Christie, Peter Kjaergaard, Jack Morrell and Roger Smith and I thank them all for prompting me to rethink many of my ideas.

My understanding of psychical research has benefited from exchanges with and welcome criticism from scholars active in this field, including Mary Rose Barrington, Bernard Carr and John Poynton. At the University of Exeter, Kristofer Allerfeldt, Jonathan Barry, Alan Booth, Tim Cooper, Jason Hall, Lucy Hilliar, Andrew Thorpe and Nicola Whyte have offered much encouragement and advice over the years. Before his untimely passing, another Exeter colleague, Nicholas Goodrick-Clarke, shared with me so much of his profound knowledge of Western Esotericism and the ancestors of psychical research. For allowing me to rehearse many lines of argument developed in this book, I also thank the numerous Exeter students who took my undergraduate course on Victorian occultism.

From 2002 to 2007 a British Academy–Royal Society Postdoctoral Fellowship in the History of Science enabled me to complete a significant amount of underlying research, while three periods of leave granted by the University of Exeter in 2010, 2014 and 2016 gave me the time to write. I want to express my heartfelt thanks to all three institutions for their considerable generosity and patience.

I have relied heavily on the wisdom, kindness and patience of numerous archivists and librarians, all of whom have gone beyond the bounds of duty in tracking down obscure materials and answering a plethora of questions. I would like to record my particular thanks to Anne Barrett at the Archives of Imperial College London; Helen Fisher and Christine Penney at the Cadbury Research Library, University of Birmingham; Susan Liberator at Ohio State University Library; Dan Mitchell at University College London Library's Special Collections; Leslie Price at the College of Psychic Studies; Jon Cable and Tim Procter at the Institution of Engineering and Technology; and Godfrey Waller and the late Peter Meadows at Cambridge University Library. In cataloguing Cambridge's SPR archive, Peter Meadows's service to the history of psychical research is immeasurable.

For their help in my search for copyright holders I thank Oliver A. W. Lodge and Philip Somervail (two of Oliver Lodge's great-grandsons) and Anita and Gwendoline Morgan.

For permission to quote from unpublished material in their collections I would like to thank the Sixth Baron Rayleigh; the Archives of Imperial College London; Boston Public Library and Rare Books; the British Library; Cadbury Research Library, University of Birmingham; Christine Fournier d'Albe; Indiana State Library, Rare Books and Manuscripts Division; the Institution of Engineering and Technology Archives; Knebworth House Archive; the Master and Fellows of Trinity College Cambridge; Niels Bohr Library and Archives, American Institute of Physics; Princeton University Library, Department of Rare Books and Special Collections; the Royal Dublin Society Library and Archives; the Royal Society Archives; the Science Museum Group Collection; Senate House Library, University of London; the Society for Psychical Research; the Syndics of Cambridge University Library; University College London Library Services, Special Collections; the United States Air Force Academy, McDermott Library; and the Wellcome Collection.

For permission to reproduce published and unpublished material in their collections I thank the Bodleian Libraries, University of Oxford; the Bill Douglas Centre, University of Exeter; Getty Images; Leeds University Library, Special Collections; Ohio State University; the National Library of Wales; the Royal Society Archives; Friederike Schriever; Andreas Sommer; the Syndics of Cambridge University Library; Telegraph Museum, Porthcurno; and the University of Bristol Library, Special Collections. I have endeavoured to identify and contact holders of the copyright of published and unpublished material used in this book and welcome any information enabling me to correct any omissions or errors in such copyright acknowledgements.

I am delighted to record my gratitude to Cambridge University Press again. My commissioning editor Lucy Rhymer has been a tremendous source of encouragement and enthusiasm, without which the bulk of the writing for the book would have been far harder to complete. My sincere thanks go to both anonymous readers of the original proposal, the series editors Simon Schaffer and Jim Secord, and to Ruth Boyes and Lisa Carter for shepherding the manuscript through the final stages. For their superb work in copy-editing and indexing I would also like to thank Catherine Dunn and Rob Sale respectively.

Portions of this book rework material that has been published in journals and essay collections: Noakes, 'World of the Infinitely Little' (part of Chapter 3); Noakes, 'Telegraphy is an Occult Art' (part of Chapter 4); Noakes, 'The Bridge which is Between Physical and Psychical Research' (parts of Chapters 3 and 4); Noakes, 'Haunted Thoughts of a Careful Experimentalist' (part of Chapter 5); and Noakes, 'Making Space for the Soul' (part of Chapter 6).

My greatest debt, however, is to my wife Jill, who has lived with and suffered from this book for vastly longer than she anticipated. I'm appalled to report that when we first met, I was 'finishing' a book that would subsequently try her patience beyond measure and give her good grounds to think that there was a third person in our marriage. I promise her that the next book will be not be a Herculean labour for either of us. This book is dedicated to her, my mother Valerie and my daughters Abigail and Emily.

Abbreviations

ARW-BL	Alfred Russel Wallace Papers, British Library (Add. 46439)
BD-NLW	Benjamin Davies Papers, National Library of Wales
GC-BL	Ghost Club Archive, British Library (Add. MSS 52258–52273)
JJT-CUL	J. J. Thomson Papers, Cambridge University Library (Add. 7654)
JSPR	*Journal of the Society for Psychical Research*
OJL-SPR	Oliver Lodge Papers, Society for Psychical Research Archive, Cambridge University Library (SPR.MS 35)
OJL-UCL	Oliver Lodge Papers, University College London (MS Add. 89)
PSPR	*Proceedings of the Society for Psychical Research*
RDS	Royal Dublin Society
R-USAF	Papers of John William Strutt (Third Baron Rayleigh) and Robert John Strutt (Fourth Baron Rayleigh), United States Air Force Academy, McDermott Library, Colorado Springs, Colorado (MS.63)
SPR	Society for Psychical Research
SPT-IC	Silvanus Philips Thompson Papers, Archives of Imperial College London
WFB-RS	William Fletcher Barrett Papers, Royal Society Archives (MS/377)
WFB-SPR	William Fletcher Barrett Papers, Society for Psychical Research Archive, Cambridge University Library (SPR. MS 3)

Introduction

On 15 October 1919, that venerable institution of British comic journalism, *Punch*, turned the focus of its regular commentaries on leading personalities of the day to Oliver Lodge, the ageing British physicist who had recently retired as Principal of Birmingham University. For "many years", the anonymous contributor explained, Lodge had

harboured the ambition of achieving distinction as a serious man of science, and was so far successful that he attained to the position of the President of the British Association. It was only comparatively late in life that he discovered that the word Physics (a science to which he had devoted so many years of patient research) by a slight rearrangement of the letters composing it and the addition of another "c", could be resolved into Psychics; and transferred his attention to a more congenial field of study.[1]

Punch had, of course, deliberately misrepresented Lodge for satirical effect. He had not "transferred" to "Psychics" – the study of psychic or psychical phenomena – simply because the word closely resembled physics; the transfer had begun much earlier in his scientific career; and it had neither been complete nor always "congenial". Yet some aspects of *Punch*'s portrait were closer to the truth. Lodge had indeed achieved scientific "distinction", and not simply as president of a major British scientific institution (the British Association for the Advancement of Science) but as someone boasting a long career in scientific research, teaching and popularisation. Much of his scientific research and writing had explored the possible similarities between what, in his later years, he termed "physics and psychics".[2] For decades he had been developing arguments that physics had the concepts, theories and practices that could illuminate the baffling psycho-physical phenomena of psychical research, and that such phenomena offered potentially fruitful directions

---

[1] [Anon.], 'Second Thoughts', *Punch*, vol. 157 (1919), p. 333. Throughout the main body of the text, I have used double quotation marks for quoted text and single quotation marks around words or phrases whose problematic nature I wish to emphasise.

[2] See, for example, Oliver Lodge, *Beyond Physics or the Idealisation of Mechanism* (London: George Allen & Unwin, 1930), pp. 19 and 114.

1

in which the scope of physics could be extended beyond its formal domains of matter and energy.

By 1919, Lodge was probably the only individual that most British reading audiences associated with connections between physics and psychics, mainly because of his staggering output of articles, books and public lectures. In the five decades before this, however, he was known as one of many eminent physicists with psychical 'connections', including four Nobel laureates, three presidents of the Royal Society of London and three other presidents of the British Association.[3] Some were well known for their role in one of the widest-reaching of all applications of the physical sciences: electrical communication (Figure 0.1). When, in the 1880s, Lodge's connection with psychical investigation started, his name jostled for attention alongside those of other, and mainly older, professional scientists in the published membership lists of an organisation that had played an important role in raising the intellectual profile of the study of psychical phenomena across the globe: the Society for Psychical Research (henceforth SPR). Founded in 1882, this predominantly British organisation aimed to subject a host of what it deemed "debatable", "remarkable" and seemingly "inexplicable" phenomena to the "exact and unimpassioned" methods of enquiry that had proven so successful in the sciences for hundreds of years.[4] The conspicuous absence of the word 'supernatural' from the SPR's manifesto was entirely consistent with this methodological ambition: like so many mesmerists and spiritualists before them, the SPR studied phenomena that it believed to be manifestations of obscure aspects of the natural order, even if they were still deemed supernatural in some quarters.

The phenomena that the SPR reclassified as 'psychical' all suggested obscure and startling powers of the human mind and body.[5] They included 'telepathy' or the capacity to communicate images, words and other impressions to other individuals independently of the known senses; the ability to

[3] The Nobel laureates were Marie and Pierre Curie, the Third Baron Rayleigh and J. J. Thomson; the Royal Society presidents were William Crookes, Rayleigh and Thomson; and the British Association presidents were Rayleigh, Thomson and Arthur Rücker.

[4] [Anon.], 'The Society for Psychical Research: Objects of the Society', *PSPR*, vol. 1 (1882–3), pp. 3–6, pp. 3–4.

[5] The *Oxford English Dictionary* suggests many alternatives to 'psychical' as collective terms relating to spiritualistic and related phenomena. 'Psychic' had been used to refer to such phenomena since the 1870s, while 'psychics' and 'psychic research' were used in the 1860s and 1880s respectively to refer to the *study* of such phenomena. 'Psychic science' came into common use as an alternative to 'psychical research' in the 1920s, partly to reflect the claimed scientific status of the enterprise. For the purposes of clarity this book will generally adopt the terms 'psychical' for the phenomena and 'psychical research' or 'psychical investigation' for the study of the phenomena.

TELEGRAPH & TELEPHONE MAGN(KT)ATES.

0.1 A semi-satirical portrait of the late-nineteenth-century telegraph and telephone businesses. Some of the individuals shown here – Latimer Clark (1), William Crookes (2), Amos Dolbear (3), Thomas Alva Edison (4), Desmond Fitzgerald (5), Silvanus Thompson (6) and Cromwell Varley (7) – were also interested in psychical phenomena. From F[rancis] C[arruthers] Gould, 'Telegraph and Telephone Magnet(at)es', *The City*, 5 May 1883. Reproduced by permission of the Telegraph Museum, Porthcurno.

see or otherwise perceive ghosts of the dead or dying; the power to induce a trance state, effect medical cures and share the sensory experiences of individuals via 'mesmerism'; and the ability to commune with the dead, materialise inhabitants of the spirit world, move objects at a distance and display the other startling powers associated with spiritualist mediumship. Some of the individuals who produced and studied these phenomena would, in the mid-1870s, launch the Theosophical Society. Modernising the ancient study of theosophy or 'divine wisdom', this organisation encouraged the development of obscure psychological powers for elucidating esoteric truths underlying all philosophies, religions and sciences relating to the origin, development and fundamental nature of mankind and the cosmos.[6]

Many of the psychical phenomena studied by the SPR had been the preoccupation of the 'occult philosophies', 'occult sciences' and 'occultisms' that had flourished for centuries. However, the SPR sought to distance itself from such enterprises on the grounds that they seemed to represent approaches to obscure or 'occult' phenomena that were fanciful, secretive and morally dubious rather than what the organisation upheld as the empirical, open and morally sound approaches of the established sciences.[7] Most of the protagonists of this book shared this anxiety and were more likely to speak of unusual, residual and psychical phenomena than the more freighted 'occult' phenomena and certainly repudiated the idea that they were trying to apply science to 'supernatural' effects or a realm beyond the natural.[8] To further project an image of "exact and impassioned enquiry", the SPR also denied prior commitment to "any particular explanation of the phenomena", among which the most notorious was undoubtedly the core belief of spiritualists that the information conveyed by entranced mediums came from personalities in the afterlife.[9] The SPR's rising membership (which had reached over 900 by

---

[6] The historical literature on modern Theosophy is enormous but see Bruce F. Campbell, *Ancient Wisdom Revived: A History of the Theosophical Movement* (Berkeley, CA: University of California Press, 1980); Joscelyn Godwin, *The Theosophical Enlightenment* (Albany, NY: State University of New York Press, 1994); K. Paul Johnson, *The Masters Revealed: Madame Blavatsky and the Myth of the Great White Lodge* (Albany, NY: State University of New York Press, 1994).

[7] This is evident in Oliver Lodge, 'In Memory of F. H. W. Myers', *PSPR*, vol. 17 (1901–3), pp. 1–12, p. 4; Frederic W. H. Myers, 'The Subliminal Consciousness', *PSPR*, vol. 7 (1891–2), pp. 298–355 and vol. 8 (1892), pp. 436–535, citation from vol. 8. For recent historical overviews of occultisms see Egil Asprem, 'Science and the Occult', in Christopher Partridge (ed.), *The Occult World* (Abingdon: Routledge, 2015), pp. 710–19; Wouter Hanegraaff, *Esotericism and the Academy: Rejected Knowledge in Western Culture* (Cambridge University Press, 2012), esp. chapter 3.

[8] For example, William F. Barrett, *Psychical Research* (London: Williams and Norgate, 1911), pp. 11–14; Oliver Lodge, *My Philosophy Representing My Views on the Many Functions of the Ether of Space* (London: Ernest Benn, 1933), pp. 300–1.

[9] [Anon.], 'Objects of the Society', pp. 4–5.

1900) suggests that this strategy clearly appealed to many Victorians looking for a more scientific approach to things ghostly and supernatural.

What sometimes surprised late-nineteenth- and early-twentieth-century commentators was that the SPR's members included many of the most distinguished scientific, literary, medical, political and religious figures of the period. Some, and especially those forging the academic discipline of psychology, were particularly baffled to find so many physicists and practitioners of other physical sciences in the SPR because the kinds of phenomena included in the organisation's remit were psychological to one degree or another and not the province of sciences that formally sidestepped questions of mind. In response, many physical scientists argued that since some psychical phenomena had *some* physical aspects then they were relevant and important to the physical sciences and should not be left solely in the hands of psychologists. By the time *Punch* was imagining his "transfer" from physics to psychics, Lodge was only one of a handful of professional physicists left willing to defend this argument. The scientific discipline to which psychical researchers now most closely associated their enterprise was psychology, although most professional psychologists – and, indeed, most professional scientists of the interwar period – denied that the methods and results of psychical research were robust enough to qualify the enterprise as a branch of *any* science. The situation would not change significantly over the course of the twentieth century, when most scientists, including many physicists, expressed grave doubts about the existence of psychical and paranormal effects and judged psychical research and its major offspring – parapsychology – as fields unworthy of their attention.[10]

This book is about the heyday of 'physics and psychics' which took place in the period circa 1870–1930 and was much more prominent in Britain than elsewhere. It argues that the study of psychical phenomena occupied a much more significant place among late-nineteenth- and early-twentieth-century physical scientists than we have assumed and that the encounters between physical and psychical enquiries stimulated a degree of theoretical, experimental and other types of scientific activity that has been largely overlooked. These activities were not limited to professional physicists, who until the late nineteenth century were rare individuals in the scientific landscape. Indeed, it is because psychical research was pursued by

---

[10] Examples of this attitude are in Georges Charpak and Henri Bloch, *Debunked! ESP, Telekinesis, other Pseudoscience* (Baltimore, MD: Johns Hopkins University Press, 2004); Robert Park, *Voodoo Science: The Road from Foolishness to Fraud* (Oxford University Press, 2000), and issues of *Skeptical Inquirer*, which publishes reports by professional scientific and other members of the American-based Committee for the Scientific Investigation of Claims of the Paranormal.

practitioners of a wider range of physical sciences that the phrase 'physics and psychics' is a handy but problematic shorthand.

The involvement of distinguished physicists and practitioners of other physical sciences in psychical research and ancestral occult enterprises has long stimulated, baffled and even titillated historians of psychical research, physics, and of nineteenth-century sciences and occultisms more generally.[11] The result is that we know a good deal about selected individuals and the links that they tried to forge between physical and psychical enterprises at conceptual and theoretical levels, but we still lack an understanding of the bigger picture.

---

[11] The most important studies are Egil Asprem, *The Problem of Disenchantment: Scientific Naturalism and Esoteric Discourse 1900–1939* (Leiden: Brill, 2014), pp. 208–25; Peter J. Bowler, *Reconciling Science and Religion: The Debate in Early-Twentieth-Century Britain* (Chicago University Press, 2001), pp. 89–101; William H. Brock, *William Crookes (1832–1919) and the Commercialization of Science* (Farnham: Ashgate, 2008), chapters 7–8 and 10–11; Geoffrey Cantor, *Michael Faraday: Sandemanian and Scientist* (London: Macmillan, 1991), pp. 146–54; Patrick Fuentès, 'Camille Flammarion et les force naturelles inconnues', in Bernadette Bensaude-Vincent and Christine Blondel (eds.), *Les savants face à l'occulte, 1870–1940* (Paris: Éditions la Découverte, 2002), pp. 105–21; Michael Gordin, *A Well-Ordered Thing: Dmitrii Mendeleev and the Shadow of the Periodic Table* (New York: Basic Books, 2004), chapter 4; Franz Ferzak, *Karl Freiherr von Reichenbach* (Munich: Franz Ferzak World and Space Publications, 1987), pp. 62–152; Jeff Hughes, 'Occultism and the Atom: The Curious Story of Isotopes', *Physics World*, September 2003, pp. 31–5; Mark S. Morrisson, *Modern Alchemy: Occultism and the Emergence of Atomic Theory* (New York: Oxford University Press, 2007); Albert E. Moyer, *A Scientist's Role in American Culture: Simon Newcomb and the Rhetoric of Scientific Method* (Berkeley, CA: University of California Press, 1992), chapter 10; Michael Nahm, 'The Sorcerer of Coblenzl and His Legacy: The Life of Baron Karl Ludwig von Reichenbach, His Work and Its Aftermath', *Journal of Scientific Exploration*, vol. 26 (2012), pp. 381–407; Richard Noakes, 'Telegraphy Is an Occult Art: Cromwell Fleetwood Varley and the Diffusion of Electricity to the Other World', *British Journal for the History of Science*, vol. 32 (1999), pp. 421–59; Richard Noakes, '"The Bridge Which Is Between Physical and Psychical Research": William Fletcher Barrett, Sensitive Flames and Spiritualism', *History of Science*, vol. 42 (2004), pp. 419–64; Richard Noakes, 'Cromwell Varley FRS, Electrical Discharge and Spiritualism', *Notes and Records of the Royal Society*, vol. 61 (2007), pp. 5–21; Richard Noakes, 'The "World of the Infinitely Little": Connecting Physical and Psychical Realities circa 1900', *Studies in History and Philosophy of Science*, vol. 39 (2008), pp. 323–34; Richard Noakes, 'Making Space for the Soul: Oliver Lodge, Maxwellian Psychics and the Etherial Body', in Jaume Navarro (ed.), *Ether and Modernity: The Recalcitrance of an Epistemic Object in the Early Twentieth Century* (Oxford University Press, 2018), pp. 88–106; Janet Oppenheim, *The Other World: Spiritualism and Psychical Research in Britain, 1850–1914* (Cambridge University Press, 1985), chapter 8; Courtenay Green Raia, 'From Ether Theory to Ether Theology: Oliver Lodge and the Physics of Immortality', *Journal of the History of the Behavioural Sciences*, vol. 43 (2007), pp. 19–43; Klaus B. Staubermann, 'Tying the Knot: Skill, Judgement and Authority in the 1870s Leipzig Spiritistic Experiments', *British Journal for the History of Science*, vol. 34 (2001), pp. 67–79; David B. Wilson, 'The Thought of Late-Victorian Physicists: Oliver Lodge's Ethereal Body', *Victorian Studies*, vol. 15 (1971), pp. 29–48; Brian Wynne, 'Physics and Psychics: Science, Symbolic Action and Social Control in Late Victorian England', in Barry Barnes and Steven Shapin (eds.), *Natural Order: Historical Studies of Scientific Culture* (Beverly Hills, CA: Sage, 1979), pp. 167–87.

How widespread was the interest in psychical investigation among physical scientists? What did this interest amount to? To what extent were 'physics and psychics' linked on experimental as well as theoretical levels? Why did physical scientists think that their skills were relevant to and productive in psychical investigation? And why did some change their approaches to psychical investigation or abandon such enquiries altogether? This book attempts to answer these and many other questions.

The hostility of today's physicists to psychical research has invariably shaped their attempts to understand why so many of their nineteenth- and early-twentieth-century professional ancestors displayed a serious interest in the subject. Echoing nineteenth-century critics of psychical research and spiritualism, they attribute the embarrassing spiritualist beliefs of Lodge, the chemist William Crookes and others to temporary lapses in otherwise formidable powers of scientific judgement. Driven by strong religious, metaphysical or emotional attachment to the idea that we survive bodily death, these lapses, it is said, blinded them to the trickery of spiritualist mediums.[12] Radically alternative interpretations have been given by many contemporary spiritualists. For them, Victorian scientists lend weighty scientific support to the spiritual and psychical beliefs for which they have already gained conclusive evidence.[13]

For all their differences, today's physicists and spiritualists share an interest in the past as a resource for criticising or defending the beliefs and practices associated with psychical research, spiritualism, modern Theosophy and other so-called occult subjects. The approaches of most academic historians have long deviated from this. They generally abstain from evaluating the strengths and weaknesses of these beliefs and practices in favour of understanding their origin, development and significance. For this reason their work sidesteps the frequent and often sterile debates about whether 'occult' beliefs and practices meet some transhistorical criteria of 'pseudo-science' in favour of understanding the historical processes by which such things were eventually demarcated from 'established', 'mainstream' or 'orthodox' sciences.[14] This literature is correspondingly sensitive to historical actors' notions of the scientific

[12] Exemplary here is Victor Stenger, *Physics and Psychics: The Search for a World Beyond the Senses* (Buffalo, NY: Prometheus Books, 1990), chapter 7.
[13] See, for example, Gordon Smith, *Beyond Reasonable Doubt: The Case for Supernatural Phenomena in the Modern World* (London: Coronet, 2018), chapter 1; Lynn G. De Swarte, *Thorson's Principles of Spiritualism* (London: Thorson's, 1999), pp. 4 and 7.
[14] For critical historical studies of 'pseudo-science' see Roger Cooter, 'The Conservatism of "Pseudoscience"', in Patrick Grim (ed.), *Philosophy of Science and the Occult* (Albany, NY: State University of New York Press, 1990), pp. 156–69; David J. Hess, *Science in the New Age: The Paranormal, Its Defenders and Debunkers and American Culture* (Madison, WI:

character of 'occult' subjects, which often challenged the ways in which the established sciences were defined. This sensitivity is particularly clear in historians' terms of analysis. By referring to mesmerism, spiritualism, modern Theosophy and psychical research as 'alternative' sciences rather than the more pejorative 'pseudo', 'marginal' or 'occult' sciences, they have better captured the considerable scientific potential that these controversial subjects had for so many nineteenth-century individuals.[15] Although many of these subjects were pursued for philosophical, religious and moral as well as scientific reasons, 'alternative sciences' remains a useful collective term for them and will be adopted here.

The need for an alternative collective term to 'pseudo-sciences' is especially pressing in the nineteenth century because it was a period when the boundaries of scientific orthodoxy, whether defined in terms of subject matter, forms of expertise, practices, audiences or sites of enquiry, were still being negotiated. The major revisionist studies of early Victorian phrenology and mesmerism are particularly instructive here because they demonstrate the significant role that these alternative sciences fulfilled in determining the boundaries of scientific orthodoxy and the social, political and cultural factors that necessarily informed this boundary work.[16] Alison Winter's study of mesmerism, for example, shows that the trajectory of this controversial medical therapy was less bound up with the question of whether a quasi-magnetic fluid really passed between mesmeric doctors and their patients than with the

---

University of Wisconsin Press, 1993); Seymour H. Mauskopf, 'Marginal Science', in R. G. Olby, G. N. Cantor, J. R. R. Christie and M. J. S. Hodge (eds.), *Companion to the History of Modern Science* (London: Routledge, 1990), pp. 869–85; Daniel P. Thurs and Ronald L. Numbers, 'Science, Pseudo-Science and Science Falsely So-Called', in Massimo Pigliucci and Maatern Boudry (eds.), *Philosophy of Pseudoscience: Reconsidering the Demarcation Problem* (Chicago University Press, 2013), pp. 121–44. Many recent historical approaches to science and pseudo-science owe a debt to Harry Collins and Trevor Pinch's classic sociological study of parapsychology. This understood the conflict between parapsychologists and their scientific adversaries as one between rival and incommensurable forms of scientific method, rationality and expertise, rather than between science and pseudo-science: H. M. Collins and T. J. Pinch, *Frames of Meaning: The Social Construction of Extraordinary Science* (London: Routledge and Kegan Paul, 1982).

[15] This is well captured in Arne Hessenbruch (ed.), *The Readers' Guide to the History of Science* (London: Fitzroy Dearborn, 2000). A critical assessment of 'alternative science' is Shiv Visvanathan, 'Alternative Science', *Theory, Culture and Society*, vol. 23 (2006), pp. 164–9.

[16] The classic accounts are Roger Cooter, *The Cultural Meaning of Popular Science: Phrenology and the Organisation of Consent in Nineteenth-Century Britain* (Cambridge University Press, 1984) and Alison Winter, *Mesmerized: Powers of Mind in Victorian Britain* (Chicago University Press, 1998). For analysis of the 'boundary work' involved in demarcating scientific from non-scientific enterprises see Thomas F. Gieryn, *Cultural Boundaries of Science: Credibility on the Line* (Chicago University Press, 1999).

challenges that it posed to early Victorian ideas about professional authority, and about the relations of class, ethnicity and gender.

The fluid boundaries of scientific orthodoxy make it equally perilous to approach spiritualism, modern Theosophy and psychical research with rigid distinctions between natural and supernatural, material and spiritual, manifest and occult. These distinctions were constantly being contested in the nineteenth century, not least because of the puzzling phenomena associated with the alternative sciences.[17] The notorious materialised spirits of seances, for example, seemed to possess all these qualities: they were clearly manifest to the senses, and had natural and even grossly material aspects, but their causes were hidden or occult, and ostensibly in the domain of the supernatural or spiritual. These kinds of phenomena were often deemed worthy of scientific investigation precisely because of these former qualities, but also because natural scientific enquiries had long proven successful in embracing phenomena that seemed to be supernatural, spiritual and occult.

Winter's study amply demonstrates the insights that can be gained into alternative sciences when situating them in their historical contexts. It is an approach that has, in the past few decades, yielded more nuanced and altogether more satisfactory historical interpretations. Spiritualist mediumship proved an attractive career move to many nineteenth-century American and British women because it conferred on them powers of speaking, writing and behaving that subverted the oppressive femininities of the Victorian patriarchy; spiritualism secured many followers among English plebeians because it helped them challenge the control that educational, religious, medical and political institutions wielded over them, and it gave bereaved men welcome opportunities to write about emotionally charged communions with loved ones on the other side, and thus challenge the oppressive ideologies of masculinity that shunned public displays of grief.[18]

---

[17] On alternative sciences and ideas of natural law see Asprem, *Problem of Disenchantment*, esp. chapter 7; Bret E. Carroll, *Spiritualism in Antebellum America* (Bloomington, IN: Indiana University Press, 1997), pp. 60–84 and Richard Noakes, 'Spiritualism, Science and the Supernatural in Mid-Victorian Britain', in Nicola Bown, Carolyn Burdett and Pamela Thurschwell (eds.), *The Victorian Supernatural* (Cambridge University Press, 2004), pp. 23–43.

[18] Ann Braude, *Radical Spirits: Spiritualism and Women's Rights in Nineteenth-Century America* (Boston: Beacon Press, 1989); Alex Owen, *The Darkened Room: Women, Power and Spiritualism in Late Victorian England* (London: Virago, 1989); Marlene Tromp, *Altered States: Sex, Nation, Drugs and Self-Transformation in Victorian Spiritualism* (Albany, NY: State University of New York Press, 2006); Logie Barrow, *Independent Spirits: Spiritualism and English Plebeians, 1850–1910* (London: Routledge and Kegan Paul, 1986); Bret E. Carroll, '"A Higher Power to Feel": Spiritualism, Grief and Victorian Manhood', *Men and Masculinities*, vol. 3 (2000), pp. 3–29.

One of the most important contexts for interpreting spiritualism and other alternative sciences has been the debates on the relationship between science and religion, or, more accurately, the sciences and Christianity. As many historians have shown, spiritualists, psychical researchers and modern Theosophists expressed more widely shared preoccupations with questions of mind, spirit, morality and cosmic purpose to which neither orthodox Christianity nor the seemingly materialistic sciences provided satisfactory answers.[19] In a period when the credibility of Christian doctrines was being challenged by historical criticism and by new scientific understandings of the earth's history and the development of organic life, many aimed to safeguard their Christian faith, or to find alternatives to Christianity, by applying the methods of rational and scientific enquiry to obscure phenomena of the mind and body that had considerable spiritual, religious and moral significance.

In taking scientific and rational enquiry in these directions, proponents of alternative sciences were both extending and challenging 'scientific naturalism'.[20] Succinctly characterised by Bernard Lightman as the "English version of the cult of science" pervading nineteenth-century Europe, this was an intellectual, cultural and political enterprise closely associated with some of the most vociferous scientists of the Victorian era, notably the biologist and prominent champion of Darwinism, Thomas Henry Huxley, and a physicist well known to many of this book's protagonists, John Tyndall.[21] Scientific naturalism held that the sciences provided the most reliable understandings of the physical world (including humanity). These understandings were based on scientifically established theories of material atoms, energy and biological evolution, and shunned scientifically unproven causes, including the spiritual and supernatural agencies at the core of religious institutions. Scientific naturalists' intellectual goals underpinned their other ambitions: they campaigned ardently for the sciences to be enterprises that were thoroughly professionalised, free from the control that the Anglican establishment had long

---

[19] Asprem, *Problem of Disenchantment*; Alan Gauld, *The Founders of Psychical Research* (London: Routledge, 1968); Oppenheim, *Other World*; Alex Owen, *The Place of Enchantment: British Occultism and the Culture of the Modern* (Chicago University Press, 2004); Frank M. Turner, *Between Science and Religion: The Reaction to Scientific Naturalism in Late Victorian England* (New Haven, CT: Yale University Press, 1974).

[20] The classic work on Victorian scientific naturalism is Turner, *Between Science and Religion*. Turner's other writings on the subject were collected in his *Contesting Cultural Authority: Essays in Victorian Intellectual Life* (Cambridge University Press, 1993), chapters 5–8. Recent perspectives are consolidated in Gowan Dawson and Bernard Lightman (eds.), *Victorian Scientific Naturalism: Community, Identity, Continuity* (Chicago University Press, 2014).

[21] Bernard Lightman, 'Victorian Sciences and Religions: Discordant Harmonies', *Osiris*, vol. 16 (2001), pp. 343–66, p. 346.

wielded over them, and practised by individuals whose profound knowledge of nature's laws gave them the right to challenge the clergy's cultural and moral authority. Spiritualists, psychical researchers and modern Theosophists, however, wanted to turn scientific naturalism against itself by showing how scientific methods could yield conclusive evidence of psychical, spiritual and other domains that challenged what they perceived to be scientific naturalism's 'materialistic' philosophy, which proclaimed that everything in the cosmos, including life, mind and spirit, could be reduced to matter and force.

Historical understandings of our alternative sciences have also been deepened by situating them in the context of *specific* scientific and even technological developments, as well as general scientific, religious and intellectual trends. Unsurprisingly, this has preoccupied many historians of psychology over the past few decades, not least because many of the architects of the academic discipline of psychology – notably Granville Stanley Hall and William James – were involved in psychical research. The work of these historians makes it clear that in many quarters, mesmerism, spiritualism and psychical research were pursued as new forms of psychology or sciences of the mind, and played significant roles in the nineteenth-century debates about the proper nature and scope of psychology.[22]

While historians of psychology have explored what, in debates about the nature of the human mind, gave plausibility and value to the claims

[22] M. Brady Brower, *Unruly Spirits: The Science of Psychic Phenomena in Modern France* (Urbana, IL: University of Illinois Press, 2010); Deborah J. Coon, 'Testing the Limits of Sense and Science: American Experimental Psychologists Combat Spiritualism, 1880–1920', *American Psychologist*, vol. 47 (1992), pp. 143–51; Adam Crabtree, *From Mesmer to Freud: Magnetic Sleep and the Roots of Psychological Healing* (New Haven, CT: Yale University Press, 1993); Alan Gauld, *A History of Hypnotism* (Cambridge University Press, 1992); Reginé Plas, *Naissance d'une science humaine, la psychologie: Les psychologues et de 'le merveilleux psychique'* (Presses Universitaires de Rennes, 2000); Graham Richards, 'Edward Cox, the Psychological Society of Great Britain (1875–1879) and the Meanings of an Institutional Failure', in G. C. Bunn, A. D. Lovie and G. D. Richards (eds.), *Psychology in Britain: Historical Essays and Personal Reflections* (Leicester: British Psychological Society, 2001), pp. 33–53; Andreas Sommer, 'Psychical Research and the Origins of American Psychology: Hugo Münsterberg, William James and Eusapia Palladino', *History of the Human Sciences*, vol. 25 (2012), pp. 23–44; Andreas Sommer, 'Normalizing the Supernormal: The Formation of the "Gesellschaft fur Psychologische Forschung" ("Society for Psychological Research"), c. 1886–1890', *Journal of the History of Behavioural Sciences*, vol. 49 (2013), pp. 18–44; Eugene Taylor, *William James on Consciousness Beyond the Margin* (Princeton University Press, 1996), esp. chapter 4; Elizabeth R. Valentine, 'Spooks and Spoofs: Relations between Psychical Research and Academic Psychology in the Interwar Period', *History of the Human Sciences*, vol. 25 (2012), pp. 67–90; Heather Wolffram, *The Stepchildren of Science: Psychical Research and Parapsychology in Germany, c. 1870–1939* (Amsterdam: Rodopi, 2009).

of our alternative sciences, other scholars have looked at the way these achievements involved engagements with recent developments in physics and other physical sciences, as well as technologies related to those enterprises. The claims of spiritualists, psychical researchers, modern Theosophists and others were often represented as subtler varieties of these more material forms of progress. Many described spiritualistic communion with the departed and telepathic exchanges between the living as only spiritual or psychological forms of telegraphy and other revolutionary forms of communication involving electricity and the ether of space.[23] The capacity to see invisible spirits or to clairvoyantly apprehend distant or hidden scenes seemed more intelligible in the context of photography with visible light and X-rays.[24] The realms bordering on and transcending the material evinced in spiritualism and psychical research seemed to converge with physicists' conceptions of a quasi-material or immaterial ether of space.[25] The higher spatial dimensions and subatomic structures apprehended by modern

[23] Carroll, *Spiritualism in Antebellum America*, chapter 4; Susan J. Douglas, *Listening In: Radio and the American Imagination* (Minneapolis, MN: University of Minnesota Press, 1999), pp. 40–54; Jill Galvan, *The Sympathetic Medium: Feminine Channelling, the Occult, and Communication Technologies, 1859–1919* (Ithaca, NY: Cornell University Press, 2010); Laura Otis, *Networking: Communicating with Bodies and Machines in the Nineteenth Century* (Ann Arbor, MI: University of Michigan Press, 2011), chapter 6; Roger Luckhurst, *The Invention of Telepathy 1870–1901* (Oxford University Press, 2002), pp. 135–47; Simone Natale, 'A Cosmology of Invisible Fluids: Wireless, X-Rays and Psychical Research Around 1900', *Canadian Journal of Communication*, vol. 36 (2011), pp. 263–75; Pascal Rousseau, *Cosa mentale: Art et télépathie au XXᵉ siècle* (Paris: Gallimard, 2015); Jeffrey Sconce, *Haunted Media: Electronic Presence from Telegraphy to Television* (Durham, NC: Duke University Press, 2000), esp. chapters 1–2; Jeremy Stolow, 'The Spiritual Nervous System: Reflections on a Magnetic Cord Designed for Spirit Communication', in Jeremy Stolow (ed.), *Deus in Machina: Religion, Technology and the Things in Between* (New York: Fordham University Press, 2013), pp. 83–113.

[24] For photography and spiritualism see Clement Chéroux et al., *The Perfect Medium: Photography and the Occult* (New Haven, CT: Yale University Press, 2004); David Harvey, *Photography and Spirit* (London: Reaktion Books, 2007); Jennifer Tucker, *Nature Exposed: Photography as Eyewitness in Victorian Science* (Baltimore, MD: Johns Hopkins University Press, 2005), pp. 159–93. For X-rays, spiritualism and other forms of occultism see Clement Chéroux, 'Photographs of Fluids: An Alphabet of Invisible Rays', in Chéroux et al., *Perfect Medium*, pp. 114–25; Allen W. Grove, 'Röntgen's Ghosts: Photography, X-Rays and the Victorian Imagination', *Literature and Medicine*, vol. 16 (1997), pp. 141–73; Linda D. Henderson, 'Vibratory Modernism: Boccioni, Kupka, and the Ether of Space', in Bruce Clarke and Linda D. Henderson (eds.), *From Energy to Information: Representation in Science and Technology, Art and Literature* (Stanford, CA: Stanford University Press, 2002), pp. 126–49.

[25] The classic study of nineteenth-century ether physics and psychical research is Wynne, 'Physics and Psychics'. Wynne argued that for late-Victorian Cambridge-based physicists, ether physics and psychical research were both forms of a "displaced, tacit moral discourse" (p. 168). The immaterial world evinced in these enterprises supported the

Theosophists using obscure psychological powers were more credible in the light of the mathematics of hypergeometry and the physics of electrons and radioactive emanations.[26]

In exploring the engagement of our alternative sciences with physical sciences, historians have tended to focus on concepts and theories, and some have treated such concepts and theories as relatively unproblematic resources that could be borrowed, adapted and otherwise mobilised for psychical and occult purposes. These approaches are limited in at least two respects. First, they overlook the considerable experimental and practical aspects of the physical sciences, even though these provided the resources for some of the most elaborate psychical investigations of the period. Second, they represent a rosy and overly simplistic view of physical theories and concepts and, moreover, of nineteenth-century physics and other physical sciences in general. This has led at least one major study of our subject to suppose that physics and psychics can be approximately mapped onto science and pseudo-science respectively.[27]

Challenging assumptions that nineteenth-century physics can be treated as a relatively unproblematic body of theories and practices are a growing number of studies demonstrating the enormous effort that went into establishing the discipline whose astonishing capacity for understanding, controlling and exploiting the phenomena of the physical world would prompt boasts that it was the 'king' of the sciences. These efforts necessarily embraced the natural and social, material and cultural.[28] Physicists had to confront a host of capricious effects, handle temperamental instruments, master abstruse mathematical methods, and carefully control spaces of

"ineffable spiritual and transcendent basis of social reality" enshrined in the predominantly Anglican, Tory and aristocratic institutions cherished by a Cambridge intellectual elite – institutions threatened by scientific naturalism's "atomic nihilism" and the excessively individualistic and utilitarian values underpinned by this worldview (pp. 174 and 180). Wynne's argument for Cambridge physicists is deeply flawed, but his general hypothesis regarding the ether's psychical and spiritual significances remains provocative and is explored later in this book. For a critique of Wynne's argument see Richard Noakes, 'Ethers, Religion and Politics in Late-Victorian Physics: Beyond the Wynne Thesis', *History of Science*, vol. 43 (2005), pp. 415–55.

[26] Mark Blacklock, *The Emergence of the Fourth Dimension: Higher Spatial Thinking in the Fin de Siècle* (Oxford University Press, 2018), chapters 4–5; Linda D. Henderson, *The Fourth Dimension and Non-Euclidean Geometry in Modern Art* (Princeton University Press, 1983), esp. pp. 186–93 and 245–55; K. G. Valente, '"Who Will Explain the Explanation?": The Ambivalent Reception of Higher Dimensional Space in the British Spiritualist Press, 1875–1900', *Victorian Periodicals Review*, vol. 41 (2008), pp. 124–49. On occultism and radioactivity see Egil Asprem, 'Pondering Imponderables: Occultism in the Mirror of Late Classical Physics', *Aries*, vol. 11 (2011), pp. 129–65; Asprem, *Problem of Disenchantment*, pp. 444–80; Morrisson, *Modern Alchemy*.

[27] Oppenheim, *Other World*, chapter 8.

[28] For an incisive synthesis of historical studies of this process see Iwan Rhys Morus, *When Physics Became King* (Chicago University Press, 2005).

research, teaching and display.[29] They also needed to develop literary and performative strategies that would persuade individuals crucially important to creating and securing the intellectual and cultural spaces of physics – for example, entrepreneurs, statesmen, university administrators, lawyers, educators, clergymen, editors and publishers – that they were supreme authorities on the physical world.[30] These efforts were particularly conspicuous in the numerous controversies in which physicists were embroiled over particular effects, theories, concepts, techniques, instruments and the very boundaries of their fledgling discipline. Physicists argued with each other and with individuals close to and well beyond the emergent discipline over such questions as the ultimate nature of matter, electricity and the ether; the virtues of mechanical force and energy as fundamental concepts of the science; the techniques, instruments and standards of physical

[29] Jed Z. Buchwald, *The Creation of Scientific Effects: Heinrich Hertz and Electric Waves* (Chicago University Press, 1994); Graeme J. N. Gooday, 'Instrumentation and Interpretation: Managing and Representing the Working Environments of Victorian Experimental Science', in Bernard Lightman (ed.), *Victorian Science in Context* (Chicago University Press, 1997), pp. 409–37; Graeme J. N. Gooday, *The Morals of Measurement: Accuracy, Irony and Trust in Late Victorian Electrical Practice* (Cambridge University Press, 2004); David Gooding, 'In Nature's School: Faraday as an Experimentalist', in David Gooding and Frank A. J. L. James (eds.), *Faraday Rediscovered: Essays on the Life and Work of Michael Faraday, 1791–1867* (London: Macmillan, 1985), pp. 105–35; Myles W. Jackson, *Spectrum of Belief: Joseph von Fraunhofer and the Craft of Precision Optics* (Cambridge, MA: MIT Press, 2000); Frank A. J. L. James, 'The Study of Spark Spectra, 1835–1859', *Ambix*, vol. 30 (1983), pp. 137–62; Frank A. J. L. James, 'The Practical Problems of "New" Experimental Science: Spectro-Chemistry and the Search for Unknown Chemical Elements in Britain 1860–1869', *British Journal for the History of Science*, vol. 21 (1988), pp. 181–94; Iwan Rhys Morus, *Frankenstein's Children: Electricity, Exhibition and Experiment in Early-Nineteenth-Century London* (Princeton University Press, 1998); Chitra Ramalingam, 'Natural History in the Dark: Seriality and the Electric Discharge in Victorian Physics', *History of Science*, vol. 48 (2010), pp. 371–98; Simon Schaffer, 'Where Experiments End: Tabletop Trials in Victorian Astronomy', in Jed Z. Buchwald (ed.), *Scientific Practice: Theories and Stories of Doing Physics* (Chicago University Press, 1995), pp. 257–99; Otto Sibum, 'Reworking the Mechanical Value of Heat: Instruments of Precision and Gestures of Accuracy in Early Victorian England', *Studies in History and Philosophy of Science*, vol. 26 (1995), pp. 73–106; Richard Staley, *Einstein's Generation: The Origins of the Relativity Revolution* (Chicago University Press, 2008), esp. chapters 3 and 6; Andrew Warwick, *Masters of Theory: Cambridge and the Rise of Mathematical Physics* (Chicago University Press, 2003).

[30] For literary and other performative strategies in Victorian physics see Bruce Clarke, *Energy Forms: Allegory and Science in the Era of Classical Thermodynamics* (Ann Arbor, MI: University of Michigan Press, 2001); Jill Howard, '"Physics and Fashion": John Tyndall and his Audiences in Mid-Victorian Britain', *Studies in History and Philosophy of Science*, vol. 35 (2004), pp. 729–58; Morus, *Frankenstein's Children*; Iwan Rhys Morus, 'Worlds of Wonder: Sensation and the Victorian Scientific Performance', *Isis*, vol. 101 (2010), pp. 806–16; Greg Myers, 'Nineteenth Century Popularizations of Thermodynamics and the Rhetoric of Social Prophecy', *Victorian Studies*, vol. 29 (1985), pp. 35–66; Staley, *Einstein's Generation*, chapter 4.

measurement; and the relationships of physics to medicine, engineering and religion.[31]

A more nuanced picture of nineteenth-century physics is critical to the way *Physics and Psychics* seeks to develop more satisfactory understandings of the psychical forays of physicists and practitioners of related physical sciences. Uncertainties about the 'material' constitution of atoms and ether posed a serious threat to the materialistic philosophy, and lent plausibility to myriad psychical phenomena that could not easily be explained in terms of matter and motion. These uncertainties were

---

[31] Conceptual and theoretical problems are explored in Jed Z. Buchwald, *From Maxwell to Microphysics: Aspects of Electromagnetic Theory in the Last Quarter of the Nineteenth Century* (Chicago University Press, 1985); Olivier Darrigol, *Electrodynamics from Ampère to Einstein* (Oxford University Press, 2000); Sungook Hong, 'Controversy over Voltaic Contact Phenomena, 1862–1900', *Archive for History of Exact Sciences*, vol. 47 (1994), pp. 233–89; Bruce J. Hunt, *The Maxwellians* (Ithaca, NY: Cornell University Press, 1991); Crosbie Smith, *The Science of Energy: A Cultural History of Energy Physics in Victorian Britain* (London: Athlone, 1998), esp. chapters 9–10; Crosbie Smith and M. Norton Wise, *Energy and Empire: A Biographical Study of Lord Kelvin* (Cambridge University Press, 1989), esp. chapters 11–13.

Experimental problems are analysed in Matthias Dörries, 'Balances, Spectroscopes and the Reflexive Nature of Experiment', *Studies in History and Philosophy of Science*, vol. 25 (1994), pp. 1–36; Gooday, *Morals of Measurement*; Sungook Hong, 'Efficiency and Authority in the "Open Versus Closed Transformer Controversy"', *Annals of Science*, vol. 52 (1995), pp. 49–76; Bruce J. Hunt, 'The Ohm Is Where the Art Is: British Telegraph Engineers and the Development of Electrical Standards', *Osiris*, vol. 9 (1994), pp. 48–63; Bruce J. Hunt, 'Scientists, Engineers and Wildman Whitehouse: Measurement and Credibility in Early Cable Telegraphy', *British Journal for the History of Science*, vol. 29 (1996), pp. 155–69; Simon Schaffer, 'Late Victorian Metrology and its Instrumentation: A Manufactory of Ohms', in Robert Bud and Susan E. Cozzens (eds.), *Invisible Connections: Instruments, Institutions, and Science* (Bellingham, WA: SPIE Optical Engineering Press, 1992), pp. 23–56; Simon Schaffer, 'Accurate Measurement Is an English Science', in M. Norton Wise (ed.), *The Values of Precision* (Princeton University Press, 1995), pp. 135–72; Loyd Swenson, *The Ethereal Aether: A History of the Michelson-Morley-Miller Aether-Drift Experiments, 1880–1930* (Austin, TX: University of Texas Press, 1972).

On physics and medicine see Iwan Rhys Morus, *Shocking Bodies: Life, Death and Electricity in Victorian England* (Stroud: History Press, 2011) and Iwan Rhys Morus, 'Physics and Medicine', in Jed Z. Buchwald and Robert Fox (eds.), *Oxford Handbook of the History of Physics* (Oxford University Press, 2013), pp. 679–97. On physics and engineering see Graeme J. N. Gooday, 'Teaching Telegraphy and Electrotechnics in the Physics Laboratory: William Ayrton and the Creation of an Academic Space for Electrical Engineering in Britain 1873–1884', *History of Technology*, vol. 13 (1991), pp. 73–111; Hunt, *Maxwellians*, chapter 7; Hunt, 'Scientists, Engineers and Wildman Whitehouse'; Smith and Wise, *Energy and Empire*, chapters 9, 12–13, 19–20. On physics and religion see Smith, *Science of Energy*, chapter 12; Smith and Wise, *Energy and Empire*, chapters 15–18; Matthew Stanley, *Huxley's Church and Maxwell's Demon: From Theistic Science to Naturalistic Science* (Chicago University Press, 2015); David B. Wilson, *Kelvin and Stokes: A Comparative Study in Victorian Physics* (Bristol: Adam Hilger, 1987), chapters 4–5; Ursula DeYoung, *A Vision of Modern Science: John Tyndall and the Role of the Scientist in Victorian Culture* (Basingstoke: Palgrave Macmillan, 2011), chapter 3.

partly prompted by experimental researches on matter, ether and electricity undertaken by our protagonists themselves.

This experimental work had two other kinds of 'psychical' significance for our protagonists. First, it gave them many of the investigative skills and material resources for probing such obscure phenomena as telekinesis, magnetic sensitivity and materialised spirits. Second, the troubles that they encountered bringing physical experiments to successful conclusions made them particularly tolerant of the more notorious difficulties of psychical investigation. In 1909, Lodge denied that Crookes's investigations into spiritualism could be sharply contrasted with the same scientist's work on the discharge of electricity through rarefied gases because there was a time when the latter 'scientific' enterprise had many of the qualities of the former 'unscientific' one: it too had been a "mistrusted region, full of danger, and strewn with the bones of former explorers".[32] It was partly because they remembered times when experimental work in other purely physical subjects – for example, telegraphic signalling, spectro-chemical analysis, sensitive flames and the detection of electromagnetic waves – had been 'mistrusted regions' that Lodge, Crookes and others were more sympathetic than most scientists of their era to the problems of communicating between this world and the next, observing ghostly effects in seances and of managing the wayward nature of those major psychical instruments: spiritualist mediums.

The status of physics during much of the nineteenth century as an emergent, rather than an established, scientific discipline creates other interpretative possibilities for this book. The most pragmatic is that the study of 'physics and psychics' needs to embrace scientific practitioners other than just physicists. Until late in the century there was no fixed training regime or career path for physicists, and the science was open to contributions from individuals from a range of scientific backgrounds including astronomy, analytical chemistry, practical electricity and medicine. It is from this wider pool of subjects that the discipline of physics was put together and, accordingly, this is the pool from which our cast of characters has been drawn.[33]

The fluidity of the boundaries of physics, and especially those with physiology, medicine and psychology, are important here because they serve as important contexts within which physics was pushed into the domain of psychical investigation. By the 1870s, physics was being defined as a science that formally sidestepped questions of life and

[32] Oliver Lodge, 'The Attitude of Science to the Unusual: A Reply to Professor Newcomb', *Nineteenth Century*, vol. 65 (1909), pp. 206–22, p. 212.
[33] This point is discussed in Morus, *When Physics Became King*, pp. 280–5.

mind, but this hardly stopped research into the borders of physics with physiology, medicine and psychology. Many physicists shared with medical doctors, electrical engineers and physiologists a preoccupation with using physical concepts (notably electricity and energy) and instruments in developing new understandings of and therapies for the human body. As we shall see in this book, some physicists saw the use of physical concepts and instruments in studying startling bodily phenomena of spiritualism and psychical research as extensions of this process.

Among those who applied energy physics to the human body were proponents of scientific naturalism who held that the domain of physics could be extended still further – to the mind. In a period when architects of psycho-physics and physiological psychology used the instruments of physics to study sensation and perception, scientific naturalists saw physics as offering the kind of material, mechanical, and empirically based explanation of mental processes that they believed a true science of psychology demanded. Based on an assumed close parallelism between neural and psychological states, this proposed connection between physics and psychology posed a serious threat to the strongly held belief in free will that in Victorian Britain could also undermine the basis of Christian morality.[34]

One of the fiercest critics of this argument was the Scottish physicist James Clerk Maxwell, who, in the 1870s, denied that the natural sciences per se, let alone the physics of molecules and energy, could be legitimately applied to questions of volition and consciousness.[35] Along with William Thomson (Lord Kelvin), Peter Guthrie Tait and others, Maxwell represented a formidable opposition among late-Victorian physicists to scientific naturalism, and in particular its materialistic readings of those subjects where physicists claimed particular authority: the nature of matter and energy. As academically trained practitioners heading specialised research and teaching laboratories, Maxwell and others symbolised the professionalised science championed by secular scientific naturalism. But these physicists were also devout Christians who maintained that professionalised physics could fulfil the religious purposes that the sciences had carried for centuries: to evidence a cosmos designed and ruled by divine agency.[36]

---

[34] Lorraine Daston, 'British Responses to Psycho-Physiology, 1860–1900', *Isis*, vol. 69 (1978), pp. 192–208; Roger Smith, *Free Will and the Human Sciences, 1870–1910* (London: Pickering and Chatto, 2013), chapters 1–2.

[35] Stanley, *Huxley's Church*, pp. 222–41.

[36] The historical literature on the religious functions of the sciences is vast, but see especially John Hedley Brooke, *Science and Religion: Some Historical Perspectives* (Cambridge University Press, 1991) and Peter Harrison, *The Territories of Science and Religion* (Chicago University Press, 2015).

The dispute between theistic physics and scientific naturalism focussed partly on the same issue that had been, and would continue to be, pursued by the physicists, electrical engineers and others in this book: to what extent could the theories and practices of physics be applied phenomena associated with mind? As we shall see in the following chapters, some types of psychical phenomena were at least as 'physical' as 'mental' in nature and lent themselves more easily to this kind of intervention. Our protagonists interpreted the faint luminosity allegedly seen around magnets, the movement of untouched objects by the power of thought, and the materialisation of otherwise disembodied spirits as partly magnetic, optical and mechanical puzzles requiring the skills and resources that had already proven successful in physical detection, isolation, control and measurement. But even less physical and more mental types of psychical phenomena were not closed to their interventions: for example, telepathy and the mechanisms by which spirits interacted with the material world seemed legitimate areas of physical theorising.

The protagonists of this book complicated rather than resolved the conflict between theistic physics and scientific naturalism. In some ways their forays led them to a position closer to Maxwell and other theistic physicists whose scientific work they generally revered. Their psychical forays helped evidence the apparent capacity of mind to exist independently of the material brain, which challenged the psycho-physiological link upheld by scientific naturalists. Many of them saw this as an argument against materialism and for the credibility of beliefs long associated with Christian theism, such as the divine guidance of the cosmos and the efficacy of prayer. But their forays also testified to the ascendancy of scientific naturalism over theistic science.[37] Many of the individuals studied here accepted that secular scientific methods, rather than metaphysics and religious belief, were the most reliable tools for studying all phenomena coming within human experience, including psychological phenomena previously studied under the rubrics of morality, philosophy and religion. Their attempts to extend theories of physics to telepathy, the afterlife and other psychical realms certainly transgressed Maxwell's limits and put their work closer to the materialism and pantheism often associated with scientific naturalism than to theistic science.

The foregoing discussion demonstrates the tantalising convergences now existing between the historiographies of the alternative sciences and physical sciences in the nineteenth and twentieth centuries and which

[37] Stanley, *Huxley's Church*, chapter 7. They represent what Asprem has called the 'open-ended naturalism' of psychical research, which sought to apply scientific methods to metaphysical and religious questions sidestepped by agnostic scientific naturalism: Asprem, *Problem of Disenchantment*, pp. 299–306.

encourage a more nuanced study of 'physics and psychics'. As historians of alternative sciences uncover more ways in which the physical sciences mattered to interpretations of mesmerism, spiritualism, modern Theosophy and psychical research, so historians of the physical sciences acknowledge the psychical and 'occult' implications of the ideas and practices in these enterprises. Building on and moving beyond this literature, *Physics and Psychics* argues that these implications were taken much more seriously than we have acknowledged. An entangled cultural history of physical and psychical sciences, it returns us to a period when our boundaries between these areas of enquiry were absent or only partially formed; when exchanges between these areas were both possible, desirable and even fruitful; and when some of the most distinguished physicists, electrical engineers, chemists and astronomers of the day regularly locked horns with psychologists, conjurors and spiritualists over the legitimacy, meanings and uses of a physical science of psychical phenomena.

This book is organised chronologically and thematically. Covering the period from around the 1770s to the 1860s, Chapter 1 explores the trajectories of spectacular new forms of what we label psychical investigation (mesmerism, the human sensitivity to magnetism, and spiritualism) and the emergence of physics as a scientific discipline boasting formidable resources for understanding and controlling the natural world. Not surprisingly, the same period saw the construction of some of the most important early arguments for the relevance and utility of physics to the study of psychical phenomena, arguments that would intensify from the 1870s. The identity of the scientific practitioners who took these arguments seriously is one of the principle burdens of Chapter 2. Covering the period from the 1780s to the 1930s, it analyses the complex and often radically different reasons why individuals that we collectively refer to as 'physical–psychical scientists' studied different kinds of psychical phenomena and why their attitudes to the subject changed.

Many of the physical–psychical scientists identified in Chapter 2 did not leave particularly illuminating insights into the relationships between 'physics and psychics', but those that did form the main subjects of analysis in the remainder of this book. Chapters 3–5 take a more thematic approach and consider three complementary perspectives on the reasons why confidence in the relationships between physical and psychical investigation reached a zenith in the period from the 1870s to the early 1900s. One reason was a host of concepts and theories in physical sciences that made psychical effects plausible, or at least challenged arguments against their impossibility (Chapter 3); another was that the physical sciences embodied material resources and skills that promised more effective solutions to the investigative problems that had beleaguered psychical investigation for decades (Chapter 4). Yet another equally striking reason examined in Chapter 5 is that physical–psychical scientists believed that their long experiences of the vicissitudes of physical experiment gave them a patient and altogether more 'scientific' attitude to the difficulties of doing psychical research, which they felt was lacking in conjurors and other kinds of psychical investigators who were among their most redoubtable adversaries.

None of the approaches to psychical phenomena examined in Chapters 3–5 proved as successful as physical–psychical scientists hoped. Taking the story forward to the late 1930s, Chapter 6 analyses the implications of these disappointing outcomes and shows that while the leading (and ageing) physical–psychical scientists turned increasingly from doing to writing about 'physics and psychics', a host of younger scientific practitioners explored other ways of realising and surpassing the original experimental ambitions of the older generation of physical–psychical scientists. The conclusion surveys the real and imagined connections that have been made between physics and the study of psychical, parapsychological and paranormal effects since the 1930s and considers the continuities and discontinuities with the period on which this book is principally focussed.

1 New Imponderables, New Sciences

In August 1862, the leading article of one British periodical was headed 'Animal "Magnetism"'.[1] The quotation marks around the word magnetism indicated the anonymous author's understanding of at least one of the many controversies that had surrounded this subject for over half a century. This was the question of whether, as animal magnetism's proponents claimed, a weightless, invisible bodily fluid, force or emanation by which the will of an individual was alleged to directly influence the mind and body of another person was analogous to the magnetism associated with minerals. By the early 1860s, many of those who had accepted the effects of animal magnetism but rejected the idea that they derived from a kind of magnetic fluid described the effect as mesmerism, in honour of the Swabian physician Franz Anton Mesmer, who, in the 1770s, had announced the discovery of this 'magnetic' form of influence and turned it into the basis of a medical therapy that proved both controversial and popular in Continental Europe, Britain and North America.

'Animal "Magnetism"' was much more positive about animal magnetism than its cautious title suggested. Anticipating disdain from some readers, it asserted that the "*quasi* science" rested on indisputable fact and urged the need to establish connections between facts of an "extraordinary character and occult nature" and those "accepted by science".[2] There were moral and intellectual motivations for this. Establishing facts about animal magnetism was no less important to the "cause of truth" than the recognition of other facts that "scientific orthodoxy" had a lamentable tendency to dismiss simply because such facts appeared to conflict with "accepted doctrine".[3] Moreover, recent developments in the physical sciences suggested the strong possibility that an obscure force, fluid or agency somehow connected with life could be related to the known physical forces. The eminent German chemist Justus von Liebig

---

[1] [Anon.], 'Animal "Magnetism"', *Electrician*, vol. 2 (1862), pp. 157–8.
[2] [Anon.], 'Animal "Magnetism"', p. 157.   [3] [Anon.], 'Animal "Magnetism"', p. 157.

had demonstrated the similarity between the vital and chemical forces; the doyen of British natural philosophy, Michael Faraday, had shown that all bodies, including those of living beings, were to one degree or another extent influenced by magnetism; and another German chemist, Karl von Reichenbach, had produced evidence indicating that magnetism influenced the nervous system in ways comparable to the mysterious agency of mesmerism. The idea that the body produced and was susceptible to an obscure force somehow associated with magnetism was not as implausible as many suggested and was certainly ripe for investigation.

One of the most striking features of 'Animal "Magnetism"' was *where* it was published. Articles on animal magnetism and mesmerism were not uncommon in British periodicals in the mid-Victorian era, not least in those medical and spiritualist titles that were respectively hostile and sympathetic to the topics.[4] The rhetoric of 'Animal "Magnetism"' – its attack on the narrowminded nature of scientific orthodoxy and appeal to recent discoveries in the physical sciences – was not untypical in spiritualist serials. Yet 'Animal "Magnetism"' appeared in *The Electrician*, a weekly technical paper that we might not expect to be interested in, let alone sympathetic to, the topics.

Founded in 1861, *The Electrician* aimed to represent the burgeoning number of individuals with interests in the scientific understanding and application of electricity, especially those connected with the expanding overland and undersea networks of electric telegraphs.[5] Although the periodical's content was mainly preoccupied with the electric telegraph, the use of electricity in medical therapies and physiological research, and other 'material' applications of electricity, the inclusion of material on animal magnetism was not inconsistent with its declared mission to show that electricity would "solve many of the important problems connected with the well-being of mankind".[6] Indeed, as a prominent platform for knowledge of the way electricity would improve human beings' capacity to communicate and to understand and heal their bodies, it is not surprising that it tolerated the possibility of some other subtle force of fluid, perhaps related to electricity and magnetism, that was at the basis of

[4] Examples from medical periodicals are [Anon.], 'Mesmeric Hospital Reports', *British Medical Journal*, vol. 2 (1862), pp. 308–9; [Anon.], 'Reviews', *Medical Times and Gazette*, vol. 14 (New Series) (1857), pp. 122–3. Examples from spiritualist serials include [Anon.], 'Benjamin Brodie on Spiritualism', *Spiritual Magazine*, vol. 1 (1860), pp. 97–103; John Ashburner, 'On the Connection Between Mesmerism and Spiritualism, with Considerations on Their Relations to Natural and Revealed Religion and to the Welfare of Mankind', *Supplement to the British Spiritual Telegraph*, vol. 3 (1859), pp. 1–96.
[5] *The Electrician* is discussed in P. Strange, 'Two Electrical Periodicals: *The Electrician* and *The Electrical Review*, 1880–1890', *IEE Proceedings*, vol. 132 (1985), pp. 574–81.
[6] [Anon.], 'Introductory', *Electrician*, vol. 1 (1861), p. 1.

other, perhaps 'occult', forms of communication and therapy typically maligned by scientific orthodoxy.

'Animal "Magnetism"' was not the only article in *The Electrician* to be interested in 'occult' phenomena and the possible benefits of investigating them, and neither was *The Electrician* the only British scientific and technical serial of the mid-Victorian period to be so.[7] This material yields instructive insights into aspects of the nineteenth-century cultures of the sciences which this chapter will study in detail. It affords a glimpse of the presence of animal magnetism and related psychical or 'occult' phenomena in scientific and technical cultures from which we might expect them to have disappeared. It also suggests that the damning verdict on such phenomena given at the time by leading medical and scientific practitioners – that the effects were due to well-known mechanisms of the mind and body rather than new, hidden forces or fluids connected with the body and mind – were as unconvincing in some scientific quarters as in sections of the general population enthralled by the performances of itinerant mesmeric lecturers and spiritualist 'mediums'. As the author of 'Animal "Magnetism"' demonstrated, the efforts of many medical and scientific practitioners to demarcate the study of such bodily emanations as a pseudo- or '*quasi*-science' were not decisive.

The 'occult' material in *The Electrician* and other scientific and technical serials of the 1850s and '60s also helps us to reassess the better-known forays of mid-Victorian scientific practitioners into similar subjects. In this period, a young professor of natural philosophy, John Tyndall, was testing Reichenbach's claim that some people saw lights around magnets and investigating the alleged capacity of spiritualist mediums to commune with professed denizens of the spirit world; an electrician working for one of the British electric telegraph firms, Cromwell Varley, was exploring his own powers of mesmeric healing and investigating the popular fascination with tables that seemed to turn under the influence of unknown forces or spirits of the dead; and a student of chemistry, William Crookes, was taking a keen interest in Faraday's attempt to explain the mystery of 'table-turning' in terms of a force unconsciously exerted by people participating in the popular pastime.[8]

---

[7] For example, [Anon.], 'Scientific Gossip', *Photographic News*, vol. 6 (1862), pp. 3–4; [Anon.], 'Gregory's "Letters on Animal Magnetism"', *Mechanics' Magazine*, vol. 54 (1851), pp. 364–70; [Anon.], 'Chemistry', *Popular Science Review*, vol. 1 (1862), pp. 382–9; [Anon.], 'Swedenborg – No. IV', *English Mechanic*, vol. 2 (1865), pp. 87–8; 'P.', 'On the Odic Principle of Reichenbach', *London Journal of Arts, Sciences and Manufactures*, vol. 38 (1851), pp. 124–32, 193–9.

[8] John Tyndall to Edward Frankland, [6 August 1856], in William H. Brock and Geoffrey Cantor (eds.), *The Correspondence of John Tyndall Volume 5* (Pittsburgh, PA: University of Pittsburgh Press, 2018), pp. 434–6; [John Tyndall], 'Science and the

In the context of 'Animal "Magnetism"', the 'occult' interests of Tyndall, Varley and Crookes no longer seem so exceptional. They seem to represent a more widely shared belief that there might exist obscure forces, fluids, powers and influences associated with the human body that could form the basis of potentially fruitful extensions of the physical sciences. This chapter explores the origins and development of this belief, which reached its culmination in the work of the late-nineteenth-century physical scientists, or, as we call them, 'physical–psychical scientists', who are the focus of this book. It studies the way that this belief was articulated, contested and defended from the late eighteenth until the mid-nineteenth centuries, and, in the contexts of animal magnetism, Reichenbach's magnetic researches, table-turning and Modern Spiritualism. These developments took place in a period when the boundaries of physics were still in flux. The subject areas that were beginning to constitute the scientific discipline were being extended in myriad directions: the study of electricity was transforming approaches to problems in engineering and medical therapy; studies of force and heat were being extended to, and enriched by, questions in physiology and medicine; and understandings of atoms, energy and ether were adding new, and often competing, perspectives on the relationships between science and religion. These were precisely the contexts within which it became possible for some to argue that the physical sciences could and should be extended to the puzzling phenomena often lumped together as "occult".

Animal Magnetism as Physics

"I dare to flatter myself that the discoveries which I have made, and which are the subject of this book, will push back the limits of our knowledge in physics, as much as the invention of microscopes and telescopes has done for the age preceding our own."[9] This 1799 declaration by Mesmer problematises the assumption that his historical significance lies solely in the fields of psychiatry and psychology. As Alan Gauld has emphasised, Mesmer sought not only to apply animal magnetism to the treatment of bodily rather than mental illnesses, but saw himself as the discoverer of a genuinely novel physical agency – an invisible, weightless and fluidic

Spirits', *Reader*, vol. 4 (1864), pp. 725–6; Cromwell Varley, 'Evidence of Mr. Varley', in *Report on Spiritualism of the Committee of the London Dialectical Society* (London: Longmans, Green, Reader and Dyer, 1871), pp. 157–72; William Crookes, *Psychic Force and Modern Spiritualism: A Reply to the 'Quarterly Review' and other Critics* (London: Longmans, Green and Co., 1871), p. 11; Michael Faraday, 'Table Turning', *Times*, 30 June 1853, p. 8.

[9] Franz A. Mesmer, *Mémoire de F. A. Mesmer, docteur en médecine, sur ses découvertes* (Paris: Fuchs, 1799), p. 6. My translation.

form of matter – that would transform the study of physiology and physics.[10]

Mesmer argued that, owing to its extraordinarily rarefied nature, the fluid penetrated, and acted as a medium of mutual influence between, all bodies in the universe, whether animate or inanimate. Moreover, he proposed that the fluid sustained tidal effects which, in the human body, produced imbalances that caused bodily illnesses. By manipulating the subtle fluid, Mesmer believed he could restore this imbalance, a procedure that induced a 'crisis' in patients – spasms and other violent physical movements – which accelerated the natural healing process. Initially, Mesmer achieved curative effects by applying mineral magnets to the body (which drew on established traditions of magnetic cures), but he later accepted that his own body was equally effective as a source of the fluid, or, as he was soon calling it, 'animal magnetism'. By employing a series of elaborate bodily gestures, notably touching and passing hands over patients, he believed he could cause the magnetic fluid in his own body to restore imbalances in those of ailing individuals (Figure 1.1).

Mesmer and the disciples he eventually attracted in Europe, Britain and North America had good grounds for believing that animal magnetism was an extension of existing scientific and medical thinking, as well as a development of ideas of a living, cosmic fluid promulgated in occult philosophies and sciences.[11] By the late eighteenth century, physical sciences divided the material cosmos into ponderable matter and a host of forces and imponderable (weightless) and invisible fluids such as gravity, mineral magnetism, frictional electricity and heat.[12] The apparent discovery of another invisible force or imponderable fluid fitted well within programmes of enquiry in these sciences. The ideas of a universal force or fluid linking the microcosm of animate and inanimate bodies on earth to the macrocosm of celestial objects and of the therapeutic benefits arising from the manipulation of such a fluid made sense within contemporary scientific and medical discourses. For popular scientific audiences in late-eighteenth-century European cities, many of these ideas were dramatised in scientific shows of electricity and other imponderables that were easier to sense than to comprehend. As Jessica Riskin has shown, when, in 1778, Mesmer arrived in Paris to market his mysterious

---

[10] Gauld, *History of Hypnotism*, pp. 11 and 247.

[11] Nicholas Goodrick-Clarke, *The Western Esoteric Traditions: A Historical Introduction* (New York: Oxford University Press, 2008), pp. 174–8.

[12] John L. Heilbron, 'Weighing Imponderables and Other Quantitative Science Around 1800', *Historical Studies in the Physical Sciences*, vol. 24 (1993), pp. 1–33, 35–277, 279–337, esp. p. 16.

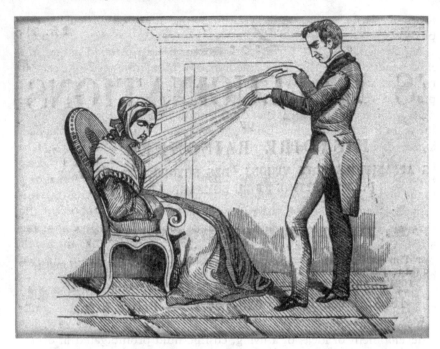

1.1 The rays of animal magnetism believed to mediate the influence of a mesmeric operator over the bodies and minds of their subjects. From Jules Baron Du Potet de Sennevoy, *Manuel de l'étudiant magnétiseur* (Paris: G. Ballière, 2é, 1851), p. 22, figure 1. Reproduced by permission of the Syndics of Cambridge University Library.

new magnetic therapy, he encountered a clientele who were "ready" for him.[13]

To give intelligibility to the bodily gestures at the heart of his 'magnetic' therapy, Mesmer proposed that animal magnetism was essentially a universal fluid akin to the medium of gravity, and which transmitted motion and produced tidal ebbs and flows. Mesmer's goal to elucidate the "unknown mechanical laws" of this fluid reflected his debt to popular Newtonianism and embodied his attempt to extend the boundaries of physics: animal magnetism involved studying a mysterious agent that seemed to share physical properties with gravity, light, heat, sound, magnetism and frictional electricity (for example, it could be reflected

[13] Jessica Riskin, *Science in the Age of Sensibility: The Sentimental Empiricists of the French Enlightenment* (Chicago University Press, 2002), p. 201. The following discussion is indebted to Riskin's incisive analysis of animal magnetism (in chapter 6).

by mirrors and accumulated in material objects), but it also promised to illuminate the ultimate nature of these better-known physical agents.[14] As the means by which the will of a 'magnetiser' appeared to affect the sensations of a patient at a distance, the animal magnetic fluid was also indebted to late-eighteenth-century physiological theories that explained sensation as the motion of an imponderable ether or fluid in the nerves. However, animal magnetism extended physiological thinking by proposing the existence of an imponderable fluid that mediated sensations seemingly inaccessible to the five ordinary senses. Individuals subject to the passes of a 'magnetiser' claimed to be able to directly experience the magnetiser's thoughts, as well as perceive hidden or distant objects (later christened 'clairvoyance'), and past and future events.

The popularity and apparent success that Mesmer's magnetic therapy enjoyed in Paris exasperated many leading French medical practitioners and led to a key development in the history of the relationship between established and occult sciences. In 1784, Louis XVI's government set up two commissions – one from the Académie Royale des Sciences and the Paris Faculté de Medécine, the other from the Société Royale de Medécine – to investigate animal magnetism. Boasting such luminaries as the American natural philosopher and diplomat Benjamin Franklin, the chemist and tax collector Antoine Lavoisier and the astronomer Jean-Sylvain Bailly, the commissions delivered an intellectually weighty and damning verdict on Mesmer's physics: while they accepted the genuineness of magnetic effects, they rejected the idea that an imponderable fluid was the cause, since all the effects could be explained in terms of the imagination. The imagination was a compelling explanation because the human subjects tested by the commissioners only experienced the effects of the fluid when they *believed* they were subjected to it.

It was hardly surprising that most of the commissioners entertained this verdict from the outset. The animal magnetic fluid was much more problematic than other imponderable agents to which it was often compared: like frictional electricity, mineral magnetism and gravity, it could only be detected by its effects, but unlike these imponderables its effects could only be exhibited on human subjects rather than inanimate objects and so there was a high probability of "moral causes" operating.[15] Given that, by the 1780s, the imagination was known to be a cause of bodily effects, the commissioners regarded their scepticism towards the animal magnetic fluid as justifiable. By declaring that sensations could be the

---

[14] Franz A. Mesmer, *Mémoire sur la découverte du magnétisme animal* (Geneva: P. F. Didot la jeune, 1779), pp. 74–83.

[15] *Rapport des commissaires de la Société Royale de Médicine, nommés par le roi, pour faire l'examen du magnétisme animal* (Paris: Chez Moutard, 1794), p. 10.

result of the imagination stimulated by verbal and other suggestions, the commissioners effectively undermined a key epistemological claim of the sciences: that sensations were a reliable basis of empirical knowledge of the world.[16]

The verdict of the animal magnetism commissions was certainly consistent with the opinion of the German-speaking physicians and natural philosophers whose hostility to Mesmer's claims and therapeutic practice had earlier prompted his decision to move to Paris. The 'official' verdict, however, was neither unanimous nor authoritative. One member of the Société Royale de Medécine commission denied that all animal magnetic effects tested by his colleagues could be put down to the imagination.[17] Plenty of Mesmer's followers, who by the 1780s could be found throughout France, attacked the commissions for sloppy experimental methods and for upholding the imagination as the main cause when this failed to explain all the evidence for animal magnetism's efficacy (notably on animals) and how, in the absence of some kind of imponderable fluid, the imagination was supposed to produce effects on the body.[18] Most mesmerists maintained that the senses and feelings of magnetised subjects could be trusted and that they remained reliable instruments of the power of the magnetic influence. The official verdict on animal magnetism certainly played a part in Mesmer's decision to leave Paris and eventually withdraw from the centre stage of animal magnetism, but this had little effect on the spread of his ideas and practices elsewhere in France and in Europe.

From the 1780s until the early 1800s, animal magnetism enjoyed less success as a possible contribution to physics than it did as a medical therapy and as a contribution to other branches of knowledge. Mesmer's vision of animal magnetism as physics was most strongly shared by several French and German physicians who sought to relate the animal magnetic fluid to electricity and other known imponderables or to a supposed atmosphere produced by the nervous fluid.[19] However, this approach competed with the more psycho-physical one of those who followed the Marquis de Puységur, a disciple of Mesmer who in 1784 announced that animal magnetism could induce a state of artificial somnambulism or 'magnetic' sleep. In this state, an individual exhibited what would become the defining characteristics of animal magnetism: they displayed a consciousness distinct from that associated with their waking self, an insensibility to pain, a capacity to see through opaque objects and to great distances, and an ability to enter into a state of 'rapport' with the

[16] Riskin, *Science in the Age of Sensibility*, pp. 222–3.
[17] This was the botanist Antoine Laurent de Jussieu.
[18] Crabtree, *From Mesmer to Freud*, pp. 31–2; Riskin, *Science in the Age of Sensibility*, p. 221.
[19] Gauld, *History of Hypnotism*, chapters 4–5.

magnetiser, whose thoughts they seemed to be able to read and whose silent commands they obeyed. Neither the imagination nor the fluid theories coped well with explaining these aspects of animal magnetism, and for this reason Puységur and his disciples largely sidestepped the question of the nature of the magnetic fluid. While they maintained that some physical influence passed from magnetiser to subject, their primary concern was with the nature of the will that mobilised this influence and, accordingly, with animal magnetism as a branch of psychology.[20]

Even less materialistic in their interpretations than Puységur and his followers were those French and German writers preoccupied with animal magnetism's spiritual and mystical significances. Some individuals in a state of magnetic sleep appeared to commune with angels and spirits and possess the powers of visionaries and prophets. The capacity of magnetised somnambules to commune with the soul of nature strengthened animal magnetism's appeal to proponents of *Naturphilosophie*.[21] This key aspect of German Romanticism emphasised the fundamental identity of nature and spirit and that a true understanding of it required special abilities to interpret external nature and the spiritual depths that it symbolised. Individuals in a state of magnetic sleep became important enquirers into these depths but also seemed to exhibit some of the polarities that *Naturphilosophen* traced in living things – in this case, between the higher animal states associated with consciousness and the lower vegetative states associated with unconsciousness.

These developments in animal magnetism embodied many of the ideas and practices that would characterise Reichenbach's 'od', spiritualist mediumship, telepathy and the other psychical phenomena in which British physical scientists would show such a strong interest. Indeed, I want to suggest that the attitudes and approaches of these latter individuals to psychical phenomena built partly on the examples set by early-nineteenth-century British practitioners of and writers on animal magnetism because, more than many other proponents of the subject, they kept alive Mesmer's hopes that his medical therapy was also a source of progress in physics.

Animal magnetism made little impact in Britain until the 1830s, when many physicians, clergymen, *littérateur* and others began to practise, publish and lecture on a subject that had impressed some of them via

[20] Gauld, *History of Hypnotism*, pp. 112–13.
[21] Gauld, *History of Hypnotism*, pp. 141–4. On *Naturphilosophie* see Robert J. Richards, *The Romantic Conception of Life: Science and Philosophy in the Age of Goethe* (Chicago University Press, 2002), esp. pp. 128–46.

the London lectures of the leading French magnetist Baron Dupotet.[22] By the time Dupotet arrived in Britain, animal magnetism in France had become, after a state of relative latency during the Revolution, one of the most controversial subjects in medical circles. In continuation of the controversy following the 1784 commissions, one of the most contentious issues remained the existence of the animal magnetic fluid. Many agreed with the French physician Alexandre Bertrand, whose *Traité du somnambulisme* (1823), a work later regarded as a foundational text in hypnotism, argued that magnetised somnambules who claimed to perceive magnetic fluids only saw what they believed or were made to believe in such fluids; others, including Dupotet, upheld the fluid theory as the only interpretation that could cope with evidence that magnetisation worked on animals and infants (who, it was supposed, could not possibly be made to believe in fluids) and when the magnetised subject was asleep or unaware of the magnetic operator's presence.[23]

Given Dupotet's significance in stimulating British mesmerism, it is not surprising that so many of its proponents should also favour the idea that the modus operandi of mesmerism was a physical influence crossing the space between the operator and subject. Nineteenth-century mesmeric texts in Britain, France, Germany, North America and elsewhere shared a strong preoccupation with mesmerism as a new form of medical therapy and as a contribution to the emergent science of the mind, but the British texts were not as uninterested in the 'philosophical' aspects as some historians have claimed.[24] In *Isis Revelata* (1836), one of the earliest English-language surveys of animal magnetism, the lawyer John Campbell Colquhoun supported his theory that the mesmeric influence was the nervous fluid flowing out of the body with a detailed exposition of a "new theory of physics" tracing the mesmeric influence and all other imponderable agents to vibrations in the space-filling medium that many natural philosophers now accepted as the carrier of light waves across empty space: the luminiferous ether.[25] The Anglican clergyman Chauncy Hare Townshend developed a similar theory in his *Facts in Mesmerism* (1840).[26]

[22] On early Victorian mesmerism see Gauld, *History of Hypnotism*, chapters 11–12; Winter, *Mesmerized*.

[23] Baron Dupotet de Sennevoy, *An Introduction to the Study of Animal Magnetism* (London: Saunders and Otley, 1838), pp. 329–46.

[24] See, for example, Gauld, *History of Hypnotism*, p. 210.

[25] J. C. Colquhoun, *Isis Revelata: An Inquiry into the Origin, Progress and Present State of Animal Magnetism*, 2 vols. (Edinburgh: Maclachlan and Stewart, 1836), vol. 2, p. 349–408.

[26] Chauncy Hare Townshend, *Facts in Mesmerism, with Reasons for a Dispassionate Inquiry into It* (London: Longman, Orme, Brown, Green and Longmans, 1840), pp. 488 and 497.

Colquhoun and Townshend exemplify a tendency among some British mesmeric writers of the 1830s and '40s to link the mesmeric and nervous fluids and to suppose that both were identical to, or at least closely related to, electricity. Yet by the 1850s, some mesmeric writers had accepted that electrical analogies for the nervous and mesmeric fluids were problematic. In 1851, for example, the physician Joseph Haddock warned that the "best physiologists" had rejected the identity of electrical and nervous fluids, so that the 'mesmeric' power of the nervous fluid outside the body could not be called electrical.[27]

Townshend shared with several leading British mesmerists a belief that establishing the relationship between the animal magnetic fluid and known imponderables would be a major step towards linking animal magnetism to the physical sciences and thus raising the intellectual credibility of the controversial practice.[28] For some British medical and scientific practitioners this was a stimulus for experimental as well as theoretical activity. An instance of the experimental approach took place in 1838 at University College Hospital, London, and involved the hospital's leading physician and medical professor John Elliotson and the Irish natural philosopher Dionysius Lardner.[29] One of the most vociferous of all English mesmerists, Elliotson had stimulated considerable publicity and professional hostility for using female patients in displays of mesmeric phenomena and treatment in the wards.[30] Elliotson regarded his work as partly the prosecution of the "physics of mesmerism" insofar as it investigated the capacity of different metals and water to carry the mesmeric influence and thereby cause muscular rigidity and other bodily reactions in patients making contact with the substances.[31] Publicised in

---

[27] Joseph W. Haddock, *Somnolism and Psycheism; Or, the Science of the Soul and the Phenomena of Nervation as Revealed by Mesmerism* (London: James S. Hodson, 2nd ed., 1851), p. 50. Haddock was probably referring to Hermann von Helmholtz, who in 1850 had shown that the speed of stimuli in sensory nerves was far lower than that of Voltaic electric currents, thus challenging their identity. See Kathryn Olesko and Frederic L. Holmes, 'Experiment, Quantification and Discovery: Helmholtz's Early Physiological Researches', in David Cahan (ed.), *Hermann von Helmholtz and the Foundations of Nineteenth Century Science* (Berkeley, CA: University of California Press, 1993), pp. 50–108.

[28] Townshend, *Facts in Mesmerism*, p. 488; John Elliotson, 'Review of an *Abstract of Researches on Magnetism and Certain Allied Subjects*', *Zoist*, vol. 4 (1846–7), pp. 104–24, p. 122.

[29] [Dionysius Lardner and Edward Bulwer Lytton], 'Animal Magnetism', *Monthly Chronicle*, vol. 1 (1838), pp. 289–306; vol. 2 (1838), pp. 11–30. The anonymous authors of this report on the experiments are identified in Winter, *Mesmerized*, pp. 52–6.

[30] See Gauld, *History of Hypnotism*, pp. 199–203; Winter, *Mesmerized*, chapters 2–4.

[31] Elliotson, 'Review of an *Abstract of Researches*', p. 123. See also John Elliotson, *Human Physiology* (London: Longman, Orme, Brown, Green and Longman, 1840), pp. 1163–94.

an anonymous article by Lardner and the English novelist Edward Bulwer-Lytton, the experiments took this physics much further and appeared to show that the mesmeric influence experienced by patients could be, like light, reflected from mirrors and metallic surfaces. The influence also seemed to penetrate opaque screens placed between the operator and patient, diminish in strength when the distance between participants was increased, and be unaffected by electric shocks given to the patients from a galvanic apparatus and Leyden jar provided by the natural philosopher and electric telegraph pioneer Charles Wheatstone.

The 1838 experiments illustrate the modest but telling overlap between early Victorian mesmerism and the cultures of electrical display and measurement.[32] Most mesmerist practitioners and writers accepted that human beings were the primary instruments of research, principally because they were sensitive to subtle influences that could not otherwise be detected. For some, however, the addition of inanimate instruments promised to illuminate the suspected connections between animal magnetism and known imponderables as well as symbolise the capacity of mesmeric phenomena to become what Lardner and Lytton called "subjects of vast importance, whether regarded as appertaining to general physics or the special science of medicine".[33]

Lardner, Bulwer-Lytton and Elliotson were not alone in recognising that their physical approach to mesmerism depended on connections with academic professors, popular lecturers and instrument makers. Thus, in the mid-1840s, the physician John Ashburner turned to the Royal Polytechnic Institution, one of London's premier venues of popular science, for a "splendid apparatus" used to determine whether magnetism induced the same physiological responses as mesmerism.[34] A few years later, the gas engineer John O. N. Rutter commissioned fellow Brighton resident, the surgeon–electrician E. O. Wildman Whitehouse, to make a "galvanoscope" that yielded quantitative support for Emil Du Bois Reymond's recent evidence that human muscular contraction generated electrical currents.[35] Although Rutter's principal concern was animal electricity, he recognised its value in a plausibility argument for animal *magnetism*: the discovery of electromagnetism suggested that as a carrier of electric currents, the human body should also be sensitive to

---

[32] On these cultures of display and measurement see Morus, *Frankenstein's Children*.

[33] Lardner and Lytton, 'Animal Magnetism', p. 28.

[34] John Ashburner, 'Observations upon the Analogies Between the Mesmeric and Magnetic Phenomena', *Zoist*, vol. 4 (1846–7), pp. 124–39, p. 130.

[35] John O. N. Rutter, *Human Electricity: The Means of Its Development* (London: John W. Parker, 1854), p. 100. Whitehouse is referred to on p. 117. Whitehouse was heavily involved in submarine cable telegraphy during the 1850s: see Bruce J. Hunt, 'Scientists, Engineers and Wildman Whitehouse'.

and the source of magnetic "currents", even if these currents were too feeble to be detected by instrumental means.[36]

British mesmerists' hopes that their physical approaches to animal magnetism would help raise the medical and scientific profile of the subject were significantly weakened by the indifference or outright hostility towards the subject shown by the British medical and scientific establishments. Few of the natural philosophers who witnessed mesmeric demonstrations (notably Faraday, Wheatstone and William Whewell) sustained any interest in the subject.[37] Worse, Elliotson, Ashburner and others found themselves regularly attacked by medical professionals and commentators for basing their claims for the reality and curative effects of the mesmeric fluid on the judgement of deceptive or potentially unreliable human subjects and on investigative methods that allowed for mesmeric effects arising from what subjects merely imagined or expected.

A further blow to claims for the reality of the mesmeric fluid was dealt by the popularity of the work of the Scottish surgeon James Braid. In the 1840s, he argued that the mesmeric sleep (which he sought to replace with a new term, 'hypnosis') was a physiological response produced when an individual fixed their gaze on a small bright object and which was entirely independent of any exterior magnetic fluid. Predictably, mesmerists challenged the capacity of Braid's theory to explain all cases of mesmerism, but they also appealed to evidence of another imponderable agency attacked by Braid and which seemed to surpass the magnetic fluid in helping to ally animal magnetism to the physical sciences.[38]

The Oddity of Od

Despite the hostility shown towards it in many quarters of the British medical and scientific establishment, mesmerism enjoyed a significant presence in early Victorian culture, from stage shows, sermons and popular tracts to dedicated hospitals, journals and treatises. It reflected and contributed to the turbulence in British political, religious and scientific as well as medical establishments, and its religious, political, scientific and medical significance changed according to the quarter in which it was experienced.[39] In the *Zoist*, a periodical launched by Elliotson in 1843, mesmerism was pushed forward as a medical therapy practised by professional elites and as a materialistic science of the mind centred on cerebral

---

[36] Rutter, *Human Electricity*, p. 165.   [37] Winter, *Mesmerized*, pp. 49–52.
[38] James Braid, *The Power of the Mind over the Body: An Experimental Enquiry into the Nature and Cause of the Phenomena Attributed by Baron Reichenbach and Others to a 'New Imponderable'* (London: John Churchill, 1846).
[39] Winter, *Mesmerised*.

physiology, and this contrasted with the more democratic and spiritual meanings of mesmerism upheld elsewhere.[40]

Elliotson well captured his journal's intellectual ambitions in an issue of 1846 when he praised a certain "philosopher" for writing a book that placed mesmerism "among the physical sciences" by showing that the mysterious animal magnetic influence obeyed laws similar to those describing other imponderables.[41] The author of the book, whose status as a "philosopher" rather than a mesmerist conferred additional weight on his claim, was Baron Karl von Reichenbach, a wealthy German industrial chemist already renowned for the discovery and manufacture of creosote and other coal-tar products.[42] For many Victorian readers, including some of the protagonists of this book, few mid-nineteenth-century natural philosophers deserved more praise than Reichenbach for struggling to bring an obscure force within the realm of physics.

The work reviewed by Elliotson was an English-language abridgement of a series of papers that had been published the year before in the *Annalen der Chemie und Pharmacie*, the prestigious scientific serial edited by the eminent German chemists Justus von Liebig and Friedrich Wöhler.[43] Later revised for publication as *Physicalisch-physiologische Untersuchungen über die Dynamide des Magnetismus* (1849–50), the papers embodied a long series of investigations into the existence and nature of a new force or power that only seemed to manifest itself to human subjects of a 'sensitive' nature and which Reichenbach, working mainly from his castle near Vienna, studied in a range of individuals, from those whose were physically healthy but melancholic to those suffering from such nervous disorders as somnambulism, catalepsy and hysteria.[44] Unable to sense the force himself,

---

[40] See Jennifer Ruth, '"Gross Humbug" or "The Language of Truth": The Case of the *Zoist*', *Victorian Periodicals Review*, vol. 32 (1999), pp. 299–323.

[41] Elliotson, 'Review', p. 122.

[42] See Ferzak, *Reichenbach* and Nahm, 'Sorcerer of Coblenzl'.

[43] Karl von Reichenbach, *Abstract of 'Researches on Magnetism and on Certain Allied Subjects'*, *including a New Imponderable*, translated and abridged from the German by William Gregory (London: Taylor and Walton, 1846).

[44] Karl von Reichenbach, *Physicalisch-physiologische Untersuchungen über Die Dynamide der Magnetismus, der Electricität, der Wärme, des Lichtes, der Krystallisation, des Chemismus in ihren Beziehungen zur Lebenskraft*, 2 vols. (Braunschweig: Friedrich Vieweg and Son, 1849–50). Two English translations of the second edition of this work were published in close succession, translated and edited by renowned British mesmerists: Karl von Reichenbach, *Researches on Magnetism, Electricity, Heat, Light, Crystallisation, and Chemical Attraction, in their Relations to the Vital Force*, translated and edited by William Gregory (London: Taylor, Walton and Moberly, 1850); Karl von Reichenbach, *Physico-Physiological Researches on the Dynamics of Magnetism, Electricity, Heat, Light, Crystallisation, and Chemism in Their Relations to the Vital Force*, translated and edited by John Ashburner (London: Hippolyte Baillière, 1850). Since the Gregory edition was the one most frequently cited by British physical-psychical scientists, it will be used throughout this book.

Reichenbach's main witnesses were a handful of women who reported having peculiar sensory responses to the physical world: they experienced a "gentle" but "unpleasant" physical sensation when magnets were passed over their bodies; they saw faint coloured luminous emanations around magnets, crystals and human bodies; they felt their hands drawn to magnets; they felt a strange coolness when exposed to the light of the sun and stars and a puzzling warmth when exposed to lunar rays; and they claimed peculiar sensations from other agents including electricity, heat, mechanical friction, artificial light and chemical activity[45] (Figure 1.2).

Reichenbach was convinced that the force, which he christened "od" but which was often called 'odyle' or the 'odic' force by his English translators, was either an entirely new one, the "modification" of an existing physical force, or a "complex" combination of existing forces.[46] Independently of these questions, it seemed to be ubiquitous in the physical world and had a complex relationship with known physical forces. On the one hand, it was closely associated with and shared many of the properties of existing physical forces: it always accompanied natural and artificial magnetic sources (including the earth's magnetism); it exhibited polar characteristics akin to magnetism and electricity; and it was propagated by radiation and conduction (like heat). On the other hand, it was not identical with any existing physical force: for example, it was not identical to magnetism or electricity because its presence was not limited to magnetic and electrical sources; and while it accompanied heat, it had no effect on thermometers and often induced opposite thermal sensations to heat.

The apparent tendency of od to follow "fixed physical laws" that matched and transcended those of existing imponderables was enormously exciting to Reichenbach because it represented a possible extension of the "domain of physics" and the opening of a "new leaf in the history of the Dynamides or Imponderables".[47] Since od was closely linked to vitality, it also had far greater potential than other imponderables to bring physiology closer to physics and to show the "unity of these imponderables in a higher form".[48]

Reichenbach's conceptions of od had obvious roots in Romantic ideas of nature that flourished in late-eighteenth- and early-nineteenth-century scientific, literary and artistic circles, particularly in the German lands of

[45] Reichenbach, *Researches on Magnetism*, p. 1.
[46] Reichenbach, *Researches on Magnetism*, p. 163.
[47] Reichenbach, *Researches on Magnetism*, pp. 3 and 59.
[48] Reichenbach, *Researches on Magnetism*, p. 164.

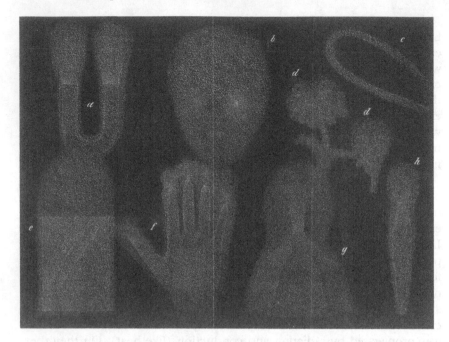

1.2 The luminous manifestations of 'od'. From *Karl von Reichenbach, Researches on Magnetism, Electricity, Heat, Light, Crystallisation, and Chemical Attraction, in Their Relation to the Vital Force,* translated and edited by William Gregory (London: Taylor, Walton and Moberly, 1850), plate III.

Reichenbach's youth and early career.[49] His belief in the close connection between physical and vital forces, the unity of imponderables in some "higher form" and the cosmic significance of polarity owed a great deal to one aspect of Romanticism: *Naturphilosophie*. It is also possible that Reichenbach's acceptance of the capacity of his sensitives to perceive occult features of the physical world owed something to the belief of *Naturphilosophen* in the power of genius and imagination to discern nature's hidden reality. Reichenbach was acutely aware of the danger of resting the case for od on human subjects, whose judgements might be impaired by poor physical and mental health. The goal of making od part of physics needed the discovery of a "universal inorganic reagent" for the force, a "means of recognising and measuring it" which could liberate

---

[49] For Romanticism and sciences see Andrew Cunningham and Nicholas Jardine (eds.), *Romanticism and the Sciences* (Cambridge University Press, 1990); Richards, *Romantic Conception of Life.*

students of the subject "from the frequently more than painful dependence on diseased persons, hospitals, and uncultivated people of every kind".[50]

Reichenbach's intellectual goals for od underpinned his ambivalence towards Mesmer and his followers. Not surprisingly, his critics and allies saw him as a latter-day Mesmer, propounding an updated theory of a universal but obscure force that was strongly associated with mineral magnetism and to which only certain individuals were sensitive. Reichenbach was certainly familiar with the literature on mesmerism, and his claims for od, especially its luminous manifestations, built partly on animal magnetism. But he was adamant that his work adopted the critical approach to the work of Mesmer and his followers that he believed would placate such formidable critics of mesmerism as the German physiologists Emile du Bois-Reymond and Johannes Müller.[51] While his researches confirmed the existence of a force that was concentrated in but not limited to magnetic sources, he deemed animal magnetism an "unfit" term because the phenomena associated with it did not exactly coincide with those "properly called Magnetism".[52]

Few aspects of Mesmer's work drew more criticism from Reichenbach than the closest thing Mesmer got to a therapeutic instrument: the *baquet*. This was a circular wooden tub containing bottles of 'magnetised' water, whose subtle influence was communicated via iron rods to patients forming a 'magnetic' chain around the perimeter of the vessel. Reichenbach was not alone in being highly circumspect about this attempt to create a collective form of magnetic therapy, not least because Mesmer sought to enhance the effect with rituals, music and darkness. As far as he was concerned, the only truth buried in this "mysterious superstructure" was the slow and continuous chemical action of the *baquet*, which, as he concluded from his own investigations, was itself a source of od which had distinct physiological effects.[53]

Reichenbach's discussion of the *baquet* in his *Physicalisch-physiologische Untersuchungen* led to telling examples of the way he sought to render od plausible in the context of recent physical discovery. Having found evidence that od was perceived in even feeble chemical reactions, Reichenbach considered it highly likely that chemical activity within the body was the source of

---

[50] Reichenbach, *Researches on Magnetism*, p. 60. His British champion William Gregory agreed. In 1851 he argued that storing od independently of the body and measuring it were important desiderata in addressing the "peculiar difficulties" surrounding the subject: William Gregory, *Letters to a Candid Inquirer on Animal Magnetism* (London: Taylor, Walton, and Moberly, 1851), p. 292.

[51] Reichenbach, *Researches on Magnetism*, pp. xxiii–xliv.

[52] Reichenbach, *Researches on Magnetism*, pp. 62–3.

[53] Reichenbach, *Researches on Magnetism*, p. 121.

its remarkable capacities for od. Not surprisingly for a work first published in the *Annalen der Chemie*, Reichenbach emphasised that a "guarantee of the essential truth" of his own "observations and deductions" was the fact that they seemed to converge with Liebig's far better-known research on the relationship between chemical activity and the vital functions, and in particular the roles of respiration and digestion in the production of heat and muscular power – the same vital functions that Reichenbach believed also yielded od.[54]

Another convergence was sought between od and Faraday's recent work on diamagnetism, which referred to the susceptibility of all bodies, whether magnetic or not, to an external magnetic field. Annoyed that the British natural philosopher had apparently ignored his researches on od, Reichenbach insisted that they were actually "drawing the same vehicle, but by different ropes".[55] Published in 1845, Faraday's research would certainly have intrigued Reichenbach because it derived from Faraday's discovery of an effect – the magnetic rotation of the plane of polarised light – lending credence to the magneto-optic connection that Reichenbach's sensitives perceived in the form of odic luminosity around magnets.[56] But Reichenbach's principal interest in diamagnetism was that, like od, it seemed to be a power shared by animate and inanimate matter, although it was not clear whether diamagnetism was a manifestation of od or whether both derived from a still-higher power. Either way, he anticipated that Faraday's "fertile genius" would unravel the mystery of universal powers.[57]

For all their intended rhetorical power, Reichenbach's connections between od research and recent physical discoveries sidestepped the critical problem of employing human beings as the principal instruments or 'reagents'. Although he appealed to the fact that patients' testimony was a necessary feature of medical discovery, he evidently felt that what his experimental subjects reported about od needed bolstering in other ways. His least successful response to this problem was to try to photograph, at some time in the late 1840s, the perceived magnetic luminosity. With the help of a Viennese photographer, he established that a daguerreotype inside a light-tight box had become fogged by a long exposure to a strong magnet. If "other causes" of the fogging could be ruled out, Reichenbach concluded, then the plate must have been

[54] Reichenbach, *Researches on Magnetism*, p. 123.
[55] Reichenbach, *Researches on Magnetism*, p. 229.
[56] On Faraday's work see David Gooding, 'A Convergence of Opinion on the Divergence of Lines: Faraday and Thomson's Discussion of Diamagnetism', *Notes and Records of the Royal Society of London*, vol. 36 (1982), pp. 243–59; Frank A. J. L. James, '"The Optical Mode of Investigation": Light and Matter in Faraday's Natural Philosophy', in Gooding and James, *Faraday Rediscovered*, pp. 137–61.
[57] Reichenbach, *Researches on Magnetism*, p. 229.

exposed to a *"real light"* flowing from the magnet.[58] Although he was satisfied that the plates had not been exposed to other sources of light, Reichenbach did not deem the result decisive and only returned to the problem in the early 1860s.[59] Nevertheless, his original photographic test would certainly pique the interest of other scientific investigators, including many British physical–psychical scientists revisiting the subject from the 1870s onwards.

Reichenbach achieved more success with a far simpler series of tests, which were designed to eliminate the possibility that his experimental subjects were deceiving him or themselves, notably by using clues in their environment rather than a genuine odic sensitivity. One test was prompted by the fact that an individual he studied particularly closely, Leopoldine Reichel, claimed to perceive the image of a magnetic 'flame' focussed onto the wall of a darkened room by a glass lens. Since the image was invisible to Reichenbach and his assistants, he decided to conduct a more stringent test of her powers, which involved her pointing to the place where the image fell as the lens was silently and repeatedly moved. Reichel's ability to correctly identify the different places where Reichenbach believed the image must have fallen confirmed her abilities "beyond a doubt".[60]

Reichenbach's most conspicuous strategy regarding his sensitives, however, responded to criticism that he had relied far too heavily on Reichel and four other young women who, despite passing crucial experimental tests, were fundamentally unreliable as observers because of the nervous disease that made them strongly sensitive to od. For this reason, *Physicalisch-physiologische Untersuchungen* presented evidence of odic sensitivity in a much larger sample of people (59), a large proportion of whom were not only physically healthy but drawn from the middling and highest social ranks of Viennese society, including baronesses, university professors and daughters of tradespeople. Not surprisingly, these individuals featured heavily in Reichenbach's more detailed studies of odic luminosity, which included examinations of the deflection of odic flames by crystals and human hands, the changing colour of the odic luminosity of a bar magnet turned in the earth's magnetic field, and the striking parallel between the odic colours of a magnetised iron sphere and the aurora. Having accepted the potential of od to unify a wide range of phenomena in nature, Reichenbach concluded from this

---

[58] Reichenbach, *Researches on Magnetism*, p. 17.
[59] Karl von Reichenbach, *Odische Begebenheiten zu Berlin in dem Jahren 1861 und 1862* (Berlin: G. H. Schroeder, 1862), pp. 6–8.
[60] Reichenbach, *Researches on Magnetism*, p. 18.

last observation that the earth's polar lights were a "vast manifestation" of odic light whose sheer power ensured that they were not just visible to sensitives.[61]

The immediate medical and scientific reactions to Reichenbach's work in the German-speaking lands were generally unfavourable. In 1846, a committee of Austrian physicians failed in their attempt to replicate Reichenbach's results and questioned the judgement and honesty of their experimental subjects; similar criticisms against Reichenbach were made by Liebig, who had been enthusiastic enough to publish the first of his papers on od, but who by 1852 was no longer convinced by the credibility of the "science of od" because it depended on observers who possessed unreliable sensory and nervous apparatuses.[62] This fundamental problem had long been aired in the British medical press, which was generally no more sympathetic to Reichenbach than it had been towards Mesmer and his followers. Reviewing the English-language *Abstract* of Reichenbach's treatise, the *Lancet* lambasted the "hysterical young women" employed by the chemist as the "most suitable subjects for the development of shams" and erroneous judgements of sensory experiences and, recalling the ferocious battle it had waged against mesmerism since the late 1830s, placed Reichenbach beside Elliotson as a perpetrator of "disguised quackery".[63]

The English translations of *Physicalisch-physiologische Untersuchungen* prompted even more trenchant criticisms from the British medical press. For many medical commentators, the case for od was undermined by the fact that Reichenbach and his chief English-language translator and champion, the Scottish academic chemist William Gregory, were physical scientists who seemed to lack the knowledge of physiology and psychology that, as the *British and Foreign Medico-Chirurgical Review* charged, was essential to a "right investigation of the phenomena".[64] For this latter reviewer, the "fatal" gap in Reichenbach's "high character as an inductive and experimental philosopher" was his failure to understand the nature and reliability of his

---

[61] Reichenbach, *Researches on Magnetism*, p. 455.

[62] Justus von Liebig, *Über das Studium der Naturwissenschaften. Eröffnungsrede zu seinen Vorlesungen über Experimental-Chemie im Wintersemester 1852/53* (Munich: Cotta, 1852), pp. 18–19. On the German reception see William H. Brock, *Justus von Liebig: The Chemical Gatekeeper* (Cambridge University Press, 1997), pp. 66–7; Ferzak, *Reichenbach*, pp. 120–8; Nahm, 'Sorcerer of Coblenzl', pp. 391–401.

[63] [Anon.], 'Reviews', *Lancet*, vol. 2 (1846), pp. 103–4.

[64] [Anon.], 'Odyle, Mesmerism, Electro-Biology', *British and Foreign Medico-Chirurgical Review*, vol. 8 (1851), pp. 378–431, p. 385. See also [Anon.], 'Gregory's Edition of Reichenbach', *Medical Times*, vol. 21 (1850), pp. 451–2.

"*instruments of research*" and to recognise his unconscious role in determining what such human apparatus sensed.[65] A good deal of this criticism appealed to Braid's work on hypnotism, which yielded powerful
evidence of the way that individuals could be made to experience the
tactile and visual sensations claimed by Reichenbach's sensitives simply
as the result of verbal suggestions and in the absence of magnets and
crystals.[66] For Braid and other critics, Reichenbach's descriptions of his
experiments simply failed to rule out the possibility that he had inadvertently led his highly suggestible subjects towards their observations.

Gregory was one of several mid-Victorian scientific and medical practitioners who, in opposition to this hostility, upheld Reichenbach's work
and used it in their writings on the credibility and therapeutic benefits of
animal magnetism.[67] Their independent studies of odic sensitivity,
coupled with Reichenbach's scientific reputation and the sheer quantity
of his empirical evidence for odic sensitivity, lent powerful support to
Mesmer's original discovery. However, British mesmerists diverged from
Reichenbach himself in the extent to which they believed od was relevant
to mesmeric phenomena: Gregory well exceeded the limits of
Reichenbach's speculations in proposing od as the possible mechanism
of two psychological powers associated with the mesmeric trance –
thought-reading and clairvoyance – while the physician John Ashburner
believed that Reichenbach had unfairly neglected using mesmerised sensitives who, as Ashburner's own experiments revealed, possessed significant odic sensitivity.[68] Gregory's use of Reichenbach was hardly
surprising given mesmerists' preoccupation with the subtle physical
means by which psychological powers extended beyond the material
brain, and it was precisely this function that would attract so many
spiritualists and psychical researchers to the question of od later in the
nineteenth century.

Reichenbach's researches were ignored rather than explicitly criticised
in the most prestigious British scientific publications, but in some scientific quarters they were certainly considered worthy of critical or sympathetic comment. The young John Tyndall was not the only scientific
practitioner who privately shared Liebig's doubts about Reichenbach's

[65] [Anon.], 'Odyle, Mesmerism, Electro-Biology', pp. 388–9.
[66] Braid, *Power of the Mind.*
[67] Gregory, *Letters to a Candid Inquirer*, pp. 247–84; Herbert Mayo, *On the Truths Contained
in Popular Superstitions with an Account of Mesmerism* (London: William Blackwood,
1851), pp. 11–16; John Ashburner, 'Preface', in Reichenbach, *Physico-Physiological
Researches*, pp. vii–xx.
[68] Gregory, *Letters*, pp. 285–319; Reichenbach, *Physico-Physiological Researches*, footnotes
on pp. 11–13, 75–7.

claims.[69] That prominent forum of intellectual and scientific comment, the *Athenaeum*, was more equivocal insofar as it questioned the scientific judgement of Reichenbach's witnesses but expressed great confidence in his abilities and in the possibility that with further investigations employing more reliable observers, an important contribution could be made to the understanding of the correlation of the physical and vital forces.[70] Scientific reputation weighed even more heavily with the *Mechanics' Magazine*, which in 1851 pointed to the scientific stature of Reichenbach, as well as of sympathisers Gregory and the Swedish chemist Jöns Jacob Berzelius, as reasons for defying the British medical profession's notorious contempt for animal magnetism. Part of the defiance was repudiating the argument that Reichenbach, like Mesmer and Elliotson, had at best inadvertently contributed to the study of physiology and psychology by showing how scientific experts and their human subjects perpetuated false beliefs: on the contrary, the *Mechanics' Magazine* had no doubt that Reichenbach had helped make animal magnetism a "new branch of physical enquiry".[71]

By the early 1860s, some British scientific commentators had fresh reasons for sharing the *Mechanics' Magazine*'s optimism. In 1861, Reichenbach seemed to be moving closer to his ambition of bringing od within the remit of physics by getting a new instalment of his research published in the distinguished German scientific journal, the *Annalen der Physik*.[72] The paper focussed on the different sources of phosphorescence, including crystallisation, fusion, fermentation and, most tellingly, the human body. His experimental evidence for human phosphorescence was clearly drawn from ongoing od researches, but references to od were strategically omitted and the work was linked to existing debates in the *Annalen* and in physics more generally on the relationship between molecular movement and imponderable forces.[73]

Reichenbach's success with German physicists, however, was short lived, and from this point until his death in 1869 he encountered stiff

[69] See note 8. In 1846 Faraday was reputedly "not disposed to place faith in the magnetic experiments of Reichenbach" or mesmerism until effects could be shown with "inorganic matter or a baby" – whose responses to new forces could not possibly derive from knowledge of what to expect: Walter White, *The Journals of Walter White* (London: Chapman and Hall, 1898), p. 69.

[70] [Anon], Untitled review of Reichenbach's *Researches on Magnetism*, *Athenaeum*, 19 October 1850, pp. 1088–90.

[71] [Anon.], 'Gregory's "Letters on Animal Magnetism"', p. 370.

[72] Karl von Reichenbach, 'Zur Intensität der Lichterscheinungen', *Annalen der Physik und Chemie*, vol. 112 (1861), pp. 459–68.

[73] On the contents of the *Annalen* in this and later periods see Christa Jungnickel and Russell McCormmach, *Intellectual Mastery of Nature: Theoretical Physics from Ohm to Einstein*, 2 vols. (Chicago University Press, 1986), vol. 2, chapter 13.

opposition from leading savants who, having accepted his invitations to witness new attempts at odic photography and related experiments, maintained their doubts about the reality of the new force.[74] As we have seen, Reichenbach enjoyed a more sympathetic hearing in some sections of the British scientific and technical press, where, in agreement with the *Electrician* author with which we began this chapter, writers directly or indirectly encouraged further investigations into od. In 1862, for instance, an anonymous writer in the *Popular Science Review* (probably William Crookes) pointed out that although readers were entitled to be circumspect about those statements of Reichenbach that had been greeted with "incredulity", he remained "one of the first chemists and physicists of the day, and his researches in this 'occult' science" were "characterised by equal philosophical acumen with his chemical experiments".[75] What made a less incredulous approach to od particularly pressing by this time was that it promised to explain aspects of another 'occult' science, but one that had become the talk of Victorian society.

Outdoing the Electric Telegraph

If there was one area of agreement between Reichenbach's supporters and critics it was that od was related to occult phenomena other than just mesmerism. It was often lumped together with the divining rod (typically a forked twig held in the hands that appeared to move near hidden sources of water independently of the will), the puzzling movements of a pendulum bob suspended from a stationary finger, and the crazes for table-turning and spirit-rapping that swept across the United States, Continental Europe and Britain from the early 1850s.

In table-turning, groups of people gathered around tables and, after placing their fingers lightly on table-tops, observed the furniture rotating seemingly independently of their volition. Electricity, magnetism, a new physical force, disembodied spirits and the Devil were offered as possible causes. Spirit-rapping was the ability of professed spirits of the dead to commune with the living via messages encoded as rapping noises on furniture and other objects. The manifestation of raps tended to require the presence of individuals, later called 'mediums', whose bodily and mental constitutions made them especially susceptible to otherworldly influences. Od offered a possible explanation of these effects insofar as it

---

[74] Reichenbach's defensive account of these battles is his *Odische Begebenheiten*.
[75] [Anon.], 'Chemistry', p. 388. See also [Anon], 'Obituary', *Chemical News*, vol. 19 (1869), p. 82. See also [Anon.], 'Scientific Gossip'; [Anon.], 'Swedenborg'.

was the imponderable channel through which the unconscious human will could cause mechanical effects beyond the body, or the invisible carrier of communication between the living and the dead.[76]

Table-turning and spirit-rapping were early phases of that iconic aspect of nineteenth-century occultism: spiritualism. Emerging in the United States in the late 1840s, spiritualism spread to Continental Europe, Britain, Russia, Australia and elsewhere and reached the peak of its popularity in the final quarter of the century, when the number of followers had swelled to several millions.[77] It was primarily a culture focussed on the production, interpretation and promulgation of evidence that the human spirit or soul survived the dissolution of the material body, communed with and manifested itself to the living, and experienced moral and spiritual progress in the next state of existence. Although communion with spirits of the dead had been practised for millennia, spiritualism approached the question via distinctive practices and increasingly startling physical and psychological phenomena.

One of spiritualism's most distinctive practices, spirit-rapping was developed in the founding events of 'Modern Spiritualism', a term used by many spiritualists to distinguish what they did from older forms of spiritual communion. The events took place in 1848 at Hydesville, New York, where three teenage sisters, Katherine, Leah and Margaretta Fox, appeared to be able to communicate with mysterious rapping noises in their family home. The source of the rapping seemed to be intelligent because it imitated the girls' finger-snapping and clapping noises and revealed information about itself by sounding one rap for 'no' and two for 'yes' in response to vocalised questions. Relatives and friends of the Fox sisters had several reasons to accept the genuineness of the girls' spiritual powers and of the communicating intelligence. The rapping noises could not be easily ascribed to any known natural cause (including

---

[76] Edward Coit Rogers, *Philosophy of the Mysterious Agents, Human and Mundane* (Boston: John P. Jewett and Company, 1853), esp. pp. 171–203; George Sandby, 'The Mesmerisation and Movement of Tables', *Zoist*, vol. 11 (1853–4), pp. 175–85.

[77] By the 1860s, the number of spiritualists in the United States and Britain was reputedly in the millions and thousands respectively: Gauld, *Founders of Psychical Research*, pp. 29 and 77. The literature on Modern Spiritualism is vast. The more analytically sophisticated studies include Barrow, *Independent Spirits*; Braude, *Radical Spirits*; Cathy Gutierrez, *Plato's Ghost: Spiritualism in the American Renaissance* (New York: Oxford University Press, 2009); Sophie Lachapelle, *Investigating the Supernatural: From Spiritism and Occultism to Psychical Research and Metapsychics in France, 1853–1931* (Baltimore, MD: Johns Hopkins University Press, 2011); John Warne Monroe, *Laboratories of Faith: Mesmerism, Spiritism and Occultism in Modern France* (Ithaca, NY: Cornell University Press, 2008); Oppenheim, *Other World*; Owen, *Darkened Room*; Diethard Sawicki, *Leben mit den Toten: Geisterglauben und die Enstehung des Spiritismus in Deutschland 1770–1900* (Paderborn: Ferdinand Schöningh, 2002).

trickery) and the intelligence revealed information unknown to anybody present and which proved to be correct (it claimed to be the spirit of a man murdered in the house years earlier, and whose remains were soon discovered in the cellar of the property).

The much-publicised displays of spirit-rapping that the Fox sisters subsequently staged in New York initiated the spread of Modern Spiritualism in the United States and promulgated two further key elements of spiritualist practice. These were the presence of 'mediums' and the staging of spirit 'circles' or seances, by which small groups of individuals gathered, typically in the presence of a known medium and in a dimly lit room, to contact the professed denizens of the other world. As spiritualism spread, it came to be associated with a plethora of other, equally striking physical and psychological phenomena. Mediums seemed to be able to move objects and play musical instruments without touching them; to levitate themselves and handle hot coals; to write, draw and speak under the guidance of spirits; to cause spirits to directly write and draw, and to communicate via wooden 'planchettes' and ouija boards; and to produce phosphorescent lights, cool breezes and images of spirits on photographic plates. Most spectacular and controversial of all, mediums seemed to partially or fully 'materialise' the bodies of spirits, and fully formed varieties were even able to walk and talk like living people (Figure 1.3).

The increasingly startling nature of spiritualistic phenomena ensured that it catered to burgeoning tastes for magical and 'supernatural' entertainment but also to medical and scientific interests in obscure powers of the human mind and body.[78] But spiritualism proved popular for a host of other and more commonly shared reasons. As a culture focussed on otherworldly interventions, it served popular fascinations with ghosts, haunted houses and other preternatural phenomena that the critical theological, philosophical and scientific arguments of the Reformation and Enlightenment had not vanquished.[79] It consoled myriad bereaved individuals with opportunities to contact deceased loved ones, and it gave others answers to questions about the existence and nature of the post-mortem state that were more satisfactory than those offered by established religions, philosophies and sciences.

The teachings and practices that came to define spiritualism built partly on the existing cultures of American Universalist religion (which preached the salvation of all, irrespective of earthly sins), Swedenborgianism (which

[78] On spiritualism and popular entertainment see Simone Natale, *Supernatural Entertainments: Victorian Spiritualism and the Rise of Modern Media Culture* (University Park, PA: Pennsylvania State University Press, 2016).
[79] See Owen Davies, *The Haunted: A Social History of Ghosts* (Basingstoke: Palgrave Macmillan, 2007).

1.3 A typical late-Victorian seance. After seating themselves at a table, people joined hands and observed such spectacular effects as untouched objects floating about and disembodied hands writing messages. From [Anon.], '"Spirits" and their manifestations. An evening séance', *Frank Leslie's Illustrated Newspaper*, 2 April 1887, p. 105. Reproduced by permission of Corbis Historical/Getty Images.

emphasised the close proximity of the earthly and spiritual realms), and mesmerism, from which spiritualism drew some of its personnel, techniques and language. Spiritualist mediumship conferred upon an individual many of the powers – notably thought-reading, clairvoyance and spiritual vision – that magnetised somnambules had long been achieving, and the means by which professed spirits influenced mediums were often believed to be the mesmeric fluid.[80]

Yet spiritualism catered to spiritual and religious needs in ways that mesmerism rarely did. It offered more powerful evidence that consciousness could exist without the material body and was accordingly embraced as an argument against materialist philosophies seeking to reduce humans and the cosmos to mere matter and motion. It presented evidence of that cornerstone of Christian faith – the existence of an

---

[80] [Anon.], *Heaven Opened; Or, Messages for the Bereaved, From Our Little Ones in Glory. Through the Mediumship of F. J. T.* (London: James Burns, 1870), p. 19; 'C. D.' [Sophia De Morgan], *From Matter to Spirit: The Result of Ten Years' Experience in Spirit Manifestations* (London: Longman, Green, Longman, Roberts and Green, 1863), p. 100.

afterlife – but connected it with beliefs and practices that many favoured over those associated with orthodox Christianity.[81] For example, spiritualism abolished hell as a distinct place of punishment and thereby responded to a moral revulsion many felt towards the idea of eternal damnation. It taught that the afterlife was an altogether happier place where the spirits of all individuals, irrespective of their earthly sins, experienced moral and spiritual progress through effort. It also emphasised individual approaches to, and sensuous forms of, spirituality, as opposed to those heavily mediated by orthodox Christian clergy or embodied in abstract theological concepts.

The source of spiritualism's appeal that is most relevant to the purposes of this study was its claim to be a new scientific approach to religion and spirituality, and the moral and ethical questions that followed from them. In 1856, Britain's first spiritualist newspaper quoted one medical follower's declaration that spiritualism was a "religion of works – not a passive, dead faith. Spiritualism is a science – a positive, practical, teachable science".[82] Spiritualists usually believed that they exuded the scientific spirit of the age and were merely applying the empirical, inductive and rational methods of enquiry that had proved so successful in understanding the material cosmos to questions of mind and spirit. As the leading English–American spiritualist Emma Hardinge put it in 1866, these questions had been answered satisfactorily neither by established religion, which required belief in God and spirit yet closed "against our spiritual eyes the realm of investigation", nor by the established sciences, which had "contentedly endured banishment to the realm of matter, dealt only with effects, and offered us systems which trace creation no farther than the visible universe conducts us".[83] By the systematic study of the psycho-physical phenomena of spirit circles, spiritualists believed they could elucidate laws of the mind and put together a "science of the soul" or a form of psychology to rival the physiological-based form being vigorously promoted in Britain with limited success.[84]

---

[81] See Georgina Byrne, *Modern Spiritualism and the Church of England, 1850–1939* (Woodbridge: Boydell Press, 2010), chapter 4.

[82] Mr Randall cited in [Anon], 'What Constitutes a Spiritualist?', *Yorkshire Spiritual Telegraph*, vol. 1 (1856), pp. 127–8, p. 127.

[83] Emma Hardinge, 'Psychology; Or, the Science of Soul', *Spiritual Magazine*, vol. 1 (New Series) (1866), pp. 385–401, p. 388. She became Emma Hardinge Britten in 1870.

[84] Hardinge, 'Psychology'. A similar argument is made in [Anon.], 'The Study of Human Nature', *Human Nature*, vol. 1 (1867), pp. 1–5. On the troubles of mid-Victorian psycho-physiology see Kurt Danziger, 'Mid-Nineteenth Century British Psycho-Physiology: A Neglected Chapter in the History of Psychology', in William R. Woodward and Mitchell G. Ash (eds.), *The Problematic Science: Psychology in Nineteenth Century Thought* (New York: Praeger, 1982), pp. 119–46.

Hardinge's ambition well reflected more widely shared spiritualist convictions that they, like other scientific enquirers, were merely trying to extend the realm of natural law – in their case, to phenomena whose 'supernatural', spiritual and psychological attributes had excluded them from the domain of the sciences.[85] But Hardinge's ambition also highlighted the anti-materialist stance that many spiritualists adopted towards natural law: spiritualistic phenomena followed laws that far transcended those of matter, force and purely physical qualities and which necessarily embraced mind and spirit.[86]

The seriousness with which spiritualists took the scientific status of their enterprise is evident in their borrowing and adaptation of the languages, concepts and theories of the established sciences. They often explained the interactions between spirit and matter in terms of electricity and the ether, both of which had long associations with religion and spirituality.[87] Seances were often described as groups of individuals whose combined 'vital' magnetism or electricity composed the 'battery' required by a spirit to manifest itself.[88] Spiritualists took a special interest in scientific achievements in the study and manipulation of imponderables because this work demonstrated the interconnectivity and power of agents far subtler than gross matter. Physiological studies of the connection between the nervous force and galvanic electricity lent plausibility to the idea that the body produced and was influenced by subtler forces that could be related to these more material forces.[89]

The immense strides made in extending the electric telegraph across continents and under oceans gave many spiritualists reason to think that communication by subtler forces would be no less successful. For one American spiritualist in the late 1840s, the Fox sisters were already doing this by showing that "God's telegraph has outdone Morse's altogether".[90] In 1860, and in the wake of the first attempts to lay submarine cables across the Atlantic, another American spiritualist could boast that

---

[85] See, for example, Robert Dale Owen, *Footfalls on the Boundary of Another World* (London: Trübner and Sons, 1860), pp. xi–xii, 42–58.

[86] Thomas Brevoir, 'The Religious Heresies of the Working Classes', *Spiritual Magazine*, vol. 6 (1865), pp. 29–32; William Howitt, 'A Letter from William Howitt', *Spiritual Magazine*, vol. 2 (1861), pp. 449–56.

[87] Ernest Benz, *The Theology of Electricity* (Eugene, OR: Pickwick Publications, 2009); Carroll, *Spiritualism in Antebellum America*, pp. 65–71.

[88] Emma Hardinge, 'Rules to Be Observed for the Spirit Circle', *Human Nature*, vol. 2 (1868), pp. 48–52; [De Morgan], *From Matter to Spirit*, p. 100. For discussion of spiritualists' electrical analogies and metaphors see Stolow, 'Spiritual Nervous System'.

[89] De Morgan, *From Matter to Spirit*, pp. 96–100.

[90] A. H. Jervis to E. W. Capron, circa 1849, cited in Emma Hardinge, *Modern American Spiritualism: A Twenty Years' Record of the Communion Between the Earth and the World of Spirits* (New York: Emma Hardinge, 1870), p. 51.

spiritualists' "modern study of the imponderables", already "productive" of astonishing results, would afford "glimpses of progress in another direction" that promised to outshine the "lightning-wire" joining the United States and Britain.[91]

By the early 1860s, however, many had reason to question the achievements of these telegraphs. The earliest attempts (in 1857–8) to span the Atlantic with submarine telegraph cables had proven costly failures, but the continued growth of overland and short undersea networks testified to the fact that public confidence in the technology per se had not been shattered.[92] Public confidence in the spiritual telegraph was far shakier. Challenging the estimated millions of spiritualist converts on both sides of the Atlantic were myriad individuals who were openly critical of or indifferent to spiritualism. The London Review echoed so many critics when, in 1862, it charged that "the mania for séances with spirits has passed like a disease from America to England", whose inhabitants had "no idea how infected American society is with spirit-rappers, spirit-mediums, spirit-orators, spirit-newspapers, and spirit-humbugs generally".[93]

The 'mania', which many critics feared had reached epidemic proportions, had serious religious, moral and philosophical implications. In sections of the British press, spiritualism was attacked as the sorry revival of old superstitions about supernatural visitations and witchcraft, and ridiculed as the displacement of traditional spiritual experiences with such vulgar, absurd and 'material' alternatives as spirits that rocked tables, gave erroneous information about the living and offered vague platitudes about the future life.[94] The revival of beliefs in the agency of spirits of the dead exasperated those who thought they were living in an age in which science and "matter-of fact" had eradicated the "love of the marvellous".[95] For these critics, the enlightened and matter-of-fact Victorians seemed to have forgotten the well-known argument against miracles proposed by the eighteenth-century Scottish philosopher David Hume: this proposed that it was more likely that testimony in favour of phenomena violating long-established natural laws was mistaken than

---

[91] Owen, Footfalls, p. 38.
[92] On mid-nineteenth-century electric telegraphy see Daniel R. Headrick, The Invisible Weapon: Telecommunications and International Politics, 1851–1945 (New York: Oxford University Press, 1991), chapters 2–3; Richard R. John, Network Nation: Inventing American Telecommunications (Cambridge, MA: Harvard University Press, 2010); Roland Wenzlhuemer, Connecting the Nineteenth Century World: The Telegraph and Globalization (Cambridge University Press, 2013).
[93] [Anon.], 'Spirit Rapping', London Review, 1 March 1862, pp. 206–7, p. 206.
[94] [Anon.], 'Howitt on the Supernatural', London Quarterly Review, vol. 21 (1863–4), pp. 27–70; [Anon.], 'Modern Spiritualism', Quarterly Review, vol. 114 (1863), pp. 179–210.
[95] [Anon.], 'The Mystery of the Tables', Illustrated London News, 18 June 1853, pp. 481–2.

that such laws needed to be abandoned. Christian-minded critics were more vexed by threats to Christian morality and authority: spiritualists had abjured scriptural warnings about exchanges with potentially deceptive spirits, abandoned hell as the ultimate source of moral sanction in the earthly life and chosen vulgar mediums rather than respectable clergymen as their spiritual guides.[96]

For many critics, the honesty of spiritualism's principal instruments weighed more heavily than these philosophical and theological concerns. Mediums had been associated with fraud almost from Modern Spiritualism's birth. In 1851, the Fox sisters were accused of faking 'spirit' rapping noises by surreptitiously cracking the joints in their knees and toes. Two years later, the American medium who brought spiritualism to Britain, Maria Hayden, was 'exposed' by the writer George Henry Lewes, who charged that her knowledge of dead persons known only to a particular seance-goer had nothing to do with her rapport with spirits and everything to do with her ability to exploit unconscious but telling hesitations of the participant as they used the alphabet method to decode spirit raps.[97]

Mediums who produced more physical effects were especially suspect because their feats more closely resembled what stage magicians claimed to be able to replicate without the agency of spirits. This was exactly the problem faced by Ira and William Davenport, two American mediums who, during their sensational tour of Britain in the mid-1860s, sparked a heated debate over public seances in which they appeared to play levitating musical instruments and cause spirits to speak whilst they were tied to chairs within a wooden cabinet. Many were baffled by these performances, but two newcomers to the world of conjuring, John Nevil Maskelyne and George Cooke, fuelled growing hostility towards the Davenports by claiming to reproduce the 'cabinet' manifestations using sleight-of-hand.[98]

Many of these 'exposures', however, were not decisive. Rumours about the Fox sisters did not deter Cromwell Varley, who, in the 1860s, clearly found it difficult to attribute to joint-cracking a deafening "chorus of raps" that he heard during a seance given by Katherine Fox.[99] Likewise,

[96] [Anon.], 'Howitt on the Supernatural'.

[97] George Henry Lewes, 'The Rappites Exposed', *Leader*, vol. 4 (1853), pp. 261–3. The 'alphabet' method involved composing messages by seance-goers calling out or pointing to, sequentially, letters of the alphabet to which 'spirits' rapped either yes or no.

[98] On nineteenth-century conjuring and spiritualism see Simon During, *Modern Enchantments: The Cultural Power of Secular Magic* (Cambridge, MA: Harvard University Press, 2002), chapter 5; Peter Lamont, *Extraordinary Beliefs: A Historical Approach to a Psychological Problem* (Cambridge University Press, 2013), chapter 4.

[99] Varley, 'Evidence of Mr. Varley', p. 165.

Lewes's damning verdict was not shared by the eminent British mathematician Augustus De Morgan, who, in seances given by Maria Hayden in 1853, was convinced that she could see "neither my hand nor my eye, nor at what rate I was going through the letters" of the alphabet, and either she or the "spirits" had been able to correctly answer questions that he had asked purely mentally.[100] The Davenports may have left Britain under a dark cloud, but some spiritualists responded to accusations of their trickery with an argument that they would make in response to the 'exposure' of other mediums: they had made more searching studies of the mediums over a longer period and these had failed to reveal fraudulent activity.[101]

What exasperated so many contemporary nineteenth-century people about mediumistic fraud was that the rules by which seances were conducted made it harder to detect. Typically, these rules or conditions were designed to optimise the vital magnetic powers in the 'circle' that spirits required to manifest themselves, although spiritualists warned novice enquirers that following the rules would not necessarily guarantee manifestations that were notoriously capricious. The rules related to physical and psychological conditions: the best circles, for example, were those held in warm, dimly lit rooms and which involved a small group of friends, family members or other individuals who could strike up an atmosphere of harmony, sympathy and mutual trust.[102] Individuals who were dogmatic, mischievous or strongly sceptical were usually excluded because they poisoned this psychological atmosphere. Professional scientists were often seen by spiritualists as particularly unpromising seance participants because their education had made them overly sceptical and altogether unable to adapt to the protocols of spiritualist scientific practice. As we shall see in the following section, however, spiritualists had many more reasons to think that 'scientific men' were more threatening than useful to their new science of imponderables.

'Scientific Men' and Spiritualism

One of the reasons why so many scientific practitioners had misgivings about spiritualism was because seance rules seemed to conflict with their ideas about fair conditions of scientific enquiry and

---

[100] 'A. B.' [Augustus De Morgan], 'Preface', in De Morgan, *Matter to Spirit*, pp. v–xlv, pp. xli and xliii.

[101] Robert Cooper, *Spiritual Manifestations, Including Seven Months with the Brothers Davenport* (London: Heywood and Son, 1867), pp. 215–19.

[102] Hardinge, 'Rules to Be Observed for a Spirit Circle'.

testing.[103] This was certainly one of the reasons why, in 1861, Faraday declined an invitation to a seance with Daniel Dunglas Home, the famous Scottish–American medium who in the 1850s had stunned Americans, Britons and Europeans with such feats as spirit-rapping, playing untouched musical instruments, self-levitation and handling hot coals.[104] For Faraday, "occult manifestations" had to be studied with the "strictest critical reasoning and the most exact and open experiment" that had yielded so many discoveries in natural philosophy. Among the conditions that Faraday insisted on being met was that Home himself "investigate as a philosopher" and have "no conceal-ments – no darkness – to be open in communication – and to aid inquiry all that he can".[105]

Faraday was not convinced that mediums and spiritualists would ever meet his conditions and this is why he later declined an invitation to a seance and asked John Tyndall, a fellow professor at London's Royal Institution, to go in his place. Tyndall's notorious account of this seance revealed that he was as dismayed as Faraday by spiritualists' apparent want of critical reasoning. He had no qualms about breaking the rules of seance conduct to highlight the self-deception at play: at one point he surreptitiously kicked the leg of the seance table and was bemused to report that one credulous participant ascribed the resulting tremor com-municated to his chair entirely to "spirits' work".[106]

Spiritualists did not think that a dogmatic attitude towards seance conditions or a mischievous approach to puzzling physical effects was becoming to the likes of Faraday and Tyndall, who should have dis-played a spirit of serious, open-minded enquiry. In 1868, for example, the leading spiritualist publisher James Burns criticised Tyndall for abandoning "his usual scientific method" in allowing himself, an "ignor-ant outsider" to spiritualism, to demand "superlative effects" to appear under conditions that more experienced investigators knew were detri-mental to the manifestation of such effects.[107] For spiritualists, the main

[103] See, for example, Henry Dircks, 'Science Versus Spiritualism', *Times*, 27 December 1872, p. 10; W. Matthieu Williams, 'Science and Spiritualism [1871]', in *Science in Short Chapters* (London: Chatto and Windus, 1882), pp. 237–51.

[104] The most balanced recent account of Home is Peter Lamont, *The First Psychic: The Peculiar Mystery of a Notorious Victorian Wizard* (London: Little, Brown, 2005).

[105] Michael Faraday to James Emerson Tennent, 14 June 1861, in Frank A. J. L. James (ed.), *The Correspondence of Michael Faraday Volume 6 November 1860–August 1867* (London: Institution of Engineering and Technology, 2012), pp. 106–7, p. 107.

[106] [Tyndall], 'Science and the Spirits', p. 725. Like Faraday, Tyndall's low opinion of spiritualists never changed: see John Tyndall to [Sved Hassan] El Medini, 23 August 1889, f. 30, Add. 41295, British Library.

[107] James Burns cited in [Anon.], 'Professor Tyndall and the Spiritualists', *Human Nature*, vol. 2 (1868), pp. 455–6, p. 455.

problem with Tyndall was that, like other scientists, he dogmatically treated spiritualism as a problem in physics, and was accordingly blind to the fact that, unlike purely physical enquiries, the success or failure of effects depended on the psychological state of participants.[108]

For all their differences, scientific and medical critics of spiritualism and spiritualists agreed on the importance of the mental state of spiritualist enquirers. But to explain why so many were falling prey to mediumistic trickery, scientific and medical critics turned increasingly to the psycho-physiological causes that had been invoked for mesmerism, Reichenbach's od and related phenomena. Few individuals represented this Victorian bulwark against Victorian spiritualism more volubly than William Benjamin Carpenter.[109] Trained as a doctor, Carpenter established a reputation as a leading authority on medicine and physiology primarily through textbooks, journalism, original research papers and academic positions in London. In the 1840s, he, the British physician Thomas Laycock and others spearheaded a physiological approach to psychology which sought to extend the material laws established to describe bodily behaviour to the mind. This approach built on earlier work finding that bodily responses to nervous stimuli could occur independently of the brain via an 'excito-motory' mechanism centred on the spinal cord. Carpenter argued that a higher level of the nervous system – the cerebrum – could also produce 'automatic' reflexes. If the directing power of the will – the highest level of nervous system – was temporarily absent then the cerebrum could reflect external impressions, sensations, ideas and emotions as 'ideo-motor' responses. As he remarked in 1852, an individual in this state had become a *"thinking automaton*, the whole course of whose ideas is determinable by suggestions operating from without"*, even if those suggestions involved false or irrational ideas normally dismissed by the power of the will.[110] An individual could reach this state in various ways: they could allow their thoughts to become dominated by particular ideas; they could give themselves up to a state of

---

[108] Emma Hardinge, 'The Scientific Investigation of Spiritualism', *Spiritual Magazine*, vol. 6 (1871), pp. 3–17.

[109] On Carpenter see Danziger, 'Mid-Nineteenth Century British Psycho-Physiology'; Vance D. Hall, 'The Contribution of the Physiologist William Benjamin Carpenter (1813–1885) to the Development of the Principles of the Correlation of Forces and the Conservation of Energy', *Medical History*, vol. 23 (1979), pp. 129–55; Roger Smith, 'The Human Significance of Biology: Carpenter, Darwin and the *vera causa*', in U. C. Knoepflmacher and G. B. Tennyson (eds.), *Nature and the Victorian Imagination* (Berkeley, CA: University of California Press, 1977), pp. 216–30; Winter, *Mesmerized*, pp. 287–305.

[110] William B. Carpenter, 'On the Influence of Suggestion in Modifying and Directing Muscular Movement, Independent of Volition [1852]', *Notices of the Proceedings of the Meetings of the Members of the Royal Institution*, vol. 1 (1851–4), pp. 147–53, p. 147.

reverie; or, following Braid's work on hypnosis, they could fix their attention on a small bright object.

The integration of mental reflexes into nerve mechanisms reflected Carpenter's ambition to turn psychology into a branch of physiology and thus to raise its scientific status. Cases of mesmerism, od, table-turning and spirit-rapping to which he devoted considerable attention furnished him with new insights into abnormal powers of the mind and further material for achieving this goal for the study of psychology. One of the earliest of many expositions of this argument appeared in his scathing anonymous review of works on mesmerism, od, table-turning and related subjects in an 1853 issue of the distinguished forum of intellectual debate, the *Quarterly Review*.[111] Carpenter argued that physiological and psychological phenomena ascribed directly to agencies *beyond* the body (for example, mesmeric fluids, od and disembodied spirits) were more likely to have been caused by ideas *within* the body – in the mind of the subject. Claims that mesmeric operators directly controlled the will of their subjects via a physical influence was vitiated by the lack of evidence for such a will being exercised independently of ideas about the supposed influence being inadvertently communicated to the subject. Mesmerised subjects were more likely to be automata of mesmeric operators because the mesmeric gaze and passes induced the very state of volitional abandonment that enabled their thoughts and actions to become directed by ideas suggested by operators. Reichenbach's subjects were no less vulnerable to delusive ideas: they were individuals whose "considerable powers of voluntary abstraction" made it possible for ideas about od inadvertently suggested by Reichenbach to produce physiological sensations ascribed to the alleged new imponderable.[112]

Table-turning and spirit-rapping were also best understood as only the latest examples of the deplorable consequences of ideo-motor action. Table-turners themselves, rather than spirits or demons, were the probable cause of the effect. The "dominant power" exerted on their minds and bodies by the very idea of tables turning caused them to push the tables via involuntary muscular action.[113] Witnesses to spirit manifestations – notably those ascribing table raps and movements to discarnate intelligences – were simply not reliable owing to the mental state into which they were

---

[111] [William B. Carpenter], 'Electro-biology and Mesmerism', *Quarterly Review*, vol. 93 (1853), pp. 501–57. Carpenter's critique also covered the topic of electro-biology, which was closely related to mesmerism. After staring at a bimetallic coin held in the hand, individuals were thrown into a state of mental and physical susceptibility to the will of an electrobiological operator. The effect was explained in terms of imbalances in the flow of electricity in the body.

[112] [Carpenter], 'Electro-biology and Mesmerism', p. 540.

[113] [Carpenter], 'Electro-biology', p. 547.

probably thrown by the "solemn expectancy" and darkness of seances.[114] In this state, they were likely to unconsciously produce the movements of tables corresponding to the dominant ideas they had about spirits or to unconsciously produce the revealing bodily gestures exploited by fraudulent mediums fishing for clues.

For Carpenter, mental education was the main remedy for the spiritualistic and other "epidemic disorders" of the mind because it could train the will to bring the power of reasoning to bear on the automatic tendencies of lower regions of the mind and nervous system.[115] But it was not just popular judgement that was a problem: some of the scientific practitioners to whom the public turned on occult matters, notably Gregory and Reichenbach, had no authority because they lacked the "philosophical discrimination" required in subjects that were "essentially physiological and psychological".[116] In the 1870s, Carpenter would have new reasons to reiterate this attack on physical scientists appearing to exceed their authority.

Not all physical scientists seemed to lack such discriminatory powers. Indeed, to support his theory of table-turning, Carpenter appealed to Faraday's recent and much-debated intervention on the subject. In the summer of 1853, and beleaguered by repeated requests for his verdict, Faraday communicated the results of his investigations to leading London newspapers. For at least one of the protagonists of this book, Faraday's decision to weigh into this early spiritualistic controversy set a powerful example of the right of natural philosophers or others with expertise in physical science to take the lead on strange physical phenomena capturing the public's attention.[117] What especially concerned Faraday was that the public mind was being exposed to explanations that struck him as either scientifically dubious or morally repugnant. He was sceptical of proposals that table-turning was due to electricity, magnetism, an unknown "physical power", the earth's rotation or some "diabolical or supernatural agency".[118] Faraday's decision to go public was prompted by a conviction that table-turners had failed to exercise proper scientific judgement and offended his deeply held Christian belief in the dangers of dabbling with potentially "unclean spirits".[119]

---

[114] [Carpenter], 'Electro-biology', p. 551.
[115] [Carpenter], 'Electro-biology', p. 556. See Winter, *Mesmerized*, pp. 294–300.
[116] [Carpenter], 'Electro-biology', p. 541.
[117] This was Crookes: see Crookes, *Psychic Force and Modern Spiritualism*, pp. 4–6.
[118] Faraday, 'Table-Turning'.
[119] Cantor, *Michael Faraday*, p. 149; Frank A. J. L. James, *Michael Faraday: A Very Short Introduction* (Oxford University Press, 2010), p. 100.

In his public intervention, however, Faraday played down his moral repugnance and represented table-turning as a regrettable public problem to which the methods of "physical investigation" could be decisively applied.[120] Employing such methods in several table-turning sessions, he reported failing to detect electrical, magnetic and 'attractive' forces. Moreover, he constructed simple mechanical devices showing that table-turners' hands moved before the tables (suggesting that they dragged the table, not vice versa) and that they involuntarily exerted a horizontal force, even when they were convinced that they only pressed downwards. The most elaborate of the devices, which Faraday strategically displayed in a leading London instrument-maker's shop window, comprised two horizontal wooden platforms that rolled on each other via glass cylinders, the relative motion of the platforms being magnified by the motion of long straw indicators attached to the platforms. When table-turners rested their hands on the platform but were prevented from seeing the indicators, the indicators moved in the direction in which they expected the table to move, but when they could see the indicators, the indicators failed to move, showing table-turners' ability to correct the muscular forces that they had been unconsciously exerting in the other scenario. Without this visual evidence of their own agency, table-turners became slaves to the illusory idea that external agencies were responsible.

Faraday linked his interpretation to Carpenter's psycho-physiological theory and, despite embodying a "physical" approach to table-turning, shared the physiologist's conclusion that the subject had more to do with psychology and morality than with natural philosophy.[121] The situation testified to the public's woeful lack of educated judgement and lack of deference to experts on the mind.[122] Yet both Faraday and Carpenter would be frustrated in their ambition to deliver decisive blows against a subject that challenged their sense of moral and intellectual propriety. Many medical and scientific practitioners, as well as critics of spiritualism, welcomed their interventions and helped give them a prominent place in mid- to late-nineteenth-century debates on epidemic delusions and unconscious powers of the mind.[123] But there were many others who, from the mid-1850s onwards, challenged the ideo-motor explanation

---

[120] Michael Faraday, 'Experimental Investigation of Table-Moving', *Athenaeum*, no. 1340 (2 July 1853), pp. 801–3, p. 801.

[121] Faraday, 'Experimental Investigation', p. 803.

[122] Michael Faraday, 'Observations on Mental Education [1854]', in Michael Faraday, *Experimental Researches in Chemistry and Physics* (London: Richard Taylor and William Francis, 1859), pp. 463–91.

[123] S. E. D. Shortt, 'Physicians and Psychics: The Anglo-American Medical Response to Spiritualism', *Journal of the History of Medicine and Allied Sciences*, vol. 39 (1984), pp. 339–55; Winter, *Mesmerized*, pp. 287–305.

upheld by Faraday and Carpenter. A common criticism was that it failed to explain how tables had moved with only light pressure being applied or with no bodily contact at all.[124] Spiritualists were not alone in expressing such doubts. A contributor to the *Mechanics' Magazine*, for example, regretted that Faraday had not determined whether the force attributed to table-turners was *sufficient* to turn the tables and had inadvertently hindered the study of a "principle" that was potentially "precious" to science.[125] Faraday's verdict would certainly persuade William Thomson, but not the British telegraph engineer Latimer Clark, who in 1857 told Faraday about his experiences of seances in which mere finger contact had caused heavy tables to tilt, and which he denied could be put down to self-delusion and trickery.[126]

The case for a better 'physical' investigation of this "precious" subject gained a far weightier advocate in the American academic chemist Robert Hare. In *Experimental Investigations into Spirit Manifestations* (1855), a work hailed by many spiritualists as the most scientific approach to their subject to date, Hare explained that in 1853 he had accepted Faraday's explanation of table-turning and repudiated the electrical theory because the human body could not produce electric currents and, even if it could, tables could not store enough electricity to cause the rotation.[127] However, Hare's confidence in the theory of unconscious muscular action was undermined after attending numerous seances where he was convinced that he had seen objects moving without being touched. The motion seems to have been initiated by disembodied spirits working through a medium and which had also persuaded him of their genuineness based on their ability to convey information to him that nobody present at the seances could have known.

To conclusively rule out trickery in these physical and psychological feats, Hare followed Faraday's example of introducing mechanical devices into a site of spiritualistic enquiry. One device was designed to

---

[124] J. H. Powell, *Spiritualism: Its Facts and Phases* (London: F. Pitman, 1864), pp. 83–4; Epes Sargent, *Planchette, Or, the Despair of Science* (Boston: Roberts Brothers, 1869), p. 15.

[125] [Anon.], 'Professor Faraday and Table-Turning', *Mechanics' Magazine*, vol. 59 (1853), pp. 23–5, p. 23.

[126] Silvanus P. Thompson, *The Life of William Thomson, Baron Kelvin of Largs*, 2 vols. (London: Macmillan and Co., 1910), vol. 2, p. 1105; Latimer Clark to Michael Faraday, 29 April 1857, in Frank A. J. L. James (ed.), *The Correspondence of Michael Faraday Volume 5 November 1855–October 1860* (London: Institution of Engineering and Technology, 2008), pp. 221–3. Cf. the science journalist William H. Harrison who later criticised Faraday's theory as "all stuff & nonsense": William H. Harrison to Balfour Stewart, 1868, PRO BJ1/25, Records of the Kew Observatory, National Archives.

[127] Robert Hare, *Experimental Investigation of the Spirit Manifestations* (New York: Partridge and Brittan, 1855), p. 35.

measure the force flowing from a medium's hands on a wooden board, direct contact being eliminated by positioning the hands in a vessel of water resting on the board. The significant result that Hare managed to replicate several times – the medium managed to produce a force of 18 pounds without appearing to experience any mechanical reaction – suggested the direct mechanical effect of spirits. While Hare's argument persuaded few scientific readers of the 1850s, it certainly made an impact on later scientific enquirers such as Crookes, Varley and the naturalist Alfred Russel Wallace.[128]

When interpreting the results of his experimental investigations, Hare shared the reservations of many spiritualists towards odic and similar theories of spirit manifestations. Od, like the mesmeric fluid and better-established imponderables, could not explain the intelligence of the force behind most of these manifestations, including that causing another of Hare's mechanical devices to spell out 'spirit' messages via a pointer moving around letters on a dial.[129] Even physical tests such as Hare's confirmed that spirit manifestations seemed to be intractably psychological in nature. As we have seen with Carpenter, the problem faced by many enquirers into spiritualism was whether the intelligence behind spirit manifestations was associated with a material brain or was entirely disembodied. For some commentators on spiritualism, including many British mesmerists, spirit manifestations were entirely under the control of an embodied mind, even if that mind exerted an influence beyond the material brain via a magnetic fluid.[130]

This was also the conclusion reached by the eminent French statesman Count Agénor de Gasparin, who, in an 1854 work later praised by Crookes, reported myriad cases of heavy objects levitating without physical contact and responding to the human will that could not be put down to fraud or ideo-motor action.[131] Yet despite impressing readers

[128] Hare, *Experimental Investigation*, pp. 49, 164–5, and text facing plate 3; William Crookes, 'Some Further Experiments in Psychic Force', *Quarterly Journal of Science*, vol. 1 (1871), pp. 471–93, p. 477; Cromwell F. Varley, 'Psychic Force', *English Mechanic and World of Science*, vol. 14 (1872), pp. 454–5; Alfred Russel Wallace, *The Scientific Aspect of the Supernatural* (London: F. Farrah, 1866), pp. 33–5.

[129] Hare, *Experimental Investigation*, pp. 392–4. Cf. William Howitt to G. H. Forbes, 9 May 1861, cited in Thomas P. Barkas, *Outline of Ten Years' Investigations into the Phenomena of Modern Spiritualism* (London: Frederick Pitman, 1862), pp. 148–60.

[130] Crabtree, *Mesmer to Freud*, pp. 245–65. These mesmerists included John Elliotson, who, a philosophical materialist, bitterly condemned the "delusion" about spirits, which he classed as "supernatural fancies" having no place in "philosophical work". See John Elliotson, 'The Departed Spirits', *Zoist*, vol. 11 (1853–4), pp. 191–201, p. 200. Seances with Home in 1861, however, persuaded him of the genuineness of disembodied spirits.

[131] Comte Agénor de Gasparin, *Des tables tournantes, du surnaturel en général et des esprits*, 2 vols. (Paris: E. Dentu, 1854). Crookes, 'Some Further Experiments', pp. 476–7. On Gasparin and the context of French table-turning see Monroe, *Laboratories*, chapter 1.

with the stringency of its tests and its measurements of the mechanical strength of levitating force, Gasparin's approach would not have satisfied thousands of seance-goers who were convinced that *disembodied* minds had to be behind the physical effects because such agents also relayed information that they were satisfied could not have been obtained by any embodied mind.

By the 1860s, sections of the British press were in no doubt that spiritualism remained an acute moral and scientific problem given the estimated millions of converts across the globe.[132] The interventions by Carpenter, Faraday and other scientific practitioners had simply failed to vanquish what many deemed a mixture of delusion and imposture. In 1861, the *North British Review* contended that spiritualism merited a "more philosophical and scientific examination than it has yet received", while over a decade later the London *Times* criticised "our scientific men" for "signally failing to do their duty by the public, which looks to them for its facts", including facts that would "decide a prejudiced controversy".[133] These pleas were partly prompted by the increased cultural profile that spiritualism had gained in Britain during the 1860s. In addition to the much-publicised activities of Home, the Davenport brothers and other visiting American mediums, spiritualism now boasted an expanding number of home-grown mediums, a handful of dedicated periodicals and organisations, a plethora of private spirit circles and a growing list of distinguished converts, including the Tory peer and archaeologist Lord Adare, the publisher and author Robert Chambers and the physician James Manby Gully.

Spiritualism continued to divide the scientific practitioners whose approach to the subject disappointed so many journalists. Two of Britain's most senior natural philosophers – Faraday and David Brewster – had no time for the subject because their negative experiences (of table-turning and Home seances respectively) persuaded them that further enquiries would only confirm the delusive, banal and morally pernicious nature of spirit manifestations.[134] Many younger scientific savants agreed: Thomas Henry Huxley notoriously refused to join a systematic enquiry into spiritualism launched by the London Dialectical Society because he had no time or inclination to investigate

---

[132] Estimate given in [Anon.], 'Modern Necromancy', *North British Review*, vol. 67 (1861), pp. 110–41, p. 110.

[133] [Anon.], 'Modern Necromancy', p. 110; [Anon.], 'Spiritualism and Science', *Times*, 26 December 1872, p. 5.

[134] Michael Faraday to Robert Cooper, 31 January 1863, in James, *Correspondence of Michael Faraday Volume 6*, p. 290; [David Brewster], 'Pretensions of Spiritualism – Life of D. D. Home', *North British Review*, vol. 39 (1863), pp. 174–206.

the banal "chatter" of spirits or the "twaddle" of mediums.[135] But a growing number of scientific savants challenged such dismissive attitudes. These included Carpenter and his closest scientific and medical allies, who, during their battles against popular delusions, accepted that spiritualism was a legitimate branch of psycho-physiological enquiry that could yield new insights into mental mechanisms.[136]

But there were other savants – notably De Morgan, Varley and Wallace – who believed that spiritualistic investigation had a still-wider scientific importance. Few savants tried harder to persuade fellow scientific savants of this possibility than Wallace. After attending seances between 1865 and 1867, the cofounder with Darwin of the theory of biological evolution by natural selection was convinced that disembodied spirits had communicated information about a dead brother that could not have been known by any living soul, and had caused heavy objects to levitate and flowers and fruit to materialise out of thin air.[137] Like many Victorians, Wallace was drawn to spiritualism because of personal experiences of the mesmeric trance, because its spiritual theories better fitted the psychological and physical facts of the seance than any other theory (including fraud), and because the future state revealed by spirits was altogether more appealing than that taught by orthodox Christianity.[138]

Unlike many of the spiritualists who championed his work, however, Wallace had much more confidence in scientists' capacity to investigate seance phenomena as long as they were humble and patient enough to learn the conditions under which phenomena could be witnessed.[139] But he also echoed the views of Mesmer and Reichenbach in recognising that the study of strange new imponderable agencies was relevant to and potentially important to the physical sciences. His argument stemmed partly from Faraday's declaration of 1854 that while table-turning was a delusion, conclusive proof of a new force enabling the fingers to attract

---

[135] Thomas Henry Huxley quoted in *Report on Spiritualism*, pp. 229–30.
[136] This is apparent from William B. Carpenter to G. W. Bennett, 24 December 1869, quoted in *Report on Spiritualism*, p. 266.
[137] On Wallace and spiritualism see Martin Fichman, *An Elusive Victorian: The Evolution of Alfred Russel Wallace* (Chicago University Press, 2004), chapter 4; Malcom J. Kottler, 'Alfred Russel Wallace, the Origin of Man and Spiritualism', *Isis*, vol. 65 (1974), pp. 145–92; James R. Moore, 'Wallace in Wonderland', in Charles H. Smith and George Beccaloni (eds.), *Natural Selection and Beyond: The Intellectual Legacy of Alfred Russel Wallace* (Oxford University Press, 2008), pp. 353–67; Charles H. Smith, 'Wallace, Spiritualism, and Beyond: "Change", or "No Change"', in Smith and Beccaloni, *Natural Selection*, pp. 391–423.
[138] Wallace, *Scientific Aspect*; Alfred R. Wallace, *My Life: A Record of Events and Opinions*, 2 vols. (London: Chapman and Hall, 1905), vol. 2, 275–350.
[139] Alfred R. Wallace, draft letter to the editor of the *Pall Mall Gazette*, circa May 1868, ff. 35–43, Add. 46439, ARW-BL.

or repel untouched objects would gain the discoverer "the attention of the whole scientific and commercial world".[140] The same remark would make an impression on at least one of our protagonists.[141] But Wallace also expected physicists to take spiritualism seriously because he, and other reputable scientific inquirers, could now "show" a force that had been "declared *impossible*".[142] For Wallace, the "rapid strides" made in "physical science" over the previous few decades was a good reason why physicists could not legitimately adhere dogmatically to ideas about the possible or impossible.[143]

Extending the Boundaries of Physics

In *The Scientific Aspect of the Supernatural*, his first substantial publication on spiritualism, Wallace argued that the spiritualist idea of the human personality transiting, more or less unchanged, to a disembodied state was rendered "more probable" by one "great law" of science.[144] This was the universal law of 'continuity', which the Welsh barrister and natural philosopher William Robert Grove had chosen as the theme of his presidential address to the British Association for the Advancement of Science, which held its 1866 annual meeting only a few months before Wallace's book appeared.[145] Grove had defined the law as a belief that the "progress of science" would reveal the "intermediate links" uniting "apparently segregated" natural phenomena with "other more familiar phenomena".[146]

Although Grove declined to discuss spiritualism, Wallace was evidently impressed by the fact that his conception of continuity allowed the seemingly "segregated" manifestations of disembodied spirits to be reconciled with the more familiar phenomena of psychology.[147] Continuity was the perfect principle for underpinning an argument, shared by Wallace and most spiritualists, that seemingly supernatural phenomena such as

---

140 Faraday, 'Mental Education', p. 471.
141 This was William Fletcher Barrett. See William F. Barrett to Oliver Lodge, 20 November [circa 1900], SPR.MS 35/155, OJL-SPR.
142 Alfred R. Wallace to Thomas Henry Huxley, 1 December 1866, f. 12, Add. 46439, ARW-BL.
143 Wallace, *Scientific Aspect of the Supernatural*, p. 10.
144 Wallace, *Scientific Aspect of the Supernatural*, p. 43.
145 On Grove see Iwan Rhys Morus, *William Robert Grove: Victorian Gentleman of Science* (Cardiff: University of Wales Press, 2017).
146 William Robert Grove, 'Address', in *Report of the Thirty-Sixth Meeting of the British Association for the Advancement of Science Held at Dundee in September 1867* (London: John Murray, 1867), pp. liii–lxxxii, p. lvi.
147 Grove's refusal to pass judgement on spiritualism was reported in [Anon.], 'Professor W. R. Grove on Spirits', *Spiritualist*, vol. 1 (1869–71), p. 95.

disembodied spirits would turn out to be part of an enlarged conception of the natural order. Wallace would not be the only Victorian man of science who found within the sciences, and especially the physical sciences, laws, principles and discoveries that lent weight to this.

Few achievements in the sciences gave Grove greater confidence in the law of continuity than those illuminating the close relationship between different physical forces or "affections of matter".[148] Since the 1840s, he had been a leading architect of the widely used idea that gravitation, inertia, heat, light, electricity, magnetism and chemical affinity could be closely correlated. Scientific investigations into these forces since the early 1800s had yielded evidence that they were not only dependent on each other, but mutually convertible, and ultimately reducible to forms of matter in motion.[149] As Iwan Morus has argued, Grove's conception of correlation was one of many strategies used by mid-nineteenth-century natural philosophers to give coherence to a range of divergent scientific enterprises and to redefine such enterprises as aspects of what would become the discipline of physics.[150]

The period between the 1830s and the 1860s witnessed many other steps towards the creation of the discipline. By the start of this period, many of the methodological features of physics had been established, especially those relating to experiment, measurement and calculation.[151] Subsequent decades saw the growth of professorships and academic research and teaching laboratories, as well as the emergence of such unifying concepts as energy, whose laws of conservation and dissipation were promoted by its architects as surpassing the long-cherished Newtonian ideas about force.[152] The middle decades of the century also saw the increasing presence of physical subjects in popular lectures, scientific and industrial exhibitions and commercial culture, and a marked rise in the place of physical

---

[148] William Robert Grove, *The Correlation of Physical Forces* (London: Longman, Green, Longman, Roberts, & Green, 1862), p. 240.
[149] Morus, *When Physics Became King*, chapter 3.
[150] Morus, *When Physics Became King*, pp. 76–7.
[151] Jed Z. Buchwald and Sungook Hong, 'Physics', in David Cahan (ed.), *From Natural Philosophy to the Sciences: Writing the History of Nineteenth-Century Science* (Chicago University Press, 2003), pp. 163–95, pp. 163–6.
[152] On the rise of physics laboratories see David Cahan, 'An Institutional Revolution in German Physics, 1865–1914', *Historical Studies in the Physical Sciences*, vol. 15 (1985), pp. 1–65; Graeme J. N. Gooday, 'Precision Measurement and the Genesis of Physics Teaching Laboratories in Victorian Britain', *British Journal for the History of Science*, vol. 23 (1990), pp. 25–51; Graeme J. N. Gooday, 'Teaching Telegraphy'; Terry Shinn, 'The French Science Faculty System, 1808–1914: Institutional Change and Research Potential in Mathematics and Physical Sciences', *Historical Studies in the Physical Sciences*, vol. 10 (1979), pp. 271–332. On the rise of the energy concept see Smith, *Science of Energy*.

1.4 Few individuals embodied the growing cultural significances of physics more than John Tyndall, here depicted giving a lecture on electricity at the Royal Institution, London's premier venue for public science. From *Illustrated London News*, 14 May 1870. Reproduced by permission of De Agostini/Getty Images.

sciences in print culture, from classroom textbooks to the blossoming periodical press (Figure 1.4).[153] This was also the period in which some of this book's protagonists built their careers, and the diversity

[153] On popular physics lectures see Sophie Lachapelle, *Conjuring Science: A History of Scientific Entertainment and Stage Magic in Modern France* (London: Palgrave Macmillan, 2015), chapter 1; Bernard Lightman, *Victorian Popularizers of Science: Designing Nature for New Audiences* (Chicago University Press, 2007), chapter 4; Morus, *When Physics Became King*, chapter 4. On physics periodicals see Graeme Gooday, 'Periodical Physics in Britain: Institutional and Industrial Contexts, 1870–1900', in Gowan Dawson, Bernard Lightman, Sally Shuttleworth and Jonathan Topham (eds.), *Constructing Scientific Communities: Science Periodicals in Nineteenth*

of their career paths illustrates the fluidity of the boundaries of physics. It was possible for individuals with backgrounds in such areas as analytical chemistry and scientific journalism (Crookes), medicine (William H. Stone), meteorology (Balfour Stewart) and telegraph engineering (Varley) to make significant contributions to the nascent discipline. It was precisely this fluidity that underpinned their conviction that psychical phenomena were a potential area into which physics could be further extended.

As one of the leading scientific popularisers of the 1840s, Grove well understood that the success of the intellectual transformation of natural philosophy hinged on persuading audiences of its ultimate cultural, social and economic utility. Natural philosophers needed to demonstrate that their ability to understand and control physical forces had more than just intellectual significance. The importance of serving the "thinking portion of mankind" and "practical minds" was precisely the reason why, in his 1866 address, Grove deemed the "greatest triumph of force-conversion" to be the recent laying of two telegraph cables under the Atlantic Ocean.[154] A large-scale application of the conversion of chemical into electrical force, the first successful transatlantic cables gave new continuity to Britain and the United States of America, whose relationship had been strained during the American Civil War.[155]

One of Grove's auditors would have been especially gratified to hear the Atlantic cable reference. This was Cromwell Varley, who, enjoying a popular image as one of the scientific 'heroes' of this engineering feat, was at the British Association to present new researches on the electrical properties of the Atlantic cable.[156] Varley's career exemplified the wider significances of natural philosophy that Grove and others had long been emphasising. The son of an artist and optical instrument maker, Varley had attended Grove's popular scientific lectures in the 1840s before building a career as an electrician for Britain's largest commercial operator of inland telegraphs, the Electric Telegraph Company, and later as

Century Britain, forthcoming. On physics textbooks see Josep Simon, 'Physics Textbooks and Textbook Physics in the Nineteenth and Twentieth Century', in Buchwald and Fox, *Oxford Handbook of the History of Physics*, pp. 651–78. For physical topics in the general periodical press see Myers, 'Nineteenth Century Popularizations'; Smith, *Science of Energy*, chapter 9.

[154] Grove, 'Address', pp. liii and lxvi.

[155] Historical studies of oceanic telegraphy are numerous but see Headrick, *Invisible Weapon*, chapter 2; Simone Müller-Pohl, *Wiring the World: The Social and Cultural Creation of Global Telegraph Networks* (New York: Columbia University Press, 2015), introduction and chapter 1; Smith and Wise, *Energy and Empire*, chapters 19–20.

[156] [Anon.], 'The Heroes of the Atlantic Telegraph Cable', *Illustrated Times*, 25 August 1866, p. 21.

a consultant for transatlantic cable businesses.[157] His ascent depended on his ability to make expertise in understanding and manipulating physical forces the key lto telegraphy's technical and commercial success. Indeed, he was one of many British electrical engineers, natural philosophers and physicists who were involved in the development of new and more accurate tools of telegraph signalling and fault detection and more robust standards of electrical measurement, all of which made telegraphy technically more reliable and commercially more attractive.[158]

From the late 1850s onwards, Varley was actively involved in campaigns to raise public confidence in and the huge financial backing needed for long-distance submarine telegraphy, an enterprise plagued by sloppy, secretive and even fraudulent engineering practices. The failures of the first transatlantic cables and the Red Sea cable of 1859–61 compounded perceptions that, as one commentator charged in 1862, telegraphy was an "art occult even to many of the votaries of electrical science".[159] Varley would have agreed with this assessment but had no doubt that the extension of electrical science to the oceanic engineering experiments would vanquish professional scientific and public misgivings about the integrity of telegraphy and its practitioners (Figure 1.5).

By 1866, Varley had accepted that his skills in correlating physical forces could be profitably extended to other, and much more troublesome, 'occult' telegraphs. In 1868, many would have been intrigued to learn that one of their cable heroes had publicly declared his belief in the phenomena and teachings of spiritualism. Nearly a decade of investigating the subject had persuaded him of the reality of "physical manifestations" not accounted for in "known laws of nature" and which opened up an "extensive field of mental and physical knowledge".[160] As he explained to Wallace a year later, however, the task of relating this and the other world would require some of the skills that had proved so successful in connecting the Old and New Worlds. Giving spiritualism an "intelligible shape" to the world required a "clever man" to "establish a clue to the relations existing between the physical forces known to us and those forces, by which the spirits are sometimes able to call into play the power by which they produce physical phenomena".[161] He had already

---

[157] On Varley see Noakes, 'Telegraphy'.

[158] Hunt, 'Ohm Is Where the Art Is'; Smith and Wise, *Energy and Empire*, chapter 19.

[159] [Anon.], 'Moral Causes', *Electrician*, vol. 2 (1862), pp. 39–40, p. 39.

[160] Varley quoted in [Anon.], 'Lyon v. Home', *Spiritual Magazine*, vol. 3 (New Series) (1868), pp. 241–54, p. 244. Varley's declaration appeared as part of his affidavit testifying to the honesty of D. D. Home, who had been accused of using fraudulent mediumistic methods to swindle money out of a wealthy widow.

[161] Cromwell F. Varley to Alfred R. Wallace, 28 January 1869, ff. 47–50, Add. 46439, ARW-BL.

1.5 The transatlantic cable breaks while being laid from the steamship *Great Eastern* in August 1865. This accident dramatised the troublesome nature of oceanic telegraphy in this period. From *Illustrated London News*, 26 August 1865. Reproduced by permission of Hulton Archive/ Getty Images.

made some headway in this quest. Two years earlier, in seances with the famous American medium Katherine Fox, he had used electrical apparatus commonly used in telegraphy to establish that professed spirits, like Reichenbach's subjects, appeared to perceive powers accompanying electricity and magnetism that were invisible to most people.[162]

Varley's approach to mysterious 'spirit' forces requiring the presence of a medium represented a bold extension of work done by physiologists and physicians over previous decades to relate physical and vital forces. It

[162] Varley, 'Evidence of Mr. Varley', pp. 165–6.

owed something to the development of the field of electrophysiology, which explored the relationships between different types of electricity, vitality and the nervous force.[163] A greater debt was to the more general shifts of physiology towards chemistry and natural philosophy. In separate studies during the 1840s, the German physician Julius Robert Mayer and his compatriot, the physician-turned-physiologist and natural philosopher Hermann von Helmholtz, analysed the close connections between the human body's vital processes (notably the oxygenation of food) and its capacity to produce heat and mechanical work.[164] Although suffering a poor initial reception, their researches were later hailed as foundation stones for two of the major generalisations in nineteenth-century physics – that the total amounts of force and energy in the cosmos were constant, although they could change form.

Another medically trained scientific practitioner whose work was seen to have helped establish these generalisations was Carpenter.[165] In 1850, and before his most public interventions on mesmerism and spiritualism, he extended Grove's concept of correlation to the vital forces that produced physiological phenomena from the transformation of physical forces. His research examined some of the psychophysical interactions that we have already discussed in this chapter, including studies of the close link between electricity and the nervous force and, more significantly, the apparently convergent discoveries by Reichenbach and Faraday of the link between magnetism and the nervous force.[166]

Carpenter's enthusiasm for Reichenbach's proposed odic link between the physical and vital forces was short-lived, and by 1853 he was accusing Reichenbach of relying on witnesses who were victims of the closely correlated forces of human physiology and psychology. However, others, including the author of *The Electrician*'s 'Animal "Magnetism"', maintained that studies of od could fuel the progress of the physical sciences. The physician and medical electrician William H. Stone agreed. In a period when the British medical establishment typically associated medical electricity with quackery, he fought hard to raise the intellectual profile of research on the

---

[163] On electrophysiology see Edwin Clarke and L. S. Jacyna, *Nineteenth-Century Origins of Neuroscientific Concepts* (Berkeley, CA: University of California Press, 1987), chapter 5.

[164] See Kenneth Caneva, *Julius Robert Mayer and the Conservation of Energy* (Princeton University Press, 1993) and Fabio Bevilacqua, 'Helmholtz's *Über die Erhaltung der Kraft*: The Emergence of a Theoretical Physicist', in Cahan, *Hermann von Helmholtz*, pp. 293–333.

[165] For discussion see Hall, 'Contribution of the Physiologist'.

[166] William B. Carpenter, 'On the Mutual Relations of the Vital and Physical Forces', *Philosophical Transactions of the Royal Society of London*, vol. 140 (1850), pp. 727–57, p. 746.

relationship between electricity and the human body.[167] As someone who had long recognised the value of turning medical and physiological problems into fruitful enquiries in physical science, Stone's interest in studying alleged connections of od to physical and vital forces is hardly surprising.

Like others establishing close links between the physical and vital realms, Carpenter tried to distance himself from charges of philosophical materialism to which the more 'physical' approaches to physiology and psychology had been subjected. For him, the mechanisms of the human body and mind, as well as those pervading the cosmos, were under the control of an immaterial will, whether human or divine.[168] This need to challenge materialist interpretations of extensions of physical principles into the vital realm was also felt by a young William Crookes in 1862. By this time, Crookes had established himself as a prominent analytical chemist and scientific journalist.[169] Between 1849 and 1854 he had studied at London's Royal College of Chemistry, and by the early 1860s he had significantly boosted his scientific reputation by editing scientific periodicals and discovering a new chemical element: thallium. This latter achievement depended critically on Crookes's mastery of the relatively new technique of spectro-chemical analysis (later christened 'spectroscopy'). A significant extension of optics into the field of analytical chemistry, this yielded evidence of the chemical composition of material bodies (both terrestrial and celestial) by analysing the light that they emitted into a spectrum of lines. For Crookes and others, mastering this technique was not easy because of such practical difficulties as observing faint and transient chemical 'spectra' and conclusively distinguishing known and unknown elements from such spectral fingerprints[170] (Figure 1.6).

Through his teacher, the German chemist August Wilhelm Hofmann, Crookes acquired an understanding of the chemical and physical bases of animal physiology, a major research area of Liebig, who had taught Hofmann and inspired Helmholtz. A good deal of this understanding is evident in an essay that Crookes contributed to an 1862 number of the *Popular Science Review*, one of a plethora of semi-popular science

---

[167] Iwan Rhys Morus, 'The Measure of Man: Technologizing the Victorian Body', *History of Science*, vol. 37 (1999), pp. 249–83; Takahiro Ueyama, 'Capital, Profession and Medical Technology: The Electrotherapeutic Institutes and the Royal College of Physicians, 1888–1922', *Medical History*, vol. 41 (1997), pp. 150–81.

[168] Smith, 'Human Significance of Biology', p. 223.

[169] On Crookes's early career see Brock, *William Crookes*, chapters 1–4.

[170] On spectroscopy see Klaus Hentschel, *Mapping the Spectrum: Techniques of Representation in Research and Teaching* (Oxford University Press, 2002); James, 'Practical Problems'; Frank A. J. L. James, 'The Study of Spark Spectra, 1835–1859', *Ambix*, vol. 30 (1983), pp. 137–62; Schaffer, 'Where Experiments End'.

1.6 A standard late-nineteenth-century compound spectroscope. Here, light from chemical substances burned in a gas flame (far right) is passed through a slit and collimator which focusses the image of the flame on a prism (centre). The prism disperses the image into a spectrum, which is observed in the telescope (left). The candle flame on the centre-right illuminates a photographic scale in a telescope lens enabling measurements of spectra. From J. Norman Lockyer, *The Spectroscope and Its Applications* (London: Macmillan and Co., 2nd ed., 1873).

periodicals established in the mid-Victorian period.[171] The essay explored an analogy between the life of a human being and of a candle flame and was partly designed to emphasise the social utility of science. Scientific analysis of human respiration revealed that human life, like that

[171] William Crookes, 'The Breath of Life', *Popular Science Review*, vol. 1 (1862), pp. 91–9. For these periodicals see Ruth Barton, 'Just before *Nature*: The Purposes of Science and the Purposes of Popularization in English Popular Science Journals of the 1860s', *Annals of Science*, vol. 55 (1998), pp. 1–33; Susan Sheets-Pyenson, 'Popular Science Periodicals in Paris and London: The Emergence of a Low Scientific Culture, 1820–1875', *Annals of Science*, vol. 42 (1985), pp. 549–72.

of the flame, was critically dependent on fresh air, and that this made the urban and industrial evils of poor ventilation more deplorable. The analogy supported more than just an argument for public health reform. It was designed to show that the conservation of force had potential spiritual implications, and that the physical sciences were relevant to religious questions. Crookes tentatively suggested that the analogy extended to the "dim shadowy realms" beyond the extinctions of life and the flame: if "philosophy" proved that the physical forces or energies of the flame were conserved, and merely changed form after the flame died, "shall not faith accept the same proof that our own spiritual life is continued after the vital spark is extinguished?"[172]

Crookes's physical "proof" of the future life was, of course, enormously presumptuous, but it was indicative of his confidence in the capacity of physical science to illuminate spiritual questions, and of "progressive" science to move beyond "stationary" theology.[173] It was a confidence that he and many other physical–psychical scientists would carry into the investigation of spiritualistic phenomena. It was also a confidence which physical–psychical scientists would share with many distinguished natural philosophers and physicists who, from the 1860s onwards, defended natural philosophy against scientific naturalism.

One of the most controversial features of scientific naturalism was that it appeared to promulgate philosophical materialism and determinism. By reducing the cosmos to matter and motion, scientific naturalism denied a place for the divine and human will in scientific interpretations of the cosmos. Most scientific naturalists were devout Christians, but, as one of its proponents argued in 1867, they separated questions of "real religion" and those of science, partly by assigning the former task to the "affections and emotions" and the latter to the "dry light of the intellect alone".[174] Not surprisingly, the targets of their numerous intellectual and cultural critiques included those who confused these approaches, from Christian theologians who appealed to the affections rather than scientific reasoning in weighing evidence of miraculous physical effects, to spiritualists whose emotional attachment to the idea of a spirit world undermined their capacity to judge seance manifestations.

Many Victorian physicists were at least as devout as their scientific naturalist peers but denied that religious and scientific questions could

---

[172] Crookes, 'Breath of Life', pp. 98–9.
[173] Crookes drew this contrast between science and theology in William Crookes to Herbert MacLeod, 22 April 1864, f. 8, Add. 5989, Department of Manuscripts and University Archives, Cambridge University Library.
[174] John Tyndall, 'Miracles and Special Providences', *Fortnightly Review*, vol. 7 (New Series) (1867), pp. 645–60, p. 649.

or ought to be separated, and often turned to the physical sciences to support their opposition to the apparent materialism and determinism of scientific naturalism.[175] For them, a proper understanding of energy, ether and matter reinforced the idea of nature created and guided by a divine mind and helped reconcile an intuitive sense of free will and the idea of a law-bound cosmos. In the 1860s, two of the most prominent architects of the science of energy, the Scottish physicists William Thomson and Peter Guthrie Tait, argued that the universal dissipation of energy testified to the divine purpose or directionality of the cosmos, as opposed to the entirely reversible and mechanistic cosmos implied by scientific naturalism. In the same decade, Balfour Stewart and Norman Lockyer offered a different argument for immaterial will within energy physics: their painstaking studies of sunspots suggested a "delicacy of construction" for our nearest star that enabled it to transform inscrutable impulses (Divine Will) into spectacularly visible outcomes on the solar surface, just as physiology indicated the capacity of the 'delicate' human body to transform the inscrutable will into tangible results.[176]

The perceived rising threat of materialism from scientific naturalism would, in the 1870s, prompt further interventions from proponents of theistic physics. James Clerk Maxwell argued that the incredible similarity of molecules of a given element, whether on earth or in the heavens, suggested that they were divinely "manufactured", while the apparently perfect continuity of the interstellar ether supported the idea of a cosmos filled with "symbols of the manifold order of His kingdom".[177] And in 1875, in their anonymous and hugely controversial *The Unseen Universe; Or Physical Speculations on a Future State*, Stewart and Tait argued that the proper interpretation of the physics of matter, energy and ether was not incompatible with Christian teachings on the soul, the afterlife and the Resurrection.[178]

Some of this book's protagonists occupied positions combining those of the scientific naturalists and their physicist adversaries. Their zealous investigation of psychical phenomena echoed scientific naturalists' bold, empiricist attitude towards the supernatural and miraculous. In 1872,

---

[175] Smith, *Science of Energy*, chapter 9; Stanley, *Huxley's Church*, esp. chapter 6; Wilson, *Kelvin and Stokes*, chapters 4–5.

[176] Balfour Stewart and J. Norman Lockyer, 'The Sun as a Type of the Material Universe', *Macmillan's Magazine*, vol. 18 (1868), pp. 246–57, 319–27, p. 257.

[177] James Clerk Maxwell, 'Molecules', *Nature*, vol. 8 (1873), pp. 437–41, p. 441; James Clerk Maxwell, 'On Action at a Distance', *Nature*, vol. 7 (1872–3), pp. 323–5, 341–3, p. 343.

[178] [Balfour Stewart and Peter Guthrie Tait], *The Unseen Universe; Or Physical Speculations on a Future State* (London: Macmillan and Co., 1875). Unless otherwise stated, all subsequent references will be to this (first) edition of the work.

during an intense public debate over the efficacy of prayer, Tyndall charged that when theologians represented prayer as a form of "physical energy" then the "scientific student" claimed the right to study it with scientific methods used to understand the physical universe.[179] Comparable declarations about the energetic physical phenomena claimed by spiritualists were made by Varley, Crookes and two of Tyndall's students: Barrett and Lodge.

Yet some of our protagonists deviated significantly from Tyndall's example in agreeing with Thomson, Tait and others that physical enquiry could reinforce rather than undermine Christian belief, even if these latter scientists did not think spiritualism was worth investigating. William Fletcher Barrett well captures this complexity. The son of a Congregationalist minister, he gained an informal science education in London before entering, in 1863, the Royal Institution, where he spent three years as an assistant in Tyndall's physical laboratory.[180] The skills he acquired in studying and publicly exhibiting the properties of hidden phenomena – molecules, invisible radiation and inaudible sound waves – owed much to Tyndall, but the theistic interpretations of Tyndall's physics that Barrett developed in popular scientific writings owed more to his mentor's physicist adversaries. In 1866, for example, he interpreted the "exquisite" crystalline structures of ice, phenomena that Tyndall believed testified solely to the material powers of molecules, as something also glorifying "One who employs His works as witnesses of His existence".[181]

Within four years Barrett was using Tyndall's physics for psychical as well as religious purposes. In an 1870 article for Crookes's *Quarterly Journal of Science*, he explained that studies of light and sound revealed realms of vibration beyond human sensitivity and suggested that some "forces unrecognised by our senses are perceptible elsewhere".[182] By "elsewhere", Barrett probably included those individuals he now believed were genuinely sensitive to a mesmeric 'force' and those he would soon accept could perceive the manifestations of od. Barrett only kept implicit what other scientific writers of the period were prepared to make explicit. The science journalist and photographic inventor William H. Harrison regularly praised Tyndall for his popular lectures on imponderable forces, but was convinced that Tyndall's example of researching and popularising physics at the Royal

---

[179] John Tyndall, 'On Prayer', *Contemporary Review*, vol. 20 (1872), pp. 763–6, p. 764.

[180] On Barrett see Noakes, 'Bridge Which Is Between Physical and Psychical Research'.

[181] William F. Barrett, 'Glaciers and Ice', *Popular Science Review*, vol. 5 (1866), pp. 41–54, p. 54. Cf. John Tyndall, 'Scientific Materialism [1868]', in John Tyndall, *Fragments of Science: A Series of Detached Essays, Addresses and Reviews* (London: Longmans, Green and Co., 7th ed., 1889), vol. 2, pp. 75–90.

[182] William F. Barrett, 'Light and Sound: An Examination of Their Reputed Analogy', *Quarterly Journal of Science*, vol. 7 (1870), pp. 1–16, p. 8.

Institution would help promote a cause for which he had become a vigorous campaigner by the early 1870s: spiritualism. Since spiritualism's phenomenal or "lower" aspects were related to "*physical* science" then it needed "experimental lectures" showing the public "marvellous physical powers of things both imponderable and invisible" and "patient research" linking law-like, "ordinary physical phenomena" and those spiritualistic manifestations deemed "miraculous" by the "uneducated".[183]

The spiritualist implications of Tyndall's physics were no less clear to Wallace. In his *Scientific Aspect of the Supernatural*, a work much admired by Harrison, Wallace appealed to the theory, closely associated with Tyndall, that different "modes of motion" in the invisible, space-filling and "almost infinitely attenuated form of matter" that was the ether produced the powerful forces acting on material nature. On this basis, it was not implausible that invisible "beings" composed of the "most diffused and subtle forms of matter" could use hitherto unknown forms of motion in the capacious ether to cause the notorious movement of "ponderable bodies" in seances.[184]

Our protagonists may not have cited Wallace's argument, but from the 1870s until the 1930s one of their most important strategies of linking 'physics and psychics' effectively developed his basic spiritualist move of appealing to the ether to render psychical effects more intelligible. The naturalist's argument only hinted at a problem whose importance would grow over the decades. Few physical scientists before the early 1900s questioned the need for an ether, but most were baffled by its constitution and structure. The problem would prompt the invention of the radical new theories of relativity but also inspire some of the most daring psychical theorising within physics.

***

By the late 1860s, many of the ingredients of late Victorian 'physics and psychics' were in place. The investigations of Mesmer, Elliotson, Reichenbach, Gregory, Faraday, Hare and others remained hugely controversial, but they had established important precedents to the idea of treating the problem of an imponderable, invisible influence, force or power flowing from the body as one for the physical sciences. In many quarters, such purely 'physical' approaches were judged either unnecessary or misguided. For most mesmerists, the reality of the mesmeric

<hr>

[183] William H. Harrison quoted in [Anon.], 'National Conference of Spiritualists in Liverpool', *Spiritualist*, vol. 3 (1872–3), pp. 291–7, p. 293. See also William H. Harrison, 'The Invisible Photographic Image Considered at Motion', *British Journal of Photography*, vol. 6 (1867), pp. 424–5.

[184] Wallace, *Scientific Aspect of the Supernatural*, p. 5.

influence was convincingly evidenced by its therapeutic efficacy and its psychological effects; for most spiritualists, the reality of the physical effects in seances and, moreover, of the invisible intelligences were matters of simple experience rather than scientific intervention; and for many medical and scientific practitioners, questions of these and other powers so closely associated with the body were questions for doctors, physiologists and physiological-psychologists alone.

A small but significant minority of scientific practitioners, however, disagreed and turned to the blossoming periodical press to reinvigorate older arguments that phenomena of an extraordinary and seemingly occult nature were ripe subjects for the physical sciences. The same platforms – the flurry of new commercial scientific and technical serials and general audience periodicals rather than the proceedings of established scientific societies – were the same spaces where physical–psychical scientists would present some of their strongest cases for new, physical approaches to phenomena whose grip on the public mind showed no sign of abating. These serials were entirely appropriate fora for this because these were the places where Victorian scientists and their allies frequently championed the supreme role of the sciences in the solution of social, moral, economic and intellectual issues of the day.[185]

Contributors to these serials upheld the more widely shared belief in the progress of the sciences and, particularly, nineteenth-century achievements in understanding, connecting and manipulating electricity, magnetism, light, the vital force and other imponderable agencies. The first generation of physical–psychical scientists well understood that these achievements often involved an enormous amount of scientific and other types of work, often in the face of professional and popular adversity. In the 1850s and '60s, the extensions of the boundaries of the physical sciences – whether electrical science to the domains of long-distance submarine telegraphy and the human body, optics to chemistry via spectro-chemical analysis, or the physics of matter, energy and ether to spiritual and religious questions – were fraught with practical, intellectual and moral difficulties whose solutions depended critically on the mastery of experimental, rhetorical and other scientific skills. Long and sometimes bitter experiences of these enterprises furnished the likes of Barrett, Crookes and Varley with formidable abilities in detecting, measuring, manipulating and interpreting often puzzling, capricious and invisible physical phenomena, as well as enormous confidence in being able to successfully apply these abilities to other, sometimes vastly more controversial phenomena. When physical scientists with such abilities and

---

[185] Turner, *Contesting Cultural Authority*, pp. 203–4.

confidence turned to mesmerism, spiritualism, thought-reading and related questions, some of the scientific ambitions of Mesmer, Reichenbach and Hare looked particularly appealing. In Chapter 2, we examine the identity of these individuals in depth and look at what sparked, sustained and undermined their psychical interests.

2 A Survey of Physical–Psychical Scientists

In 1912, the British psychiatrist Ivor Tuckett observed that "the interest which physicists in different countries have taken in psychic research is a really striking fact".[1] Eight years later, a very different observer, the American Episcopal clergyman, spiritual healer and psychical researcher Elwood Worcester, considered it "curious that so many physicists should have interested themselves in Psychical Research".[2] Tuckett's and Worcester's perplexity reflected a more widely shared view that the phenomena investigated by the relatively young field of enquiry, psychical research, were predominantly psychological and therefore the province of the emergent science of psychology.

Tuckett's comment was partly a response to *New Light on Immortality* (1908) by the British physicist and popular science writer Edward E. Fournier d'Albe. For Fournier d'Albe it was not "presumptuous of a physicist to venture an opinion" on human immortality, a question "usually associated with psychology and theology".[3] Human immortality concerned the "relations between mind and matter" and this required an "extensive acquaintance with what is actually known about matter and what is *not* known about it".[4] Since physicists confronted these "ultimate questions" more often than chemists and physiologists, they could justify their fascination with a subject – psychical research – that studied phenomena of great relevance to the questions of the existence of mind independent of matter and of the postmortem existence of the human soul.[5]

Fournier d'Albe, Tuckett and Worcester were probably aware that the most striking evidence for this fascination came from the membership of the Society for Psychical Research. This chapter uses the SPR as one of

[1] Ivor Tuckett, 'Psychical Investigators and "The Will to Believe"', *Bedrock*, vol. 1 (1912–13), pp. 180–204, p. 192.
[2] Elwood Worcester, 'The Intrepid Pioneer', *Journal of the American Society for Psychical Research*, vol. 14 (1920), pp. 501–10, p. 504.
[3] E. E. Fournier d'Albe, *New Light on Immortality* (London: Longmans, Green, and Co., 1908), p. vii.
[4] Fournier d'Albe, *New Light on Immortality*, p. vii.
[5] Fournier d'Albe, *New Light on Immortality*, pp. vii–viii.

the main ways of critically examining the trend that so surprised Tuckett and Worcester. After exploring the SPR's foundation and early development, it examines the identities of the physicists and practitioners of other physical sciences who joined the organisation, as well as the nature of their commitment to psychical research. This picture of what, for the purposes of brevity, we call 'physical–psychical scientists', cannot be comprehensive because many of them pursued psychical interests independently of the SPR or died before it was founded. For this reason, much of the following analysis depends on source materials beyond the SPR's publications.

The mainly British connections that we will establish between late-nineteenth- and early-twentieth-century 'physics and psychics' are both stronger and more complex than historians have suggested. Not only were more professional physicists interested in psychical phenomena than previously assumed, but they were joined by a host of practitioners of other physical sciences, and the personal, professional and other links between many of these individuals were far closer than hitherto supposed. Furthermore, the phrase 'physics and psychics' smooths over the different kinds of connection that physical scientists had with psychical investigation, whether this related to what motivated their inquiries, the kinds of phenomena that primarily interested them or the strength and duration of their interest.

Inventing Psychical Research

In December 1881, several Victorians received an invitation from the British physicist William Fletcher Barrett to attend a conference in London for "friends interested in Spiritualism and Psychological Research", whose aims included discussing the "present condition of affairs, suggesting lines of work, and discussing the advisability of having a select Central Society, organised under some such name as the London Psychical Society".[6] Barrett's invitation was the outcome of several discussions he had had over the previous few years with Edmund Dawson Rogers and William Stainton Moses, two leading spiritualists, and Frederic William Henry Myers, a well-known classicist, essayist and government schools' inspector who had spent much of the 1870s investigating spiritualism. These discussions persuaded Barrett that some spiritualists, academics and scientists would be willing to collaborate on more scientific approaches to spiritualism and related subjects.

---

[6] William F. Barrett, 'The Early Years of Psychical Research', *Light*, vol. 44 (1924), p. 395.

Barrett was on familiar territory. Only eight years earlier he had helped launch a 'Society of Physical Research' that would exhibit and publish original researches in the physical sciences in which existing scientific societies did not appear to be interested. Eventually founded in 1874 as the Physical Society of London, the organisation recruited a host of Victorian natural philosophers, electrical engineers and physical chemists and it symbolised the disciplinary boundaries of physics that Barrett and other early members were negotiating.[7] Barrett had good reason to think that existing scientific societies were even more blinkered about the spiritualistic and psychological researches he had been pursuing for over a decade. He had provoked the ire of many fellow scientists for presenting, at the 1876 meeting of the British Association, his positive evidence for thought-reading and spirit-rapping.[8] His subsequent request that the British Association form a committee for studying mesmerism and spiritualism more scientifically than previously done was refused, and this merely compounded the fears of Barrett and others that the scientific authorities typically associated with these subjects (notably medical practitioners, physiologists and psychiatrists) were either misleading the public with unsatisfactory psycho-physiological explanations or failing in their duty to conduct satisfactory investigations.

Barrett's struggles to raise the scientific profile of psychical investigation took place in the context of what Graham Richards has identified as a transitional period in British psychology.[9] Unlike the situations in Germany, the United States and France, British psychology lacked an academic institutional base, and those medical practitioners, physiologists and psychiatrists spearheading efforts to provide one did so by trying to shift psychology's subject matter away from questions of the will, soul and spirit towards the application of evolutionary theories to the human mind and body. The questions sidestepped by the emerging elite of British psychology were precisely those that many Victorian spiritualists explored in their numerous 'psychological' societies and which formed the focus of the short-lived but largely non-spiritualist Psychological Society of Great Britain (1875–9). Most of the members of these organisations shared the belief that Christian theology, 'metaphysical' philosophy and what they often perceived to be materialistic science were failing to deliver satisfactory answers to questions about human nature, morality and cosmic purpose, and some would pursue their questions in the

---

[7] Gooday, 'Periodical Physics'; John L. Lewis (ed.), *Promoting Physics and Supporting Physicists: The Physical Society of London and the Institute of Physics, 1874–2002* (Bristol: Institute of Physics Publishing, 2003), chapter 1.

[8] See Noakes, 'The Bridge Which Is Between Physical and Psychical Research', pp. 438–41.

[9] Richards, 'Edward Cox'.

Metaphysical Society, an exclusive British forum of intellectual debate that ran from 1869 to 1880, and later in the SPR.[10]

Among those individuals who link the Metaphysical Society and SPR is the eminent British moral philosopher Henry Sidgwick. As several historians have shown, Sidgwick was at the centre of a group of intellectuals who were colleagues, students or associates of his at Cambridge University, and who would dominate the SPR's activities in its first decades.[11] They included Myers and the psychological writer Edmund Gurney, and in the 1870s they joined Sidgwick and thousands of other Victorians in visits to spiritualistic seances, which promised to yield more satisfactory answers to questions about the human soul and its future state than those given by 'materialistic' science and the orthodox Christianity on which most seance attendees had been raised. Although its investigations of spiritualism (and especially of the physical phenomena of seances) proved disappointing, the Sidgwick group's hopes for an alternative science of psychology were raised by Barrett's controversial paper of 1876, his collection of testimony to various psychical phenomena, and his intervention in the 'thought-reading' craze that swept across Britain and the United States in the 1880s. Indeed, in 1881, they joined the physicist in tests suggesting that several young female performers of the extraordinary psychological feat were genuinely able to read words and images in the minds of others, independently of sensory clues and to a degree that ruled out simple chance coincidence (Figure 2.1).

The support that Barrett enjoyed from Myers, Gurney and Sidgwick extended to his London conference on spiritualistic and psychological research which, when held in January 1882, resolved to found the SPR. The principal objective of the new society was to make an "organised and systematic attempt to investigate that large group of debatable phenomena designated by such terms as mesmeric, psychical and Spiritualistic" phenomena for which there seemed to be plenty of credible testimony but which were "*prima facie* inexplicable on any generally recognised

[10] On the Metaphysical Society see Alan Willard Brown, *The Metaphysical Society: Victorian Minds in Crisis* (New York: Columbia University Press, 1947). Metaphysical Society members who joined the SPR included Arthur Balfour, William Ewart Gladstone, Richard Holt Hutton, John Ruskin, Henry Sidgwick and Alfred Lord Tennyson.

[11] Gauld, *Founders of Psychical Research*; Oppenheim, *Other World*, chapter 4; Turner, *Between Science and Religion*, chapters 3 and 5; John Peregrine Williams, 'The Making of Victorian Psychical Research: An Intellectual Elite's Approach to the Spirit World', unpublished doctoral thesis, University of Cambridge, 1984. More focussed studies are Gordon Epperson, *The Mind of Edmund Gurney* (Madison, NJ: Fairleigh Dickinson University Press, 1997); Trevor Hamilton, *Immortal Longings: F. W. H. Myers and the Search for Life After Death* (Exeter: Imprint Academic, 2009); Bart Schultz, *Henry Sidgwick: Eye of the Universe. An Intellectual Biography* (Cambridge University Press, 2004), esp. chapter 5.

2.1 The thought-reading craze in late-Victorian Britain. This popular
pastime appeared to show the ability of some people to correctly 'read'
words and images in the minds of others, even though blindfolds and
silence seemed to deprive them of obvious sensory clues. From [Anon.],
'Amateur Thought-Reading', *Illustrated London News*, 19 October 1889.
Reproduced by permission of Hulton Archive/Getty Images.

hypothesis".[12] The SPR founders' preoccupation with projecting an
image of a credible scientific organisation informed many aspects of its
operation. They strategically chose 'psychical' for the title to prevent
misconceptions that they were just another pro-spiritualist or 'occult'
organisation, and elected as their first president Sidgwick, whose reputation for open-minded but cautious enquiry helped symbolise the organisation's approach to controversial questions. The SPR's structure
imitated many of the scientific societies to which many of its professional
scientific and medical members already belonged.[13] In the first few years,
its research was organised into committees – on thought-reading, mesmerism, Reichenbach's od, apparitions, the physical phenomena of

---

[12] [Anon.], 'Society for Psychical Research', p. 3.
[13] [Eleanor] Sidgwick, 'The Society for Psychical Research. A Short Account of Its History
and Work on the Occasion of the Society's Jubilee, 1932', *PSPR*, vol. 41 (1932–3), pp. 1–
26, pp. 4–5.

spiritualism, and literary testimony of psychical effects – and submitted papers were refereed by the SPR's governing Council before being published in the organisation's publicly circulated *Proceedings*.

A conspicuous difference between the SPR and earlier Victorian psychological societies was the social status of its members. Most hailed from the middle and upper classes of British society and included eminent figures in the sciences, medicine, politics, philosophy, religion and literature. Its vice-presidents and Council members included the Tory statesmen Arthur and Gerald Balfour, the Anglican bishop William Boyd Carpenter, the logician John Venn, the journalist Richard Holt Hutton, the surgeon Charles Lockhart Robertson, the American psychologist and philosopher William James, the astronomers John Couch Adams, Lord Lindsay (the 26th Earl of Crawford and 9th Earl of Balcarres) and Samuel P. Langley, the naturalist Alfred Russel Wallace, and the physicists Lodge, Stewart, the Third Baron Rayleigh and J. J. Thomson.[14] Many senior members also had close family ties to the Sidgwick group: Eleanor Balfour, who became one of the most powerful figures in the SPR after 1900, was Arthur and Gerald's sister, the Third Baron Rayleigh's sister-in-law, and in 1876 she married Henry Sidgwick.[15]

Even the lower-ranking positions of honorary, corresponding and ordinary member boasted such significant draws from British, American and European intellectual, literary and scientific life as Crookes, the Liberal prime minister William Ewart Gladstone, the psychologists Hipployte Bernheim, Granville Stanley Hall, Ambroise-Auguste Liébeault and Pierre Janet, the physiologist Charles Richet, the literary critic John Ruskin, the poet Lord Tennyson and the painter George Frederic Watts. The intellectual lustre that such members conferred on the SPR partly explains why the size of its membership grew steadily from 286 in 1883 to 925 in 1901 and why an American branch of the society launched in Boston in 1885 attracted so many scientific members, including the psychologists Stanley Hall and William James and the astronomers Simon Newcomb and Edward C. Pickering.[16]

---

[14] For simplicity 'Rayleigh' will subsequently be used to refer to the Third Baron Rayleigh (John William Strutt) but, where necessary, full aristocratic titles will be used to distinguish him from his son, the Fourth Baron Rayleigh (Robert John Strutt).

[15] On Eleanor Sidgwick see Ethel Sidgwick, *Mrs. Henry Sidgwick: A Memoir by Her Niece* (London: Sidgwick and Jackson, 1938).

[16] Figures established from membership lists in the SPR's *Proceedings* for volumes 1 (1882–3) and 15 (1900–1). The American branch of the SPR had a chequered early history. Leading members' frustration with inconclusive enquiries and financial difficulties forced it to become taken over by the main (British) SPR in 1889 and it was not until 1907 that an independent American Society for Psychical Research was relaunched. See Seymour H. Mauskopf and Michael R. McVaugh, *The Elusive Science: Origins of Experimental Psychical Research* (Baltimore, MD: Johns Hopkins University Press, 1980), p. 16.

For many Britons, Americans and Europeans who joined the British or American branches, the main source of the SPR's appeal lay in the results of its painstaking investigations into psychical effects that they had long believed required further study. A large proportion of these investigations was undertaken by those with the financial independence and time to do so – notably the Sidgwick group. Although the SPR's administrative headquarters was in London, it was this largely Cambridge-based faction that was responsible for the society's important early achievements. This included the production, collection and critical analysis of a vast amount of evidence for 'telepathy'.[17] Coined by Myers in 1882 to describe the communication of ideas, images and other impressions between minds, independently of the recognised sensory channels, this term was partly intended to avoid the assumptions involved in such ancestral terms as 'mind-reading', 'thought-reading' and 'thought-transference', although all four terms were used interchangeably well into the early 1900s. The evidence for telepathy was put together from experimental tests of thought-reading and from unsolicited or 'spontaneous' cases of individuals who appeared to perceive apparitions of distant relatives and friends who were dying or critically close to death. On this theory, ghosts and, much to the annoyance of spiritualists, disembodied spirits were brought down from their supernatural and spiritual planes respectively by being interpreted as telepathically projected images of embodied minds. In opposition to those who maintained fluid theories of mesmerism, telepathy was also invoked as an explanation of recent evidence, closely analysed by Myers, that individuals had been hypnotised at a distance.

Myers, Gurney and the London Post Office clerk Frank Podmore presented the evidence for telepathy in *Phantasms of the Living* (1886), a monumental work that helped raise the SPR's cultural profile but which certainly did not achieve the scientific and medical impact for which its authors hoped. Many scientific critics attacked the SPR's methods, which, in experimental telepathy, did not appear to safeguard against observational error, fraud and the use of sensory clues, and which, in spontaneous cases, showed poor understanding of statistical analysis and relied too heavily on the potentially flawed judgement of those excited or distressed by seeing ghosts of loved ones.[18] One of the most prominent critics was G.Stanley Hall, whose misgivings about the SPR were shared by several other academic psychologists and other scientists who, like Hall, had resigned from the American SPR by the early 1890s on the

---

[17] On telepathy and its impact see Luckhurst, *Invention of Telepathy.*
[18] Shane McCorristine, *Spectres of the Self: Thinking About Ghosts and Ghost-Seeing in England, 1750–1920* (Cambridge University Press, 2010), chapter 4.

grounds that psychical research was failing to adopt the methods that they were vigorously upholding as critical to their idea of the true academic science of psychology.[19]

By the mid-1890s, the SPR leadership had responded to the attacks of Hall and others by applying more robust methods of investigation to a large-scale study of hallucinations, a proportion of which they claimed as evidence for the reality of telepathically projected apparitions. This 'Census of Hallucinations' certainly helped give the SPR leadership the professional scientific profile it desired, since they were invited to present papers on it at the International Congresses of Experimental Psychology in 1889 and 1892.[20] The emphasis that the Census placed on large-scale, analytical and dispassionate approaches to psychical phenomena diverged from the emphasis that most spiritualists placed on a more emotionally engaged approach to a smaller number of powerful seance 'facts'. Indeed, as Perry Williams has argued, profound methodological and epistemological differences underpinned the bitter conflict between the SPR's academic leadership and spiritualist members in the mid-1880s.[21]

The conflict came to a head in 1886 when Eleanor Sidgwick, who exemplified the more academic approach to psychical research, accused the British 'slate writing' medium William Eglinton of fraud. Drawing heavily on the testimony of the amateur magician and SPR member S. J. Davey, Sidgwick concluded that Eglinton had employed "clever conjuring" in producing writing between enclosed slates that the medium and his spiritualist supporters attributed to the agency of disembodied spirits.[22] For many spiritualists inside and outside the SPR, this was not only a slur on a valued medium but another example of an overly cautious and hostile approach to spiritualism that they had resented for years. Disgusted by the SPR's refusal to distance its corporate position from Sidgwick, many spiritualists members resigned.

The SPR's leadership was hardly dismayed by this exodus, but their confidence in spiritualism as a fruitful area of enquiry, already lower than that in other areas of psychical enquiry, was not completely shattered: the late 1880s and 1890s would see the SPR undertake its most elaborate studies of trance mediumship and the physical phenomena of spiritualism. Much of this work informed one of the SPR's most elaborate theoretical achievements: Myers's theory of the 'subliminal self'. Welcomed

[19] Coon, 'Testing the Limits'.   [20] McCorristine, *Spectres*, chapter 5.
[21] Williams, 'Making of Victorian Psychical Research', chapter 8. See also John J. Cerullo, *The Secularization of the Soul: Psychical Research in Modern Britain* (Philadelphia, PA: Institute for the Study of Human Issues, 1982), chapter 4.
[22] Eleanor Sidgwick, 'Mr. Eglinton', *JSPR*, vol. 2 (1886), pp. 282–334, p. 332.

by many fellow psychical researchers as a stimulating if tentative hypothesis, but rejected by many academic psychologists as overly speculative, it proposed that the human self comprised multiple layers of consciousness, including many whose activities fell below the "ordinary *threshold (limen)*" of consciousness.[23] Some psychological phenomena, however, including telepathy, clairvoyance and spiritualist mediumship, seemed to evidence the capacity of these activities to emerge into normal conscious experiences.

The strengths and weaknesses of Myers's theory proved to be one of many sources of conflict within the SPR, and the organisation's leading physical scientists found themselves torn by such conflicts. They were generally not as wealthy or as well connected as the Sidgwick group and their bourgeois social status put them closer than this socially elevated circle to most spiritualists. This may have been one reason why they were more tolerant of spiritualists whom the Sidgwick group typically regarded as their social inferiors as well as unscientific in their approaches to seances. There was also a good pragmatic reason for this greater tolerance: physical scientists within the SPR needed the cooperation of spiritualists to gain access to the kinds of phenomena that they believed would best exploit their expertise: the physical manifestations in seances.[24] However, most of the SPR's physical scientists shared some of the Sidgwick group's misgivings about spiritualists' methods and believed that the physical sciences could be drawn upon to forge approaches that fulfilled the requirements of robust scientific enquiry and the conditions of seances. This was also part of their attempt to limit what they perceived to be the drift of the SPR towards telepathy and purely 'psychological' subjects and to return it to the broader range of research topics that they felt lacked satisfactory scientific investigation and that the organisation had partly been established to probe.[25]

Identifying Physical–Psychical Scientists

Who were the practitioners of physical sciences in the SPR? The organisation's membership lists, regularly published in its *Proceedings*, are the obvious source here. Besides easily recognisable individuals such as

[23] Frederic W. H. Myers, *Human Personality and its Survival of Bodily Death*, 2 vols. (London: Longmans, Green, and Co., 1903), vol. 1, p. xxi. On the subliminal self and its reception see Gauld, *Founders of Psychical Research*, chapter 12.

[24] William F. Barrett, 'The Society for Psychical Research and Spiritualism', *Light*, vol. 6 (1886), pp. 51–2.

[25] See, for example, William F. Barrett to Oliver Lodge, 21 October 1912, SPR.MS 35/73, OJL-SPR; Oliver Lodge, 'On the Scientific Attitude to Marvels', *Fortnightly Review*, vol. 79 (1906), pp. 460–74, esp. p. 471.

Crookes, Rayleigh and J. J. Thomson, others can be identified from the academic titles, professional affiliations and other information provided alongside names. Over seventy astronomers, chemists, electrical and civil engineers, and physicists can be identified in the period between 1882 and about 1940, and they are shown in Table 2.1, which also gives years of birth and death, membership dates, positions reached within the SPR, primary fields of expertise and nationality. It does not constitute an exhaustive list, since there are undoubtedly some physical scientists whose identities are not obvious from the membership information provided. Nevertheless, this table serves to both strengthen and complicate contemporary observers' perceptions of the unusual presence of physicists in psychical research. In addition to the number of physicists, it shows that in the SPR's first thirty years, five of its twenty-nine presidents were professional physicists (Barrett, Lodge, the Third and Fourth Barons Rayleigh and Stewart), four of whom also fulfilled roles as Council members and as vice-presidents.

Senior ranks were filled by plenty of practitioners of other physical sciences too. Crookes, the astronomers Camille Flammarion and Frederick J. M. Stratton served as president, while Council members included John Couch Adams, the chemists William R. Bousfield and William Ramsay, the electricians Walter H. Coffin and Desmond G. Fitzgerald, and the medical electrician William H. Stone. Some additional patterns regarding physical sciences as a whole can be gleaned from Table 2.1. More joined as ordinary rather than associate members and roughly equal numbers of members hailed from astronomy, chemistry, electrical and civil engineering and physics.[26] As we might expect in an organisation based in Britain and whose founders were all Britons, most were British. We can also see that they held onto their membership for widely different lengths of time. Some, such as Barrett, Lodge, Rayleigh and J. J. Thomson, clung on for more than three decades, but others, such as Cuthbert B. Horwood and Joseph P. Hall, lasted only a year. The number of women among physical scientist members was tiny (only Marie Curie and Eleanor Sidgwick), which was proportionately smaller than the number of women in the SPR as a whole, and a reflection of the widespread gender inequality in nineteenth- and twentieth-century professional sciences and engineering.

---

[26] In the 1880s, ordinary members paid two guineas per annum, received the SPR's publications, and were eligible to occupy an official position within the organisation, vote in its Council elections and borrow books from the organisation's library and reading rooms in London. Associate members paid one guinea per annum but did not have voting or book-borrowing rights. See [Anon.], 'Constitution and Rules', *PSPR*, vol. 1 (1882–3), pp. 331–6, p. 332.

2.1 *Physical–psychical scientists in the SPR, 1882–C. 1940*

Name	Dates	In SPR	SPR position	Primary field(s) of expertise	Nationality
John Couch Adams[a]	1819–92	1891–2	H, CI	Astronomy	British
William Fletcher Barrett[b]	1844–1925	1882–1925	M, CI, VP, P	Experimental physics	British
George Thomas Beilby[c]	1850–1924	1915–24	M, CI	Industrial chemistry	British
Henry Percy Boulnois[d]	1846–1927	1894–1901	A	Civil engineering	British
William Robert Bousfield[e]	1854–1943	1924–43	M, CI	Physical chemistry and physics	British
Guy Burniston Brown[f]	1902–89	1929–31	A	Physics	British
Walter Rayleigh Browne[g]	1842–84	1882	M, CI	Civil engineering and mathematical physics	British
Alexander Mikhaylovich Butlerov[h]	1828–86	1885	Cor	Chemistry	Russian
William Lant Carpenter[i]	1841–90	1886–7	M	Electrical engineering and medical electricity	British
Arthur Prince Chattock[j]	1860–1934	1890–1934	Life A	Experimental physics	British
Walter Harris Coffin[k]	1853–1916	1882–1916	M, CI	Electrical engineering and medical electricity	Anglo-American
Daniel Frost Comstock[l]	1883–1970	1907–9	A	Electrical engineering and physics	American
Sherard Osborn Cowper-Coles[m]	1866–1936	1902–15	M	Electrometallurgy	British
John Cox[n]	1851–1923	1884–92	A	Experimental physics	British
James Mason Crafts[o]	1839–1917	1891–8	A	Chemistry	American
Quentin Charles Alexander Craufurd[p]	1875–1957	1912–29	M	Wireless telegraphy	British

Name			H, Cl, VP, P		
William Crookes[q]	1832–1919	1882–1919		Analytical chemistry and experimental physics	British
Charles Robert Cross[r]	1848–1921	1890	A	Electrical engineering	American
Marie Curie[s]	1867–1934	1913–33	Cor	Chemistry and experimental physics	Polish–French
Benjamin Davies[t]	1863–1957	1894–1939	M	Electrical engineering and physics	British
Alfred Wilks Drayson[u]	1827–1901	1882–7	M	Astronomy	British
George William Fisk[v]	1882–1972	1913–73	M	Wireless telegraphy	British
Desmond Gerald Fitzgerald[w]	1834–1904	1882–1904	M, Cl	Electrical engineering	British
Camille Flammarion[x]	1842–1925	1923	P	Astronomy	French
Edmund Edward Fournier d'Albe[y]	1868–1933	1908–10	A	Electrical engineering and physics	British
Emile Oscar Garcke[z]	1856–1930	1901–4	A	Electrical engineering	German–British
Robert M. Gordon[aa]	fl. 1880s	1884–5	A	Civil engineering	British
Christopher Clive Langton Gregory[ab]	1892–1964	1935–64	M	Astronomy	British
Joseph Platt Hall[ac]	1864–1934	1908	M	Electrical engineering	British
John Baboneau Nickterlien Hennessey[ad]	1829–1910	1905–9	M	Astronomy and surveying	British
Christian Victor Charles Herbert (6th Earl of Powis)[ae]	1904–88	1932–74	M	Astronomy	British
John Herschel[af]	1837–1921	1898–1915	A	Civil engineering and astronomy	British
Heinrich Rudolf Hertz[ag]	1857–94	1892–4	Cor	Experimental and mathematical physics	German
Harold Edward Sherwin Holt[ah]	1862–1932	1883–94	M	Civil engineering	British
Cuthbert Baring Horwood[ai]	b. 1877	1925	M	Mining engineering	British
William Benjamin Hutchinson[aj]	1863–98	1896–8	A	Civil engineering and astronomy	British
Benjamin Jordan-Smith[ak]	fl. 1900	1899–1933	M	Chemistry	British

2.1 (*cont.*)

Name	Dates	In SPR	SPR position	Primary field(s) of expertise	Nationality
Samuel Joyce[al]	1829–1906	1884–9	M	Electrical engineering	British
Charles George Lamb[am]	1867–1941	1909–39	A	Electrical engineering	British
St George Lane Fox-Pitt[an]	1856–1932	1899–1921	M, CI	Electrical engineering	British
Samuel Pierpont Langley[ao]	1834–1906	1890–1906	M, CI, VP	Astrophysics	American
Léon George Harold Lee[ap]	1883–1954	1910	A	Meteorology	British
James Ludovic Lindsay (26th Earl of Crawford and 9th Earl of Balcarres)[aq]	1847–1913	1890–1913	M, CI	Astronomy	British
Oliver Joseph Lodge[ar]	1851–1940	1884–1940	M, CI, VP, P	Experimental and mathematical physics	British
Hiram Stevens Maxim[as]	1840–1916	1910–15	A	Engineering	Anglo-American
Francis Henry Neville[at]	1847–1915	1884–95	A	Physical chemistry and metallurgy	British
Simon Newcomb[au]	1835–1909	1890–1909	M	Astronomy	American
Ernest Payne[av]	1859–1936	1899–1915	M	Electrical engineering	British
Edward Charles Pickering[aw]	1846–1919	1884–1915	Cor	Astronomy	American
James R. Pickering[ax]	b. 1857	1892–1901	A	Electrical engineering	British
William Ramsay[ay]	1852–1916	1891–1901	M, CI	Physical chemistry	British
Alec Harley Reeves[az]	1902–71	1939–71	M	Electrical engineering	British
Arthur Richardson[ba]	1858–1912	1895	A	Physical chemistry	British
Frederic William Richardson[bb]	1860–1950	1893–1933	A	Analytical chemistry	British
Edward Robinson[bc]	fl. 1900	1898–1901	A	Astronomy	British
Arthur Rücker[bd]	1848–1915	1891–1913	M	Experimental physics	British
Walter Raymond Schoeller[be]	d. 1947	1909–10	A	Analytical chemistry	British

Name					
Arthur Schuster[bf]	1851–1934	1884–8	M	Experimental physics	British
Alfred Richard Sennett[bg]	1860–1932	1885–7	M	Civil and electrical engineering	British
Alfred John Shilton[bh]	1860–92	1884–90	A	Chemistry	British
Eleanor Mildred Sidgwick[bi]	1845–1936	1884–1936	M, Cl, P, HS	Experimental physics	British
Balfour Stewart[bj]	1828–87	1882–7	M, Cl, VP, P	Experimental physics	British
William Henry Stone[bk]	1830–91	1882–5	M, Cl	Medical electricity	British
John William Strutt (3rd Baron Rayleigh)[bl]	1842–1919	1884–1919	M, Cl, VP, P	Experimental and mathematical physics	British
Robert John Strutt (4th Baron Rayleigh)[bm]	1875–1947	1925–47	M, P	Experimental physics	British
Joseph Wilson Swan[bn]	1828–1914	1884–93	M	Electrical engineering	British
Harold Dennis Taylor[bo]	1862–1943	1908–25	A	Astronomy and instrument-making	British
Joseph William Thomas[bp]	1846–1914	1892–1909	M	Analytical chemistry	British
Joseph John Thomson[bq]	1856–1940	1883–1940	M, Cl, VP	Experimental and mathematical physics	British
William Mundell Thornton[br]	1870–1944	1899–1913	A	Electrical engineering	British
George Richard Tweedie[bs]	1857–1937	1891–2	M	Analytical chemistry	British
Arthur Mannering Tyndall[bt]	1881–1961	1904–7	A	Experimental physics	British
George Nugent Merle Tyrrell[bu]	1879–1952	1908–52	M, P	Wireless telegraphy	British
Henry Addenbrook Wassell[bv]	1838–1918	1883–1910	A	Astronomy	British
Walter Weldon[bw]	1832–85	1882–5	M	Industrial chemistry	British
William Cecil Dampier Whetham[bx]	1867–1952	1904–15	M	Experimental physics	British
D. L. Leedham White[by]	1838–1905	1897–1904	M	Industrial chemistry	British
Charles John Young[bz]	fl. 1890	1888–94	A	Astronomy	British

Key to columns

A = Associate Member; Cl = Council Member; Cor = Corresponding Member; H = Honorary Member; HS = Honorary Secretary; M = Ordinary Member; P = President; VP = Vice-President

Abbreviations in references

DSB: *Dictionary of Scientific Biography*
JCS: *Journal of the Chemical Society*
JIEE: *Journal of the Institution of Electrical Engineers*
MNRAS: *Monthly Notices of the Royal Astronomical Society*
ODNB: *Oxford Dictionary of National Biography*
Venn: J. A. Venn, *Alumni Cantabrigenses: A Biographical List of all Known Students, Graduates and Holders of Office at the University of Cambridge from the Earliest Times to 1900, Part II from 1752–1900*, vols. 1–6 (Cambridge University Press, 1944–54).

[a] *ODNB.*
[b] *ODNB.*
[c] *ODNB.*
[d] [Anon.], 'Henry Percy Boulnois', *Engineer*, 14 January 1927, p. 14; H. Percy Boulnois, *Reminiscences of a Municipal Engineer* (London: St. Bride's Press, 1920).
[e] *ODNB.*
[f] www.ancestry.com
[g] Obituary Notice, *Minutes of Proceedings of the Institution of Civil Engineers*, vol. 29 (1884–5), pp. 362–6.
[h] *DSB.*
[i] Obituary Notice, *Nature*, vol. 43 (1891), p. 230.
[j] Arthur M. Tyndall, 'Arthur Prince Chattock, 1860–1934', *Obituary Notices of the Fellows of the Royal Society*, vol. 1 (1932–5), pp. 293–8.
[k] Obituary Notice, *British Dental Journal*, vol. 37 (1916), pp. 316–17.
[l] *Who Was Who in America* (1970–1), vol. 5.
[m] *ODNB.*
[n] Venn.
[o] Charles R. Cross, 'James Mason Crafts 1839–1917', *National Academy of Sciences of the United States of America, Biographical Memoirs*, vol. 9 (1919–20), pp. 159–77.
[p] Obituary Notice, *Times*, 10 May 1957, p. 15.
[q] *ODNB.*
[r] H. M. Goodwin, D. H. Dewey and H. W. Tyler, 'Charles Robert Cross', *Technology Review*, vol. 14 (1922), pp. 22–4.
[s] *DSB.*
[t] [Anon.], 'Benjamin Davies, Birmingham', *Czas y Byd*, vol. 17 (1902), pp. 49–52; R. G. Roberts, 'The Training of an Industrial Physicist: Oliver Lodge and Benjamin Davies', unpublished doctoral thesis, University of Manchester, 1984.

u Obituary Notice, *MNRAS*, vol. 62 (1902), pp. 241–2.

v Donald J. West, 'Obituary. G. W. Fisk', *JSPR*, vol. 47 (1973–4), pp. 21–3.

w [Anon.], 'Obituary. Desmond G. Fitzgerald', *Electrician*, vol. 54 (1904), p. 21; www.ancestry.com.

x *DSB*.

y 'G. A. S.', 'Dr. E. E. Fournier D'Albe', *Nature*, vol. 132 (1933), p. 125; H. H. Stephenson (ed.), *Who's Who in Science International* (London: J. & A. Churchill, 1914).

z *ODNB*.

aa *Report of the Seventieth Meeting of the British Association for the Advancement of Science, Held at Bradford in September 1900* (London: John Murray, 1901).

ab E. Margaret Burbidge, 'Christopher Clive Langton Gregory', *Quarterly Journal of the Royal Astronomical Society*, vol. 7 (1966), pp. 81–2.

ac Obituary Notice, *JIEE*, vol. 75 (1934), p. 836.

ad *ODNB*.

ae Charles Mosley (ed.), *Burke's Peerage, Baronetage and Knightage* 3 vols (Wilmington, Delaware: Genealogical Books, 107th ed., 2003).

af Obituary Notice, *MNRAS*, vol. 82 (1922), pp. 250–1.

ag *DSB*.

ah *Who's Who 2008*.

ai www.ancestry.com; *Who's Who in Engineering* (London: Compendium Publishing Company, 1939).

aj Obituary Notice, *MNRAS*, vol. 59 (1899), p. 226.

ak 'University Intelligence', *Times*, 16 August 1902, p. 5.

al [Anon.], 'Professor Samuel Joyce', *JIEE*, vol. 39 (1907), p. 789.

am [Anon.], 'C. G. Lamb', *Electrical Review*, 128 (1941), p. 656.

an [Anon.], 'Mr. Fox Pitt. Inventor and Psychic Student', *Times*, 7 April 1932, p. 14.

ao *DSB*.

ap www.ancestry.com; *Quarterly Journal of the Meteorological Society*, vol. 30 (1904), p. 343.

aq *ODNB*.

ar *ODNB*.

as *ODNB*.

at Venn.

au *DSB*.

av Venn.

aw *DSB*.

ax www.ancestry.com.

ay *ODNB*.

az ODNB.

ba J. N. Collie, 'Arthur Richardson', *JCS, Transactions*, vol. 103 (1913), pp. 766–7.

bb C. H. Manley, 'Frederic William Richardson', *Analyst*, vol. 76 (1951), p. 190.

bc *Royal Astronomical Society. List of Fellows and Associates. June 1895* (London: Spottiswoode and Co., 1895).

bd ODNB.

be www.ancestry.com.

bf ODNB.

bg www.ancestry.com.

bh [Anon.], 'Death of Mr. Shilton', *Reading Mercury, Oxford Gazette, Newbury Herald and Berks County Paper*, 6 February 1892, p. 5; [Anon], Proceedings at Annual General Meeting, *Proceedings of the Chemical Society*, vol. 8 (1892), pp. 59–66, p. 60.

bi ODNB.

bj ODNB.

bk 'W. M. O.', 'William Henry Stone', *Saint Thomas's Hospital Reports*, vol. 20 (1892), pp. xxvii–xxxii.

bl ODNB.

bm ODNB.

bn ODNB.

bo T. Smith, 'Harold Dennis Taylor', *Proceedings of the Physical Society of London*, vol. 55 (1943), pp. 508–11.

bp John Greenaway, 'Joseph William Thomas', *JCS, Transactions*, vol. 107 (1915), pp. 588–9.

bq ODNB.

br 'G. W. O. H.', 'Professor William Mundell Thornton', *JIEE*, vol. 91 (1944), pp. 475–6.

bs M. G. Tweedie, 'George Richard Tweedie', *JCS, Part I*, 1938, pp. 164–5.

bt ODNB.

bu ODNB.

bv Obituary Notice, *MNRAS*, vol. 79 (1919), pp. 233–4.

bw ODNB.

bx ODNB.

by Obituary Notice, *Minutes of Proceedings of the Institution of Civil Engineers*, vol. 160 (1905), pp. 403–4.

bz *MNRAS*, vol. 47 (1887), p. 443.

It is instructive to compare the presence of physical scientists with those from other scientific fields. In 1882, physical scientists comprised roughly 10 per cent of members, but there was the same percentage in the early 1900s when the size of the membership had more than doubled.[27] Members whose reputations, professional titles and qualifications suggest connections with physiology, zoology, botany and natural history constitute roughly 3 per cent of all members in 1882 and this had fallen to about 1 per cent in the early 1900s. Members coming from psychology, psychiatry, physiology and medicine also constituted about 10 per cent of the total membership in 1882, but this had swelled to about 30 per cent in the early 1900s. While this upward trend confirms what we might expect about the professional scientific composition of enquiries into phenomena with a strong psychological component, there remains the intriguing question of why, in its early decades, the SPR's scientific members were drawn from 'physical' and 'psychological' fields in roughly equal proportion.

Valuable as the SPR's membership is as a step towards a more comprehensive understanding of the relationship between psychical investigation and the physical sciences, it needs to be handled with as much caution as any other official membership. Length of membership is not necessarily a reliable indicator of the degree of commitment (or lack of commitment) to the subject represented by the organisation. Although Fournier d'Albe was only a member from 1908 to 1910, we know from many other sources that he sustained his interest in psychical research long after this. Conversely, in their comparatively long stints Ernest Payne, Benjamin J. Smith, Arthur Rücker, Charles J. Young and even Council member Lord Lindsay contributed practically nothing to the SPR's activities. They are among a significant number of individuals in Table 2.1, and, for that matter, in the SPR's membership as a whole, who were relatively inert and whose reasons for joining are typically unknown.[28]

J. J. Thomson contributed more to the SPR's activities than the likes of Lindsay and Rücker, but this was still dwarfed by what Barrett, Crookes, Lodge and Stewart managed. Thomson's case well illustrates the extent to which the SPR's management could tolerate members' relative inactivity if their reputations conferred much-needed intellectual lustre on the organisation. In 1921, he considered resigning from the SPR Council because he had been unable to attend its meetings. The organisation's Honorary Secretary, Eleanor Sidgwick, wanted him to stay on as a vice-president

---

[27] The percentages in this discussion are based on analysis of the membership lists published in the *PSPR* for volumes 1 (1882–3) and 15 (1900–1).

[28] This inertness is noted in Mauskopf and McVaugh, *Elusive Science*, p. 13.

instead because this was a "purely honorary" position that did not involve attending meetings.[29] After reminding him that her scientifically distinguished brother-in-law Rayleigh was kept on as vice-president although he "practically never attended meetings", she explained that having Thomson's "name on our papers [...] implying a general approval of our objects and methods" would be of "very great value to our Society".[30] Thomson was evidently persuaded, since he continued to be the largely symbolic member that he had been since the 1880s.[31]

Thomson's case is illuminating in two other respects. First, he illustrates the need to understand the different levels of interest that SPR members had in different *kinds* of psychical phenomena. There were some members, notably Barrett, Crookes and Lodge, who were reasonably catholic in their psychical tastes, but Thomson was more selective: he expressed particular views about telepathy and water-divining, and helped investigate the physical phenomena of spiritualism, but seems to have had little or no interest in apparitions, haunted houses, mesmerism and Reichenbach's od.[32] Second, he illustrates the need to understand the extent to which different SPR members became convinced by the evidence for different psychical phenomena. By 1900, Barrett, Crookes and Lodge had become convinced of telepathy and 'telekinesis', Myers's term for the movement of untouched objects by the power of thought or some unknown force. In his 1933 autobiography, Thomson expressed continued reservations about both types of psychical effect, despite declaring them eminently worthy of further investigation.[33]

The most obvious limitation of the SPR's membership is that there were many practitioners of the physical sciences who showed an interest in psychical phenomena but had either (a) died before the SPR was founded, (b) largely abandoned this interest before 1882, or (c) did not join. We have already met three individuals who fall into these categories: Reichenbach (a), Tyndall (b) and Varley (c). Between the 1850s and the 1930s it is possible to identify 75 such individuals from unpublished and published source materials, including the correspondence of SPR physical scientists. They are listed in Table 2.2, which, like Table 2.1, provides

[29] Eleanor Sidgwick to J. J. Thomson, 6 January 1921, JJT H.22/1, J. J. Thomson Papers, Trinity College Library, Cambridge.

[30] Sidgwick to Thomson, 6 January 1921.

[31] Heinrich Hertz also fulfilled a symbolic function. Despite having only an armchair interest in psychical research, he was still elected as a foreign member of the SPR and earned an obituary notice in the organisation's *Journal*: Oliver Lodge, 'Professor Heinrich Hertz', *JSPR*, vol. 6 (1893–4), pp. 197–9.

[32] J. J. Thomson, *Recollections and Reflections* (London: G. Bell, 1936), pp. 147–63 and 379–83.

[33] Thomson, *Recollections and Reflections*, pp. 147–58.

2.2 *Physical–psychical scientists outside the SPR, c.1850–1940*

Name	Dates	When interested in psychical effects	Primary field(s) of scientific expertise	Nationality
Dominique François Jean Arago[a]	1786–1853	1850s	Natural philosophy	French
Jacques-Arsène d'Arsonval[b]	1851–1940	1900s	Physics and physiology	French
Francis William Aston[c]	1877–1945	1910s	Chemistry and experimental physics	British
Jacques Babinet[d]	1794–1872	1840s–50s	Natural philosophy	French
John Logie Baird[e]	1888–1946	1910s–30s	Electrical engineering	British
Kristian Olaf Bernhard Birkeland[f]	1867–1917	1910s	Experimental physics	Danish
Éduoard Eugène Désiré Branly[g]	1844–1940	1900s	Electrical engineering	French
David Brewster[h]	1781–1868	1810s–60s	Natural philosophy	British
Frank Playfair Burt[i]	1879–1938	1900s	Physical chemistry	British
James Rand Capron[j]	1829–88	1870s	Astronomy and photography	British
Richard Chenevix[k]	1774–1830	1820s	Chemistry and mineralogy	Irish
Josiah Latimer Clark[l]	1822–98	1850s–60s	Electrical engineering	British
Reginald Charles Clinker[m]	1876–1931	1890s	Electrical engineering	British
William Walter Coblentz[n]	1873–1962	1880s–1910s	Experimental physics	American
William Roy Cecil Coode-Adams[o]	1895–1961	1920s	Chemistry and electrical engineering	British
William Jackson Crawford[p]	1881–1920	1910s	Civil engineering	Irish
Pierre Curie[q]	1859–1906	1890s	Experimental physics and chemistry	French
George Howard Darwin[r]	1845–1912	1870s	Mathematics and geophysics	British
Amos Emerson Dolbear[s]	1837–1910	1880s–1900s	Experimental physics	American

2.2 (cont.)

Name	Dates	When interested in psychical effects	Primary field(s) of scientific expertise	Nationality
Thomas Alva Edison[t]	1847–1931	1870s–1920s	Electrical engineering	American
Norman Edwards[u]	fl.1920s	1920s	Wireless telegraphy	British
Giovanni Battista Ermacora[v]	1858–98	1890s	Experimental and mathematical physics	Italian
Michael Faraday[w]	1791–1867	1850s	Chemistry and natural philosophy	British
George Francis FitzGerald[x]	1851–1901	1880s–90s	Experimental and mathematical physics	Irish
Cyril Charles James Frost[y]	1894–1960	1920s–30s	Wireless telegraphy	British
Reginald Gibbs[z]	1898–1997	1930s	Experimental physics	British
Charles Henry Gimingham[aa]	1853–90	1870s	Experimental physics	British
William Gregory[ab]	1803–58	1840s–50s	Chemistry and medicine	British
Fritz Grunewald[ac]	fl. 1900s	1900s	Electrical engineering	German
Edwin Herbert Hall[ad]	1855–1938	1880s	Experimental physics	American
William Hallock[ae]	1857–1913	1900s	Experimental physics	American
Herbert Anthony Hankey[af]	1884–1961	1930s	Electrical engineering	British
Robert Hare[ag]	1781–1858	1850s	Chemistry	American
William Henry Harrison[ah]	1841–97	1860s–90s	Photography and telegraphic engineering	British
Eduard Haschek[ai]	1875–1947	1910s	Experimental physics	Austrian
John Hettinger[aj]	1880–1956	1930s	Wireless telegraphy	Romanian
Bernard Joseph Hopper[ak]	1876–1947	1930s	Experimental physics	British
Edwin James Houston[al]	1847–1914	1880s–1930s	Electrical engineering	American
William Huggins[am]	1824–1910	1870s	Astronomy	British

Name	Dates	Period	Field	Nationality
Walter John Kilner*am*	1847–1920	1890s–1910s	Medical electricity	British
William Kingsland*ao*	1855–1936	1880s–1930s	Electrical engineering	British
Paul Langevin*ap*	1872–1946	1990s	Experimental and mathematical physics	French
Edwin John Godfrey Lewis*aq*	1903–77	1920s	Wireless telegraphy	French
William Lynd*ar*	1845–1911	1890s–1900s	Telegraphic engineering	British
Dmitrii Ivanovich Mendeleev*as*	1834–1907	1870s	Chemistry	Russian
Stefan Meyer*at*	1872–1949	1920s	Experimental physics	Austrian
A. E. Mundy*au*	fl. 1920s	1920s	Chemistry and electrical engineering	British
Hans Christian Oersted*av*	1777–1851	1810s	Natural philosophy	Danish
Jean Baptiste Perrin*aw*	1870–1942	1900s	Experimental physics	French
William Henry Pickering*ax*	1858–1938	1880s	Astronomy	American
Francesco Porro de' Somenzi*ay*	1861–1937	1900s	Astronomy	Italian
Samuel Tolver Preston*az*	1844–1917	1890s	Civil engineering and physics	British
Richard Anthony Proctor*ba*	1837–88	1880s	Astronomy	British
Karl Przibram*bb*	1878–1973	1920s	Experimental physics	Austrian
Alexander Oliver Rankine*bc*	1881–1956	1930s	Experimental physics	British
Karl Freiherr von Reichenbach*bd*	1788–1869	1830s–60s	Industrial chemistry	German
Herbert Stanley Redgrove*be*	1887–1943	1880s–1940s	Chemistry	British
James Emerson Reynolds*bf*	1844–1920	1870s	Chemistry	Irish
Philip James Risdon*bg*	1879–1947	1920s	Wireless telegraphy	British
John Obadiah Newell Rutter*bh*	1799–1888	1850s	Civil engineering	British
Giovanni Virginio Schiaparelli*bi*	1835–1910	1890s	Astronomy	Italian
John Hawkins Simpson*bj*	fl. 1860	1860s	Telegraphic engineering	British

2.2 (cont.)

Name	Dates	When interested in psychical effects	Primary field(s) of scientific expertise	Nationality
John Toby Sprague[bk]	1824–1906	1840s–1900s	Electrical engineering	British
John Traill Taylor[bl]	1827–95	1890s	Photography	British
Silvanus Phillips Thompson[bm]	1851–1916	1870s–1910s	Electrical engineering and experimental physics	British
Hans Thirring[bn]	1888–1976	1920s	Experimental physics	Austrian
Augustus Trowbridge[bo]	1870–1934	1900s	Experimental physics	American
George William De Tunzelmann[bp]	b. 1856	1910s	Electrical engineering and experimental physics	British
John Tyndall[bq]	1820–93	1860s	Experimental physics	Irish
Francis Lawry Usher[br]	1885–1969	1900s	Physical chemistry	British
Cromwell Fleetwood Varley[bs]	1828–83	1850s–80s	Telegraph engineering	British
René Warcollier[bt]	1881–1962	1920s–60s	Physical chemistry	French
R. A. Watters[bu]	fl. 1930s	1930s	Experimental physics	American
Wilhelm Eduard Weber[bv]	1804–91	1870s	Experimental and mathematical physics	German
Edmund Basil Wedmore[bw]	1876–1956	1890s	Electrical engineering	British
Adolf Ferdinand Weinhold[bx]	1841–1917	1870s–80s	Experimental physics	German
Robert William Wood[by]	1868–1955	1900s	Experimental physics	American
Harry Edward Yerbury[bz]	1871–1955	1930s	Electrical engineering	British
Johann Karl Friedrich Zöllner[ca]	1834–82	1870s	Astronomy	German

Abbreviations in references

ANB: *American National Biography*

COURTIER: Jules Courtier, *Documents sur Eusapia Palladino. Rapport sur les séances d'Eusapia Palladino:à l'Institut général psychologique en 1905, 1906, 1907 et 1908 (Paris:Institut général psychologique, 1908).*

DSB: *Dictionary of Scientific Biography*

JCS: *Journal of the Chemical Society*

JIEE: *Journal of the Institution of Electrical Engineers*

JSPR: *Journal of the Society for Psychical Research*

MNRAS: *Monthly Notices of the Royal Astronomical Society*

ODNB: *Oxford Dictionary of National Biography*

Venn: J. A. Venn, *Alumni Cantabrigienses: A Biographical List of All Known Students, Graduates and Holders of Office at the University of Cambridge from the Earliest Times to 1900, Part II*, vols. 1–6 (Cambridge University Press, 1944–54).

[a] *DSB*; Jacques Babinet, 'Les sciences occultes au XIXe siècle, les tables tournantes et les manifestations prétendus surnaturelle considérées au point de vue de la science de l'observation', *Revue des deux mondes*, vol. 6 (1854), pp. 510–32.

[b] *DSB*; Courtier.

[c] *ODNB*; Jeff Hughes, 'Occultism and the Atom: The Curious Story of Isotopes', *Physics World*, vol. 16 (2003), pp. 31–5.

[d] *DSB*; Jacques Babinet, 'Sciences des tables tournantes au point de vue de la mécanique et de la physiologie', *Revue des deux mondes*, vol. 5 (1854), p. 410.

[e] *ODNB*; Antony Kamm and Malcolm Bird, *John Logie Baird: A Life* (Edinburgh: National Museums of Scotland Publishing, 2002), pp. 382–8.

[f] *BDS*; Lucy Jago, *The Northern Lights: How One Man Sacrificed Love, Happiness and Sanity to Unlock the Secrets of Space* (London: Penguin, 2002), pp. 198–200.

[g] Lance Day and Ian McNeil (eds.), *Biographical Dictionary of the History of Technology*; Courtier.

[h] *ODNB*; [David Brewster], 'Pretensions of Spiritualism – Life of D. D. Home', *North British Review*, vol. 39 (1863), pp. 174–206.

[i] H. H. Stephenson (ed.). *Who's Who in Science International* (London: J. & A. Churchill, 1914); www.ancestry.com; F. L. Usher and F. P. Burt, 'Thought Transference (some Experiments in Long Distance Thought-Transference)', *Annals of Psychical Science*, vol. 8 (1909), pp. 561–600.

[j] Obituary Notice, *MNRAS*, vol. 49 (1889), pp. 159–61; J. R. Capron, *Aurorae: Their Characters and Spectra* (London: E. and F. N. Spon, 1879), pp. 165–6.

[k] *ODNB*; Richard Chenevix, 'Observations and Experiments on Mesmerism', *London Medical and Physical Journal*, vol. 61 (1829), pp. 491–501; vol. 62 (1829), pp. 114–25, 210–20, 315–24.

[l] *ODNB*; Frank A. J. L. James (ed.), *The Correspondence of Michael Faraday Volume 5 November 1855–October 1869 Letters 3033–3872* (London: Institution of Engineering and Technology, 2008), pp. 221–3.

[m] Obituary Notice, *JIEE*, vol. 69 (1931), pp. 1319–20; Arthur Chattock, 'Experiments in Thought-Transference', *JSPR*, vol. 8 (1897–8), pp. 302–5.

[n] *ANB*; W. W. Coblentz, *Man's Place in the Superphysical World* (New York: Philosophical Library, 1951).

[o] 'G. H. R. M.', 'William Roy Cecil Coode-Adams', *JIEE*, vol. 7 (1961), p. 752; W. R. C. Coode-Adams, *A Primer of Occult Physics* (London: Theosophical Publishing House, 1927).

[p] Allan Barham, 'Dr. W. J. Crawford, His Work and Legacy in Psychokinesis', *JSPR*, vol. 55 (1988), pp. 113–38; W. J. Crawford, *The Reality of Psychic Phenomena* (London: John M. Watkins, 1919).

[q] *DSB*; Courtier.

[r] *ODNB*; George H. Darwin and others, Notes on seances in 1874 and 1876, Society for Psychical Research Archive, Cambridge University Library, SPR.MS.14.

[s] [Anon.] 'Amos Emerson Dolbear', *Popular Science Monthly*, vol. 56 (1910), pp. 415–16; A. E. Dolbear, 'Implications of Physical Phenomena', *The Psychical Review: A Quarterly Journal of Psychical Science and Organ of the American Psychical Society*, vol. 1 (1893), pp. 7–15, 211–14.

[t] *ANB*; Dagobert D. Runes, *The Diary and Sundry Observations of Thomas Alva Edison* (New York: Philosophical Library, 1968), pp. 205–44.

[u] Norman Edwards, *Through a Young Man's Eyes* (London: Heath Cranton, 1928), pp. 62–6.

[v] [Anon.], 'G. B. Ermacora', *Rivista di Studi Pschici*, vol. 4 (1898), pp. 103–8; G. B. Ermacora, *La Telepatia* (Padova: L. Crescini, 1898).

[w] *ODNB*; Michael Faraday, 'Experimental Investigation of Table-Moving', *Athenaeum*, 2 July 1853, pp. 801–3.

[x] *ODNB*; [Anon.], 'General Meeting', *JSPR*, vol. 5 (1891–2), pp. 167–72.

[y] www.ancestry.com; Q. C. A. Craufurd and Jack Frost, 'Psychic Communication and Wireless: A New Instrument', *Light*, 48 (1928), p. 305. Bill Fox, 'Reginald Gibbs 1898–1997', *Physics World*, November 1997, p. 56; Harry Price, *Fifty Years of Psychical Research: A Critical Survey* (London: Longmans, Green and Co., 1939), p. 262.

[aa] Obituary notice, *Electrician*, 3 October 1890, p. 625; E. E. Fournier d'Albe, *The Life of Sir William Crookes* (London: T. Fisher Unwin, 1923).

[ab] *ODNB*; William Gregory, *Letters to a Candid Enquirer on Animal Magnetism* (London: Taylor, Walton and Moberly, 1851).

[ac] Obituary notice, *Psychic Science: Quarterly Transactions of the British College of Psychic Science*, vol. 4 (1925–6), pp. 238–9; Fritz Grunewald, *Physikalisch-mediumistische Untersuchungen* (Pfullingen: Johannes Baum, 1920).

[ad] *DSB*; Edwin H. Hall, 'Sir Oliver Lodge's British Association Address', *Harvard Theological Review*, vol. 8 (1915), pp. 238–51.

[ae] H. H. Stephenson (ed.), *Who's Who in Science International* (London: J. & A. Churchill, 1912–14); Charles L. Dana et al., 'Report of an Investigation of the Phenomena Connected with Eusapia Palladino', *Science*, vol. 31 (1910), pp. 776–80.

[af] Obituary Notice, *Wireless World*, July 1961, p. 349; H. Anthony Hankey, 'Ether Vibrations in Television', *Psychic Science: Quarterly Transactions of the British College of Psychic Science*, vol. 13 (1934–5), pp. 149–53.

ag DSB; Robert Hare, *Experimental Investigations of the Spirit Manifestations* (New York: Partridge & Brittain, 1855).

ah [Anon.], 'The Late Mr. W. H. Harrison', *British Journal of Photography*, vol. 44 (1897), p. 539; www.ancestry.com; William H. Harrison, *Spirit People: A Scientifically Accurate Description of Manifestations Recently Produced by Spirits* (London: W. H. Harrison, 1875).

ai [Anon.], 'Haschek, Eduard', in *Deutsche Biographie*, www.deutsche-biographie.de/sfz063_00008_1.html; Eduard Haschek, 'Über Leuchterscheinungen des menschlichen Körpers', *Sitzungsberichte der Kaiserlichen Akademie der Wissenschaften. Mathematisch-Naturwissenschaften Klasse*, vol. 123 (1914), pp. 523–32.

aj 'F. H. D.', 'John Hettinger', *JIEE*, vol. 2 (1956), pp. 627–8; John Hettinger, *The Ultra-Perceptive Faculty* (London: Rider and Co., 1940).

ak www.ancestry.com; B. J. Hopper, *Enquiry into the Cloud-Chamber Method of Studying the 'Intra-Atomic Quantity'* (London: International Institute for Psychical Research, 1936).

al *ANB*; Edwin J. Houston, 'Cerebral Radiation', *Journal of the Franklin Institute*, vol. 133 (1892), pp. 488–97.

am *ODNB*; William Crookes, 'Experimental Investigation of a New Force', *Quarterly Journal of Science*, vol. 1 (1871), pp. 339–49.

an Venn; Walter J. Kilner, *The Human Atmosphere or The Aura Made Visible by the Aid of Chemical Screens* (London: Rebman Limited, 1911).

ao 'C. J. S. M.', 'William Kingsland', *JIEE*, vol. 79 (1936), p. 696; William Kingsland, *The Physics of the Secret Doctrine* (London: The Theosophical Publishing Society, 1910).

ap *DSB*; Courtier.

aq www.ancestry.com; E. J. G. Lewis, 'Spirit Communication and the Ether', *English Mechanics and the World of Science*, 12 March 1926, p. 134.

ar Obituary notice, *Electrical Review*, vol. 68 (1911), p. 929; William Lynd, 'Thought Transference and Wireless Telegraphy', *Surrey Magazine*, vol. 1 (1900), pp. 24–7.

as *DSB*; Dmitri I. Mendeleev, *Materialy dlia suzhdeniia o spiritisme* (St Petersburg: Obshchestvennaia Pol'za, 1876).

at Helmut Rechenburg, 'Meyer, Stefan', in *Deutsche Biographie*, www.deutsche-biographie.de/sfz62835.html; *Rudi Schneider: The Vienna Experiments of Professors Meyer and Przibram*, Bulletin 5 of the National Laboratory of Psychical Research (London: National Laboratory of Psychical Research, 1933).

au A. E. Mundy, 'Vibrations in the Air and Ether', *British Journal of Psychical Research*, vol. 2 (1928), pp. 8–12.

av *DSB*; Erik Bjelfvenstam, 'Hypnotism in Scandinavia 1800–1900', in Eric J. Dingwall (ed.), *Abnormal Hypnotic Phenomena: A Survey of Nineteenth Century Cases*, vol. 2 (London: J. and A. Churchill, 1967), pp. 203–46.

aw *DSB*; Courtier.

ax *DSB*; H. P. Bowditch et al., 'Report of the Committee on Thought-Transference', *Proceedings of the American Society for Psychical Research*, vol. 1 (1885–9), pp. 6–49.

ay Sandra Ciccone, 'Francesco Porro de' Somenzi', in T. Hockey et al. (eds.), *Biographical Encyclopaedia of Astronomers*, 2 vols. (New York: Springer, 2012), vol. 2; Hereward Carrington, *Eusapia Palladino and Her Phenomena* (London: T. Werner Laurie, 1910).

az Obituary Notice, *Nature*, 3 May 1917, p. 190; www.ancestry.com; S. Tolver Preston, 'On the Physics of Thought-Reading [1893]', Add. MS. 7654, P45a, J. J. Thomson Papers, Cambridge University Library.

ba *ODNB*; Richard A. Proctor, 'Thought-Reading', *Knowledge*, vol. 2 (1882), pp. 51, 68–9, 106–7, 128, 161–2.

bb Helmut Rechenburg, 'Przibram, Karl', in *Deutsche Biographie*, www.deutsche-biographie.de/sfz97580.html; *Rudi Schneider: The Vienna Experiments of Professors Meyer and Przibram*, Bulletin 5 of the National Laboratory of Psychical Research (London: National Laboratory of Psychical Research, 1933).

bc George P. Thomson, 'Alexander Oliver Rankine', *Biographical Memoirs of the Fellows of the Royal Society*, vol. 2 (1956), pp. 248–55; Harry Price, *Rudi Schneider: A Scientific Examination of His Mediumship* (London: Methuen and Co., 1930), p. 38.

bd Michael Engel, 'Reichenbach, Karl Freiherr von', in *Deutsche Biographie*, www.deutsche-biographie.de/gnd107050978.html#ndbcontent; Karl von Reichenbach, *Researches on Magnetism, Electricity, Heat, Light, Crystallization, and Chemical Attraction, in Their Relation to the Vital Force*, translated by William Gregory (London: Taylor, Walton and Moberly, 1850).

be www.ancestry.com; Herbert S. Redgrove and I. M. L. Redgrove, *Joseph Glanvill and Psychical Research in the Seventeenth Century* (London: W. Rider and Son, 1921).

bf *ODNB*; E. E. Fournier d'Albe, *The Life of Sir William Crookes* (London: T. Fisher Unwin, 1923), p. 185.

bg www.ancestry.com; P. J. Risdon, 'Psychic Phenomena and Wireless', *Popular Wireless Weekly*, 29 August 1922, pp. 237–8.

bh Obituary Notice, *MNRAS*, vol. 49 (1889), pp. 168–9; J. O. N. Rutter, *Human Electricity: The Means of Its Development, Illustrated by Experiments* (London: John W. Parker, 1854).

bi *DSB*; Camille Flammarion, *Mysterious Psychic Forces: An Account of the Author's Investigations in Psychical Research, Together with Those of other European Savants* (Boston: Small, Maynard and Company, 1907).

bj [Anon.], 'Single-Wire Typo-Telegraph by J. Hawkins Simpson', *Engineer*, 15 November 1867, pp. 421–2; [Anon.], 'Lyon v. Home', *Spiritual Magazine*, vol. 3 (1868), pp. 241–54.

bk [Anon.], 'The Late Mr. J. T. Sprague', *English Mechanic and World of Science*, 9 February 1906, pp. 3–4; John T. Sprague, 'Psychology', *English Mechanic and World of Science*, vol. 56 (1892), pp. 171–2, 219–20, 263–4, 307–8, 351–2, 393–4, 487–8, 532–4, 579–80; vol. 57 (1893), 23–4.

bl [Anon.], 'In Memoriam: The Late J. Traill Taylor', *British Journal of Photography*, vol. 42 (1895), p. 725; John Traill Taylor, '"Spirit Photography", with Remarks on Fluorescence', *British Journal of Photography*, 17 March 1893, pp. 167–9.

bm *ODNB*; Silvanus P. Thompson, 'A Physiological Effect of an Alternating Magnetic Field', *Philosophical Transactions of the Royal Society of London*, Series B, vol. 82 (1909–10), pp. 396–8.

bn [Anon.], 'Thirring, Hans', *Deutsche Biographie*, www.deutsche-biographie.de/sfz132004.html#indexcontent; Hans Thirring, 'The Position of Science in Relation to Psychical Science', *British Journal of Psychical Research*, vol. 1 (1927), pp. 164–81.

bo *ANB*; Charles L. Dana et al., 'Report of an Investigation of the Phenomena Connected with Eusapia Palladino', *Science*, vol. 31 (1910), pp. 776–80.

bp *Electrical Trades' Directory and Handbook for 1894* (London: 'The Electrician', 1894); George W. De Tunzelmann, *A Treatise on Electrical Theory and the Problem of the Universe* (London: Charles Griffin and Co., 1910), pp. 626–32.

bq ODNB; [John Tyndall], 'Science and the Spirits', *Reader*, vol. 4 (1864), pp. 725–6.

br E. G. Cox, 'Dr. F. L. Usher', *The University of Leeds Review*, vol. 2 (1950), pp. 167–8; www.ancestry.com; F. L. Usher and F. P. Burt, 'Thought Transference (some Experiments in Long Distance Thought-Transference), *Annals of Psychical Science*, vol. 8 (1909), pp. 561–600.

bs ODNB; Cromwell F. Varley, 'Evidence of Mr. Varley', in *Report on Spiritualism of the Committee of the London Dialectical Society* (London: Longmans, Green, Reader and Dyer, 1871), pp. 157–72.

bt Helene Pleasants (ed.), *Biographical Dictionary of Parapsychology with Directory and Glossary, 1964–66* (New York: Garrett Publications, 1964); René Warcollier, *La télépathie recherches expérimentales* (Paris: Félix Alcan, 1921).

bu R. A. Watters, *The Intra-atomic Quantity* (Reno, NV: Dr. William Bernard Johnson Foundation for Physiological Research, 1933).

bv DSB; Johann K. F. Zöllner, *Transcendental Physics. An Account of Experimental Investigations. From the Scientific Treatises*, tr. C. C. Massey (London: W. H. Harrison, 1880).

bw J. Greig, 'Mr. E. B. Wedmore, C. B. E', *Nature*, vol. 178 (1956), pp. 238–9; Arthur P. Chattock, 'Experiments in Thought-Transference', *JSPR*, vol. 8 (1897–8), pp. 302–5.

bx Dagmar Szöllösi, 'Adolf Ferdinand Weinhold', *Sächsische Biografie*, http://saebi.isgv.de/biografie/Adolf_Ferdinand_Weinhold_(1841–1917); Adolf F. Weinhold, *Hypnotische Versuche. Experimentale Beiträge zur Kenntniss des sogenannten thierschen Magnetismus* (Chemnitz: Martin Bülz, 1879).

by DSB; Charles L. Dana et al., 'Report of an Investigation of the Phenomena Connected with Eusapia Palladino', *Science*, vol. 31 (1910), pp. 776–80.

bz 'J. A. W.', 'Harry Edward Yerbury', *JIEE*, vol. 2 (1956), p. 49; H. E. Yerbury, 'Precis of Lecture Delivered Before the Reading Society for Psychical Investigation', *Psychic Science: Quarterly Transactions of the British College of Psychic Science*, vol. 14 (1935–6), pp. 165–70.

ca DSB; Johann K. F. Zöllner, *Transcendental Physics. An Account of Experimental Investigations, From the Scientific Treatises*, translated by C. C. Massey (London: W. H. Harrison, 1880).

years of birth and death, fields of expertise and nationality, but also includes the period during which individuals were known to have shown significant interest in psychical phenomena and references to principal sources of biographical information and of their psychical interest. Again, like Table 2.1, Table 2.2 is not intended as an exhaustive list, and further research will inevitably turn up more names, but, as it stands, it further strengthens and complicates perceptions that a curiously large number of physicists engaged in psychical investigation, the main complication being that physicists were joined by practitioners from a range of other physical sciences. It is because physicists were not the only practitioners of the physical sciences inside or outside the SPR who showed an interest in psychical investigation that, for the sake of brevity, we shall refer to them collectively as 'physical–psychical scientists'.

Although there are roughly the same number of physical–psychical scientists in Tables 2.1 and 2.2, the latter table covers more nationalities and a wider range of attitudes towards psychical phenomena. This completely male-dominated table includes individuals such as the American physicists Edwin H. Hall and Robert W. Wood, and the Russian chemist Dmitrii Mendeleev who, while being notorious for their scepticism towards psychical effects, qualify for inclusion among physical–psychical scientists because, unlike those who blankly refused to investigate such effects, they devoted at least some time and effort to them.[34] In contrast, it also includes individuals whose psychical enquiries led them to a firm conviction in one or more psychical phenomena. The American physicist William W. Coblentz and the British radio engineer Cyril 'Jack' Frost, for example, became devout spiritualists, while the British engineers William Coode-Adams and William Kingsland became followers of modern Theosophy, which had been launched in the United States by Helena Petrovna Blavatsky, Henry Steele Olcott and William Quan Judge in the 1870s. It was probably the long-running conflict between the SPR and modern Theosophy that dissuaded Coode-Adams and Kingsland from joining the organisation: the conflict started in 1884–5 when the SPR published damning evidence that Blavatsky had forged letters allegedly written and materialised out of thin air by the invisible bearers of esoteric wisdom with whom this co-founder of modern Theosophy claimed to be in communion.[35]

---

[34] On Mendeleev and spiritualism see Gordin, *Well-Ordered Thing*, chapter 4. An idiosyncratic study of Wood is William Seabrook, *Doctor Wood: Modern Wizard of the Laboratory* (New York: Harcourt Books, 1941). For Hall's involvement in psychical research see Edwin H. Hall, 'Sir Oliver Lodge's British Association Address', *Harvard Theological Review*, vol. 8 (1915), pp. 238–51.

[35] Oppenheim, *Other World*, pp. 174–8; J. Barton Scott, 'Miracle Publics: Theosophy, Christianity, and the Coulomb Affair', *History of Religions*, vol. 49 (2009), pp. 172–96. The debacle does not appear to have affected the allegiances of SPR physical scientists

Physical–psychical scientists in Table 2.2 seem to have been more selective in their psychical interests than those in Table 2.1. Some, such as Frank P. Burt, Reginald C. Clinker, Edwin J. Houston, William Lynd, S. Tolver Preston, Francis L. Usher and Edmund B. Wedmore, appear only to have been interested in telepathy, while Jacques-Arsène d'Arsonval, Kristian Birkeland, Latimer Clark, William J. Crawford, Fritz Grunewald, Augustus Trowbridge and Johann K. F. Zöllner focussed solely on the physical phenomena of spiritualism.

Many of the physical–psychical scientists in Table 2.2 were good deal more open about their psychical interests than others. Harrison edited a newspaper on spiritualism, and the chemist Herbert S. Redgrove published numerous articles and books on alchemy and other occult topics, but Francis Aston was only prepared to make subtle allusions to modern Theosophy and William Huggins kept his spiritualistic interests largely private, the latter prompting a frustrated Crookes to accuse him of being "a coward with his pen" but "bold as lion in talking" about the subject.[36] Huggins shared many scientists' anxiety that public associations with spiritualism could harm one's professional reputation, and this was probably one of several undisclosed reasons why he declined to attend Barrett's psychical research conference in 1882.[37]

Connecting Physical–Psychical Scientists

One of the reasons why we can compile Table 2.2 is because so many physical–psychical scientists within the SPR shared their psychical interests with plenty of physical–psychical scientists outside and inside the organisation. Indeed, physical–psychical scientists constitute several informal networks of individuals who corresponded, had face-to-face conversations about, collaborated on, and otherwise interacted on psychical matters. Many of these networks centred on the most active (and best-known) physical–psychical scientists, such as Barrett, Crookes and Lodge, and were simply the result of pre-existing networks. For example, it is hardly surprising that Lodge, Oliver Heaviside and George F. FitzGerald interacted on psychical matters given that by the 1880s they already defined a network of physicists

---

with strong Theosophical interests, such as Henry Boulnois, Lindsay, St George Lane Fox Pitt and Arthur Richardson.

[36] On Redgrove see Morrisson, *Modern Alchemy*, chapter 1. For Aston see Hughes, 'Occultism and the Atom'. William Crookes to D. D. Home, 18 July 1871, quoted in [Julie] Dunglas Home, *D. D. Home: His Life and Mission*, ed. by Arthur Conan Doyle (London: Kegan Paul, Trench, Trübner and Co., 1921), p. 201.

[37] William Huggins to William F. Barrett, 23 December 1881, SPR.MS 3/A4/39, WFB-SPR.

interpreting Maxwell's electromagnetic theory.[38] And Barrett, Stewart and the physicist and electrical engineer Silvanus Thompson were corresponding about their research and teaching in physics long before they started exchanges on psychical subjects.[39]

The entanglement of physical and psychical matters in these networks was particularly prominent in the 1890s. This was the decade when Crookes, Rayleigh and Ramsay collaborated on experimental evidence for the inert gas argon and often discussed old and new investigations into spiritualism; when Crookes, FitzGerald, Lodge, Heaviside, S. Tolver Preston and J. J. Thomson exchanged speculations on electromagnetic theories of optics and of telepathy; when Lodge and FitzGerald combined discussions of the fundamental laws of mechanics and the baffling mechanical powers of the American stage magician Annie Abbott; and when Lodge helped Crookes with the physical and psychical parts of Crookes's presidential address to the 1898 meeting of the British Association.[40]

A more compelling illustration of this entanglement comes from Lodge's diary for 1892. During one particularly busy week spent in London in April, he lectured on a Friday at the Royal Institution on the ether of space, lunched with fellow physicist and SPR member Arthur Rücker the following Sunday, had lunch dates with Crookes on the Monday and Arthur Balfour (another SPR member) on the Tuesday, lunched with the Rayleighs and dined with Myers on the Thursday, and ended his visit by attending meetings at the Physical Society of London and the SPR.[41] The conversations between Lodge and his dining companions are likely to have ranged across a variety of physical and psychical questions. They probably included the relationship of ether to gross matter and to psychical phenomena (topics that would have interested Crookes, Myers and Rayleigh), Nikola Tesla's recent London lectures on

[38] Hunt, Maxwellians.

[39] William F. Barrett to Silvanus P. Thompson, 10 November 1879, f. 20, SPT-IC; Balfour Stewart to William F. Barrett, 22 December 1871, f. 120, WFB-RS.

[40] Copy of a letter from Lord Rayleigh to William Ramsay, 4 January 1896, Argon Correspondence Microfilm, Box #24, Series 5, R-USAF; Lord Rayleigh, John William Strutt, Life of Lord Rayleigh (London: Edward Arnold and Co., 1924), p. 265; Mrs C. W. Earle, Memoirs and Memories (London: Smith, Elder and Co., 1911), pp. 381–2; William Crookes to Oliver Lodge, 19 August 1898, SPR.MS 35/341, OJL-SPR; Oliver Heaviside to Oliver Lodge, 26 August 1896, f. 100, MS Add. 89/50(ii), OJL-UCL; S. Tolver Preston to J. J. Thomson, 18 May 1893, P47, JJT-CUL; S. Tolver Preston to George F. FitzGerald, 3 September 1890, 11/63, George Francis FitzGerald Papers, RDS Library and Archives; George F. FitzGerald to Oliver Lodge, 17 December 1891, MS Add. 89/35, OJL-UCL; William Crookes to Oliver Lodge, 25 June 1898, SPR.MS 35/328, OJL-SPR.

[41] Oliver Lodge, Diary for 1892, OJL/2/3/9, Oliver Lodge Papers, Cadbury Research Library, University of Birmingham.

the mysterious phenomena of electrical discharge (in which Crookes had strong interests), Lodge's recent tests of telepathy (a topic enormously interesting to Balfour and Myers) and Myers's analysis of the evidence for telekinesis (an effect that Crookes and Rayleigh had investigated).[42]

Lodge's metropolitan movements illustrate the fact that psychical investigation was a topic that physical–psychical scientists tended to share with more than just fellow physical scientists. It was precisely because it was a subject with obvious connections with psychology, philosophy, religion and spirituality that physical–psychical scientists often found it as fruitful to discuss their psychical interests with philosophers and clergymen as fellow physicists, chemists, astronomers and engineers.[43]

In 1881, after studying, researching and teaching physics in London, Lodge had taken up the post of Professor of Physics at University College Liverpool and this made it more difficult for him to travel to London, where many of the distinguished scientific and other learned societies had their homes, and to Cambridge, where the SPR's dominant Sidgwick group was based. The difficulty that Lodge faced in sharing physical and psychical interests with Crookes, Myers and other like-minded individuals based in London or Cambridge was also felt by Stewart, who from 1870 until his death in 1887 was Professor of Natural Philosophy at Owens College Manchester, and, *a fortiori*, by Barrett, who, from 1873 to 1910, was Professor of Experimental Physics at the Royal College of Science in Dublin. In some ways, the problem of distance explains why some physical–psychical scientists were inclined to cultivate an interest in psychical research locally as well as share it with distant peers. Indeed, in

---

[42] The prominence of Tesla as a topic is suggested by Arthur Rücker to Oliver Lodge, 2 February 1892, MS Add. 89/91, OJL-UCL, and William Crookes to Oliver Lodge, 12 March 1892, SPR.MS 35/313, OJL-SPR. Lodge and Rayleigh were corresponding on the question of the motion of the earth relative to a stationary ether in 1891: see Oliver Lodge to Lord Rayleigh, 8 February 1891, Folder 2, Box #19, Series 4, R-USAF. Publications indicating the interests of Lodge and his dining companions include: Arthur J. Balfour, 'Address by the President', *PSPR*, vol. 10 (1894), pp. 2–13; William Crookes, 'Some Possibilities of Electricity', *Fortnightly Review*, vol. 51 (New Series) (1892), pp. 173–81; Oliver Lodge, 'Some Recent Thought-Transference Experiments', *PSPR*, vol. 7 (1891–2), 374–82, Oliver Lodge, 'The Interstellar Ether', *Fortnightly Review*, vol. 53 (New Series) (1893), pp. 856–62; Frederic W. H. Myers, 'On Alleged Movements of Objects, Without Contact, Occurring Not in the Presence of a Paid Medium', *PSPR*, vol. 7 (1891–2), pp. 383–94. Lodge probably conversed with Crookes, Robert Strutt (later the Fourth Baron Rayleigh), Thomson and Rayleigh about psychical topics at dinner parties hosted by Rayleigh in his London home in the 1890s: see Hannah Gay, 'Science, Scientific Careers and Social Exchange: The Diaries of Herbert McLeod, 1885–1900', *History of Science*, vol. 46 (2008), pp. 457–96, p. 466.

[43] Lodge's interactions with philosophers and clergymen are touched on in Bowler, *Reconciling Science and Religion*, chapter 7; W. P. Jolly, *Sir Oliver Lodge* (London: Constable, 1974).

1897, the Liverpool physicist Arthur Chattock praised his senior colleague Lodge for producing writings on "psychics" that had inspired "centres of interest" in the topic in "many colleges" and had "affected" Lodge's as well as Chattock's own students, a trend that he believed would enormously benefit the subject in the long run.[44]

There is some truth in Chattock's claims. Many of the physical–psychical scientists in Tables 2.1 and 2.2 had some connection with Lodge. Francis Aston, Chattock, Benjamin Davies, Fournier d'Albe, Stratton and William Thornton were all colleagues or students of Lodge when he was at Liverpool (1881–1900) or was Principal of Birmingham University (1900–19).[45] Some of them appear to have been inspired by Lodge's own studies of psychical phenomena or directly encouraged by him.[46] Davies served as Lodge's laboratory assistant from 1880 to 1908, while Chattock knew Lodge in the 1870s when the older physicist taught at University College London, and served under him as an assistant lecturer at Liverpool from 1887 to 1889.[47] Both Davies and Chattock maintained particularly close correspondences with Lodge that blended discussion of electrical physics, telepathy and spiritualism.[48] Fournier d'Albe's exposure to the examples of physics mentors with psychical interests was particularly strong because in 1910, when he took up his post as an assistant lecturer at Birmingham, he had just spent four years assisting Barrett in the physics laboratory at the Royal College of Science, Dublin.[49]

[44] Arthur P. Chattock to Oliver Lodge, 11 April 1897, MS Add. 89/23, OJL-UCL.

[45] On Lodge's career at Liverpool see Nani N. Clow, 'The Laboratory of Victorian Culture: Experimental Physics, Industry and Pedagogy in the Liverpool Laboratory of Oliver Lodge, 1881–1900' (unpublished doctoral dissertation, Harvard University, 1999) and Peter Rowlands, *Oliver Lodge and the Liverpool Physical Society* (Liverpool University Press, 1990). On Lodge at Birmingham see Eric Ives, Diane Drummond and Leonard Schwarz, *The First Civic University: Birmingham 1880–1980. An Introductory History* (University of Birmingham, 2000), chapters 7–10.

[46] Frederick J. M. Stratton, 'Psychical Research – A Lifelong Interest', *PSPR*, vol. 50 (1954), pp. 135–52, p. 136 and James Chadwick, 'Frederick John Marrian Stratton', *Biographical Memoirs of Fellows of the Royal Society*, vol. 7 (1961), pp. 280–93, p. 283. Lodge's increasingly conspicuous psychical investigations and writings probably inspired the foundation of a 'Students' Psychical Society' at the University of Birmingham around 1905. In this year, Lodge lectured to the society on Charles Richet's evidence of supernormal 'lucidité' (Richet's alternative term for clairvoyance): see 'C.W.', 'Students' Psychical Society', *The Mermaid: The Journal of the Guild of Undergraduates of the University of Birmingham*, vol. 3 (1906), pp. 68–72.

[47] For Chattock see [Anon.], 'Arthur Prince Chattock 1860–1934', *Obituary Notices of Fellows of the Royal Society*, vol. 1 (1932–5), pp. 293–8. For Davies see R. G. Roberts, 'The Training of an Industrial Physicist: Oliver Lodge and Benjamin Davies, 1882–1940' (unpublished doctoral thesis, University of Manchester, 1984).

[48] See the letters from Chattock to Lodge from 1888 to 1927 in MS Add. 89/23, OJL-UCL and the letters from Davies to Lodge from 1885 to 1940 in Box 3, BD-NLW.

[49] Barrett and Fournier d'Albe were also chairman and honorary secretary respectively of the Dublin Section of the SPR established in 1908: [Anon.], 'Report of the Council for

Chattock's reference to "centres of interest" could well have been one that he was trying to forge himself. He was probably exposed to the psychical interests of scientific colleagues before he went to Liverpool. Between 1885 and 1887 he had been a Demonstrator in the physics department at University College Bristol (1885–7), where he worked under Silvanus Thompson, a close friend and colleague of Barrett, Crookes and Lodge who, despite not joining the SPR, read its publications and believed, like many leading SPR members, that communications from disembodied spirits were probably telepathic impressions from the living (Figure 2.2).[50]

Chattock's colleagues also included the chemist William Ramsay, who, during the mid-1880s, was interested in spirit photography and Zöllner's hyperdimensional theory of how matter appeared to pass through matter in spiritualistic seances.[51] When, in 1889, Chattock left Liverpool for a post as a lecturer in physics at Bristol, he no longer had Ramsay and Thompson as colleagues with whom to ponder physical and psychical questions, but he doubtless inspired his student (and later colleague) Arthur Tyndall to join the SPR. He did not manage to persuade any other students to do so, but in the 1890s he involved his charges Reginald Clinker and Edmund Wedmore in experimental tests of their own telepathic powers. This yielded some evidence of drawings and suits of cards being transmitted between the minds of individuals seated near each other.[52] Clinker and Wedmore do not appear to have pursued their interest much further, but two other students of Chattock – Frank Burt and Francis Usher – did, and later published what would become frequently cited evidence of telepathy over hundreds of miles[53] (Figure 2.3).

Most physical–psychical scientists did not have Lodge's or Chattock's success in creating the "centres of interest" or otherwise encouraging psychical interest among others in their workplaces. By the mid-1870s,

1908', *JSPR*, vol. 14 (1909–10), pp. 36–40. On Fournier d'Albe see Ian B. Stewart, 'E. E. Fournier d'Albe's Fin de siècle: Science, Nationalism and Philosophy in Britain and Ireland', *Cultural and Social History*, vol. 14 (2017), pp. 599–620.

[50] Jane Smeal Thompson and Helen G. Thompson, *Silvanus Philips Thompson: His Life and Letters* (London: T. Fisher Unwin, 1920), pp. 331–2.

[51] William Ramsay to George F. FitzGerald, 21 February 1886, 8/102, George F. FitzGerald Papers, RDS Library and Archives; Arthur P. Chattock to Oliver Lodge, 7 April 1888, MS Add. 89/23, OJL-UCL.

[52] Arthur P. Chattock, 'Experiments in Thought Transference', *JSPR*, vol. 8 (1897–8), pp. 302–7.

[53] F. L. Usher and F. P. Burt, 'Thought Transference (Some Experiments in Long Distance Thought-Transference)', *Annals of Psychical Science*, vol. 8 (1909), pp. 561–600. See also Arthur P. Chattock to Oliver Lodge, 7 June 1907, MS Add. 89/23, OJL-UCL and William Ramsay to E. E. Fournier d'Albe, 13 December 1908, E. E. Fournier d'Albe Papers, Private Collection, Christine Fournier d'Albe.

2.2 The physical laboratory at University College Bristol in the mid-1890s. Arthur Chattock (second from right), Reginald Clinker (second from left) and other students are seen amidst standard apparatus for teaching electricity and magnetism. The image also testifies to the growing number of women admitted to university practical physics courses in the late nineteenth century. Reproduced by permission of University of Bristol Library, Special Collections (DM2765).

Stewart had developed interests in mesmerism, thought-reading, apparitions and psychic force and, in *Unseen Universe* and other works, upheld scientific arguments for the existence of an invisible universe intimately connected with the visible one.[54] This was the very period when the physicists Arthur Schuster and J. J. Thomson were students of his at Owens College.[55] Both joined the SPR in 1884, although Schuster terminated his membership after only four years and suspected that Stewart had been duped in his tests of thought-reading.[56] Far less successful than

---

[54] See Balfour Stewart, 'Mr. Crookes on the "Psychic Force"', *Nature*, vol. 4 (1871), p. 237. His interest in spiritualism evidently predated this, as suggested by his correspondence with Harrison in 1868–9: PRO BJ1/25, Records of the Kew Observatory, National Archives.

[55] On Owens College see Robert H. Kargon, *Science in Victorian Manchester: Enterprise and Expertise* (Manchester University Press, 1977), chapter 5.

[56] Arthur Schuster, *Biographical Fragments* (London: Macmillan and Co., 1932), p. 215.

Agents: PROF. CHATTOCK and R. C. CLINKER.   Percipient: E. B. WEDMORE.

All in same room at Harrow, September, 1897. E. B. W. about 3 yards from agents with lamp and table between.

No. 1.   *First Sitting.*
ORIGINAL.   REPRODUCTONS.

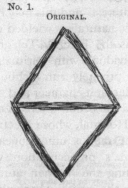

(a)   (b)

Obtained while the drawing was being made. The middle line was curved to show perspective as the outline suggested a toilet tidy. This and all further outlines and numbers were seen light on a dark background.

Remarks: "He's got it."   E. B. W.
PROF. C.

No. 2.   REPRODUCTIONS.

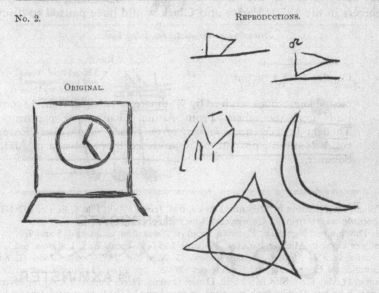

ORIGINAL.

2.3 Some of the drawings that Arthur Chattock and his students Reginald Clinker and Edmund Wedmore used in telepathy experiments. The left-hand pictures are those at which Chattock and Clinker (the 'agents') stared. Those on the right are the drawings of the

Stewart was William Thornton, a lecturer in electrical engineering at the Durham College of Science, who in the early 1900s grumbled to fellow Lodge associate Benjamin Davies that "there is a Psychical Research here somewhere not yet found. The people are very level headed & full of engineering as you might expect in the home of the Stevensons [sic]".[57]

In London there were very few "centres of interest" and this may have owed something to the power that the scientific naturalists wielded over metropolitan science.[58] The closest it gets is the Electric Telegraph Company, which employed at least three individuals with spiritualistic interests (Clark, Harrison and Varley); the privately run School of Submarine Telegraphy, whose founder, the telegraph engineer and spiritualist Desmond Fitzgerald, probably inspired the occult interests of at least one pupil (Kingsland); and St Thomas's Hospital, whose Electrical Department brought together William H. Stone and his junior colleague Walter J. Kilner, who later built on Stone's preoccupation with Reichenbach's od in his technique of imaging the 'human aura'.[59] However, there were plenty of other places in the capital where physical–psychical scientists could have, and were known to have, interacted. It is not unlikely that in the early meetings of the Society of Telegraph Engineers in the 1870s Varley and Clark would have paused to discuss

---

Caption for 2.3   (cont.)

mental impressions received by Wedmore, a 'percipient' sitting some 3 yards from the others. From Arthur Chattock, 'Experiments in Thought-Transference', *Journal of the Society for Psychical Research*, vol. 8 (1897–8), pp. 302–5. Reproduced by permission of Andreas Sommer.

---

[57] William M. Thornton to Benjamin Davies, n.d. [circa 1900], File 1, Box 2, BD-NLW. Thornton was referring to George and Robert Stephenson.

[58] See Dawson and Lightman, *Victorian Scientific Naturalism*, chapters 4–5 and 7.

[59] Latimer Clark to Michael Faraday, 29 April 1857, in Frank A. J. L. James (ed.), *The Correspondence of Michael Faraday Volume 5 November 1855–October 1860* (London: Institution of Engineering and Technology, 2008), pp. 221–3; Latimer Clark to D. D. Home, 11 May 1864, SPR.MS 28/82, Daniel Dunglas Home Papers, Society for Psychical Research Archive, Cambridge University Library; [Anon.], 'The Presentation of the Harrison Testimonial', *Spiritualist*, vol. 8 (1876), pp. 53–7; [Anon.], 'Obituary. Desmond G. FitzGerald', *Electrician*, vol. 54 (1904–5), p. 21; William Kingsland, *The Art of Life and How to Conquer Old Age* (London: C. W. Daniel Company, 1934), p. 93; William H. Stone and Walter J. Kilner, 'On Measurement in the Medical Application of Electricity', *Journal of the Society of Telegraph Engineers*, vol. 11 (1882), pp. 107–28; Walter J. Kilner, *The Human Atmosphere; Or the Aura Made Visible by Means of Chemical Screens* (London: Rebman, 1911).

spiritualism, and perhaps continued the conversation with such members as Crookes, Walter Coffin, Desmond Fitzgerald, Lord Lindsay and Stone. Later in the decade, Varley, Clark and Stone could have continued the dialogue at the Physical Society of London, at whose meetings they would also have encountered the Society's co-founder Barrett, as well as Crookes, Lodge, Rayleigh, Robert Angus Smith and Stewart.[60]

In London and, indeed, elsewhere, physical–psychical scientists were as likely to share psychical interests in private domestic spaces as in public institutions, not least because the former were the commonest types of space for seances and performances of other psychical phenomena. Crookes, for example, held regular 'at homes' for scientific colleagues at his Regent's Park residence and domestic laboratory, and it was here that Rayleigh, Huggins and others were known to have discussed and collaborated on spiritualistic investigation.[61]

The close association of the Sidgwick group with Cambridge University ensured that the institution was one of the most thriving centres of psychical interest, and we can easily place many physical–psychical scientists within it. A measure of Cambridge's importance derives from the fact that in 1884 a local branch of the SPR was established there, and this was graced by many of the university's most distinguished academics.[62] Equally telling was the presence that psychical research had elsewhere in university life, including specialist magazines, debating societies and dinner parties.[63] An insight into the overlap of psychical research and academic sociability comes from the diaries of John Couch Adams, the Cambridge astronomer who joined the SPR in 1884 and who was a regular attendee at the local branch meetings and an avid reader of publications relating to psychical research and modern Theosophy.[64] Psychical research was undoubtedly a conversation topic at a dinner party that he attended at Henry and Eleanor Sidgwick's Cambridge residence in 1885. He was joined by not only Frederic Myers and his wife Eveleen, but Rayleigh and his wife Evelyn, and

[60] Rollo Appleyard, *The History of the Institution of Electrical Engineers 1871–1931* (London: The Institution of Electrical Engineers, 1931), pp. 34–7.

[61] Crookes's 'at homes' are mentioned in Thompson and Thompson, *Silvanus Philips Thompson*, p. 153.

[62] [Anon.], 'Cambridge Branch of the S.P.R.', *JSPR*, vol. 1 (1884–5), pp. 52–3, 180–1.

[63] See for example [Anon.], 'Mrs. Verrall on Telepathy', *Cambridge Magazine*, vol. 1 (1912), p. 111; [Anon.], 'Telepathy Again', *Cambridge Magazine*, vol. 7 (1917–18), pp. 97–8. For a recollection of a Cambridge Union debate on spiritualism see Walter Leaf, *Walter Leaf 1852–1897. Some Chapters of Autobiography* (London: John Murray, 1932), p. 93.

[64] On Adams see H. M. Harrison, *Voyager in Time and Space: The Life of John Couch Adams* (Lewes: The Book Guild, 1994).

another Cambridge astronomy professor, George Howard Darwin, who had attended spiritualist seances in the 1870s.[65]

It was through the Cambridge Branch of the SPR that many of its members gained opportunities to test and debate psychical phenomena. J. J. Thomson recalled that in 1884 he participated in a seance given by Blavatsky in the King's College rooms of Oscar Browning, the Cambridge Branch's secretary. Blavatsky had accepted an invitation to visit Cambridge as part of the SPR's investigation of the psycho-physical phenomena associated with modern Theosophy. Like so many attendees at Victorian seances, Thomson was to be disappointed by his experiences. Having been informed by Blavatsky that an invisible Tibetan 'Mahatma', one of the mysterious sources of esoteric wisdom, would "precipitate a message, a cushion and a bell", he sat for over an hour and nothing happened.[66] However, as we shall see later in this book, Thomson's attendance at spiritualist seances at Myers's Cambridge residence over ten years later proved much more stimulating and perplexing.

Myers was, of course, one of Henry Sidgwick's many colleagues and students whose interests in psychical research were encouraged by the Fenland varsity's distinguished moral philosopher. There is some evidence to suggest that Rayleigh and Thomson – two of the leading physicists at the university – managed to follow the example of Sidgwick, whose college (Trinity) they both belonged to as undergraduates and fellows, and who became Rayleigh's brother-in-law when he married into the Balfour clan. By 1879, when he was appointed the Cavendish Professor of Experimental Physics and director of Cambridge's Cavendish physics laboratory, Rayleigh was convinced of the power of one mind to influence another by suggestion, and had investigated spiritualism, mainly in collaboration with the Sidgwick group.[67] Although he judged the results of his spiritualistic investigations of the 1870s to be inferior to Crookes's of the same period, he rejected the commonly held theory that the effects were due to hallucination and trickery. Rayleigh largely gave up active psychical investigation after the 1870s, but it is possible that his example inspired one research student at the Cavendish – Schuster – to take the question seriously. Given that Schuster had studied physics under Stewart and, in the early 1880s, collaborated with Rayleigh and Eleanor

[65] John Couch Adams, entry for 13 June 1885, Diary No. 18 (1885), Box 21, John Couch Adams Papers, St John's College, Cambridge. On George Darwin's spiritualistic experiences see Adrian Desmond and James Moore, *Darwin* (London; Michael Joseph, 1991), pp. 607–8.

[66] Thomson, *Recollections and Reflections*, p. 153.

[67] Lord Rayleigh, 'Presidential Address', *PSPR*, vol. 19 (1919–20), pp. 276–90; Rayleigh, *Third Baron Rayleigh*, pp. 65–8.

Sidgwick on painstaking measurements of the standard of electrical resistance, it would have been difficult for him to have entirely avoided the subject of psychical research.[68]

J. J. Thomson, who shared with Schuster connections with both Rayleigh and Stewart, seems to have been much more successful in setting an example of a physical scientist who at least dabbled in psychical research. In 1884, he succeeded Rayleigh as Cavendish Professor and laboratory director and among the students that undertook research at his laboratory were many who either joined the SPR or developed some kind of psychical interest. These included Francis Aston, a physicist later renowned for his work on isotopes, and whose psychical interests may have already been nurtured by Lodge; William R. Bousfield, who had careers as a barrister and a physical chemist, and who in the 1920s appealed to the evidence of telepathy in an attack on philosophical materialism; William Cecil Dampier Whetham, who in later life was best known as an agriculturalist and historian of science; and Robert John Strutt (later the Fourth Baron Rayleigh), the physicist and future SPR president whose psychical interests probably owed more to his father and Sidgwick and Balfour relatives.[69]

Another of Thomson's students, Daniel Comstock, an American who had careers in theoretical physics and electrical engineering, left a rare insight into the reasons why Thomson's approach to physics was relevant to psychical investigation. In 1965, the ageing Comstock recalled one of several "mental awakenings" that he had had as a Thomson student.[70] One such awakening was Thomson's insistence on clearly distinguishing fact from theory, and particularly the dangers of privileging theories of

[68] On the Cavendish and electrical measurement see Schaffer, 'Late Victorian Metrology'; Simon Schaffer, 'Rayleigh and the Establishment of Electrical Standards', *European Journal of Physics*, vol. 15 (1994), pp. 277–85. On Eleanor Sidgwick's scientific collaborations with Rayleigh see John N. Howard, 'Eleanor Mildred Sidgwick and the Rayleighs', *Applied Optics*, vol. 3 (1964), pp. 1120–2.

[69] The Cavendish work of Aston, Comstock and Whetham is discussed in Dong-Won Kim, *Leadership and Creativity: A History of the Cavendish Laboratory, 1871–1919* (Dordrecht: Kluwer Academic Publishers, 2002), chapter 5. Bousfield's and Rayleigh's major 'psychical' publications included William R. Bousfield, 'Telepathy', *Hibbert Journal*, vol. 20 (1921–2), pp. 497–506 and Lord Rayleigh, 'Presidential Address: The Problem of Physical Phenomena in Connection with Psychical Research', *PSPR*, vol. 45 (1939), pp. 1–18. For Bousfield see W. C. D. Dampier, 'William Robert Bousfield 1854–1943', *Obituary Notices of Fellows of the Royal Society*, vol. 4 (1942–4), pp. 570–6. For the Fourth Baron Rayleigh see Guy R. Strutt, 'Robert John Strutt, Fourth Baron Rayleigh', *Applied Optics*, vol. 3 (1964), pp. 1105–12.

[70] Daniel F. Comstock, 'Autobiography', typescript dated 1963–5, p. 19, MB 2014-1439; 112, Niels Bohr Library and Archives, American Institute of Physics. For Comstock's psychical investigations see Brian Inglis, *Science and Parascience: A History of the Paranormal 1914–1939* (London: Hodder and Stoughton, 1984), pp. 161–9.

puzzling phenomena (in this instance, those connected with the electrical discharge through gases) over strong evidence for such phenomena. For Comstock there was an important lesson in humility here that was immediately applicable to the puzzling but often well-documented phenomena of psychics. It warned against the "prejudice of the human mind which requires an explanation before believing in the phenomena, no matter how well authenticated".[71]

Comstock may well have been present when, at a Christmas dinner at the Cavendish Laboratory in the early 1900s, research students sung a comic song praising the dynamical laws of physics for forcing "Dame Nature" to forsake her mysteries and anticipating Oliver Lodge's ability to "investigate all spooks and their relations, / Then give them n-coordinates and use / Lagrange equations".[72] For many of those joining in, it was hard to avoid the extraordinary ambitions that Lodge, Thomson and other physicists had for the investigation of psychical phenomena, even if they thought such ambitions could be satirised via the far-fetched proposal of applying to such unpredictable effects as "spooks" the sophisticated mathematical techniques deployed by Cambridge-trained physicists to analyse vastly more predictable systems of material particles.[73]

The satirical approach that many Cambridge physicists adopted towards "spooks" points to one of the problems with the argument that 'physics and psychics' was a predominantly Cambridge-based phase in Victorian science.[74] There were probably at least as many Cambridge physicists who were ambivalent about or hostile towards psychical investigation as interested in it. Neither John Henry Poynting nor Joseph Larmor had much interest in psychical research even though they praised Lodge's courage in pursuing, and in helping to change scientific attitudes towards, the subject.[75] More decided in their opinions were two physicists who taught J. J. Thomson: George Gabriel Stokes and James Clerk

[71] Comstock, 'Autobiography', p. 20.
[72] A[lfred] A[rthur] R[obb], 'A Function of Time', in *Post-Prandial Proceedings of the Cavendish Society* (Cambridge: Bowes and Bowes, 6th ed., 1926), pp. 17–19, p. 17. Two of Thomson's students who poked fun at psychical research were J. A. McClelland and J. S. Townsend. In the early 1900s they reputedly "hoaxed" other Cavendish researchers (including Ernest Rutherford) into believing that they had genuine telepathic powers: Lord Rayleigh, *The Life of J. J. Thomson* (Cambridge University Press, 1942), p. 132.
[73] On Cambridge mathematical physics see Warwick, *Masters of Theory*.
[74] This is one of the contentions of Wynne, 'Physics and Psychics'.
[75] Joseph Larmor to Oliver Lodge, 7 January 1901, MS Add. 89/65(iii), OJL-UCL; Frederic W. H. Myers to Oliver Lodge, 19 February 1895, SPR.MS 35/1439, OJL-SPR. John Henry Poynting, 'Biographical Sketch', in Oliver Lodge, *Mind and Matter: An Address Delivered in the Town Hall, Birmingham* (Birmingham: Birmingham and Midland Institute, 1904), pp. 29–35. Larmor was evidently interested in Lodge's investigations into the physical phenomena of spiritualism but denied that the SPR's evidence for

Maxwell. A devout evangelical Anglican who adhered closely to the Bible, Stokes warned in 1893 that there was no foundation "in reason or Scripture" of the spiritualist claim that the future state and immortality of the soul could be established on a natural rather than a supernatural basis.[76] For this reason, he questioned the moral "lawfulness" of enquiring into "occult manifestations".[77]

One of Stokes's most distinguished students, Maxwell, was pouring scorn on spiritualism and animal magnetism as an undergraduate at Cambridge in the 1850s. In an essay written in 1853, around the time that he praised Faraday's intervention on table-turning, he lambasted phases of "dark science", all of which involved "speciously sounding laws" that lacked sound "experimental proof".[78] Animal magnetism was one of those dark sciences that merely "pretended to be physical sciences" by imitating the language of "popular physics" in its fluid theories, while spirit-rapping, despite offering the intellectually fruitful idea of a "spiritual medium" connecting minds, was no better than older dark sciences owing to the vulgar and absurd practices of "money-making media".[79]

Beyond Cambridge, there were many other practitioners of the physical sciences who expressed grave doubts about psychical investigation and who further challenge impressions of a generally harmonious relationship between 'physics and psychics' suggested in Tables 2.1 and 2.2. In the 1890s, the electrical engineer John Perry attacked Barrett and Wallace for not maintaining a "critical, cautious, unbelieving state" of mind and thereby falling for the "quackery" in psychical phenomena, whose reality Perry did not think was sufficiently probable to justify his attention.[80] In the same period, William Robert Grove, the chemist James Dewar and mathematician Thomas Archer Hirst reputedly agreed that Crookes was "once more 'off his head'" for praising Lodge's recent endorsement of

apparitions of dying persons was conclusive. See also Joseph Larmor to Oliver Lodge, 27 September 1905, Lm 1281, MS/603, Joseph Larmor Papers, Royal Society Archives.

[76] George G. Stokes, quoted in W. T. Stead, 'The Response to the Appeal. From Prelates, Pundits and Persons of Distinction', *Borderland*, vol. 1 (1893), pp. 10–23, p. 18. Stokes made a similar point in a warning to Barrett about disembodied spirits: George G. Stokes to William F. Barrett, 11 September 1880, f. 127, WFB-RS. On Stokes's religion see Wilson, *Kelvin and Stokes*, pp. 74–99.

[77] Stokes, quoted in Stead, 'Response to the Appeal'.

[78] James Clerk Maxwell, 'Idiotic Imps', in Lewis Campbell and William Garnett, *The Life of James Clerk Maxwell* (London: Macmillan and Co., 2nd ed., 1884), pp. 341–3, p. 341. See also James Clerk Maxwell to Charles Benjamin Tayler, 8 July 1853, in P. M. Harman (ed.), *The Scientific Letters and Papers of James Clerk Maxwell*, 3 vols. (Cambridge University Press, 1990–2002), vol. 1, pp. 220–1.

[79] Maxwell, 'Idiotic Imps', pp. 341–2.

[80] John Perry to Oliver Lodge, 29 August 1891, MS Add. 89/82, OJL-UCL. See also John Perry to Oliver Lodge, 25 June 1894, MS Add. 89/82, OJL-UCL.

psychical research in an address to the 1891 meeting of the British Association.[81]

The problem with Lodge's address was that it gave a scientific blessing to subjects that many scientists believed were, as William Thomson curtly expressed it in 1894, mostly "imposture and the rest bad observation".[82] It was probably with this in mind that Huggins, who was president of the British Association meeting in 1891, warned Lodge against discussing psychical research in his address to Section A (physical sciences) of the Association. While agreeing that the subject was worthwhile, Huggins advised Lodge to drop it for the sake of his "own reputation & peace of mind" because the leading physical scientists would feel that the Association had been compromised.[83] Lodge ignored Huggins's advice, and little that he did subsequently in psychical research allayed the astronomer's doubts that he had rendered the subject fit for scientific audiences. By the early 1900s, for example, Huggins could ask Lodge's fellow Maxwellian Joseph Larmor: "are Lodge and Science, in respect of the occult, convertible terms?"[84]

In the long run, psychical matters did seem to harm reputations. In 1895, Rücker told Lodge that Crookes's "relations to spiritualism" had aroused negative "comment" among those considering him for the distinguished position of president of the 1896 meeting of the British Association.[85] Eight years later, Lodge would have empathised with Crookes when Silvanus Thompson told him that his "dealings" with spiritualism had been a "continual stumbling block" to his nomination for a Nobel Prize.[86] Neither Crookes nor Lodge secured their respective scientific accolades but their advocacy of psychical research continued. Indeed, despite admitting in 1914 that his "occasional psychic utterances" harmed his scientific reputation, Lodge's interest in and advocacy

---

[81] The Grove anecdote was recalled by Thomas Archer Hirst, entry for November 1891, Journal XXII, p. 2831, in William H. Brock and Roy M. Macleod (eds.), *Natural Knowledge in Social Context: Journals of Thomas Archer Hirst (1830–1892)* (London: Mansell, 1980).

[82] William Thomson quoted in Stead, 'Response to the Appeal', p. 17. In recommending Varley to the Royal Society's Fellowship, Thomson (who had collaborated with Varley on patenting and working cable telegraphic inventions) opined that his want of scepticism weakened his scientific judgement but that his interests in mesmerism and spiritualism were not a "fatal objection" to his election: William Thomson to Edward Sabine, 23 March 1871, MC/9 f. 182, Royal Society Archives. Varley was elected later in 1871.

[83] William Huggins to Oliver Lodge, 9 August 1891, MS Add. 89/56, OJL-UCL.

[84] William Huggins to Joseph Larmor, 1 February 1909, in Barbara J. Becker (ed.), *Selected Correspondence of William Huggins*, 2 vols. (London: Routledge, 2014), vol. 2, pp. 474–6, p. 475.

[85] Arthur Rücker to Oliver Lodge, 3 March 1895, MS Add. 89/91, OJL-UCL.

[86] Silvanus P. Thompson to Oliver Lodge, 4 January 1913, MS Add. 89/104, OJL-UCL.

of the subject hardly diminished after this date.[87] Lodge's scientific critics may have underestimated the extent to which his psychical enquiries were driven by a host of scientific, philosophical, religious and other preoccupations and for that reason were more difficult to relinquish. This was hardly rare among physical–psychical scientists.

Gold Mine of Science, Handmaid to Faith

A simplistic interpretation of the informal networks analysed above is that individuals turned to psychical investigation because they were taught or otherwise 'influenced' by the likes of Lodge, Crookes and Barrett. This is obviously inadequate, since many who came within the orbits of these prominent physical–psychical scientists were either uninterested in or positively hostile to psychical subjects. To understand why some were 'influenced' we need to consider their other sources of motivation, which in the case of many physical–psychical scientists in Tables 2.1 and 2.2 are frustratingly difficult to establish.

One of the commonest motivations was the startling experience of a trusted friend, colleague, relative or other individual. In December 1869, for example, Crookes told Tyndall that such an experience prompted him to study spiritualism, having had only a limited, second-hand knowledge of the subject. About six months earlier, a Fellow of the Royal Society standing "in the foremost rank of experimental philosophers" had assured him that he had "witnessed phenomena alleged to be spiritual, which he was unable to explain by any known physical force".[88] The distinguished scientific colleague, who was probably the sanitation chemist Robert Angus Smith, persuaded Crookes to witness the phenomena for himself.[89] A few years earlier, Barrett's interest in mesmerism had been stimulated by John Wilson, a respected Irish member of the establishment employing both him and Tyndall, the Royal Institution, and who persuaded the incredulous young laboratory assistant to witness his mesmeric performances near his home in County Westmeath.[90] In 1881, not long after

[87] Lodge quoted in J. Arthur Hill, *Letters from Sir Oliver Lodge: Psychical, Religious, Scientific and Personal* (London: Cassell and Company, 1932), p. 49.

[88] William Crookes to John Tyndall, 22 December 1869, published in R. G. Medhurst (ed.), *Crookes and the Spirit World: A Collection of Writings by or Concerning the Work of Sir William Crookes, O.M., F.R.S. in the Field of Psychical Research* (London: Souvenir Press, 1972), pp. 232–4, p. 232.

[89] Smith's identity is suggested by a letter Crookes received from him about spiritualism in April 1869: E. E. Fournier d'Albe, *The Life of Sir William Crookes* (London: T. Fisher Unwin, 1923), p. 182.

[90] William F. Barrett, 'Some Reminiscences of Fifty Years' Psychical Research', *PSPR*, vol. 34 (1923–4), pp. 275–97, p. 282.

taking up his professorship at University College Liverpool, Lodge decided to intervene in the thought-reading craze when Malcolm Guthrie, a partner in a local drapery firm, asked him and other local scientists to witness the psychological power as developed by two of his employees. As a prominent figure in civic life and a "severe student of philosophy" whose own tests demonstrated "care and systematic vigilance", Guthrie's invitation was evidently difficult to refuse.[91]

George W. De Tunzelmann was rare among physical–psychical scientists because his introduction to psychical phenomena derived partly from his own powers. He seems to have become more interested in telepathy after convincing himself that he had correctly determined, independently of any other sensory channels, suits of cards in the minds of relatives.[92] In the 1850s, fellow electrical engineer Cromwell Varley claimed to possess mesmeric powers and used them to entrance his wife, Ellen, who in this state answered questions that he had put to her purely mentally and reacted to silent mesmeric passes that he made through solid objects.[93] Ellen Varley was rare among the wives of physical–psychical scientists because she, like many mesmerised individuals, developed powers of spiritualist mediumship.[94] The professed spirits, who spoke through her and rapped out messages, impressed Varley with their ability to correctly predict the timing and nature of critical developments in her state of health. These were decisive in converting Varley from being a "hard-headed unbeliever" in spiritualism.[95]

Varley's argument for the genuineness of the communicating spirits hinged on his confidence that known explanations of their predictive power (including fraud and self-deception) were simply inadequate. Comparable appeals to the inadequacy of known explanations were expressed by Crookes, Barrett and Lodge in their early experiences of psychical effects. As an unofficial participant in the London Dialectical Society's enquiry into spiritualism in 1869, Crookes "saw phenomena

---

[91] Oliver Lodge, 'An Account of Some Experiments in Thought-Transference', *PSPR*, vol. 2 (1884), pp. 189–200, p. 189.

[92] George W. De Tunzelmann, *A Treatise on Electrical Theory and the Problem of the Universe* (London: Charles Griffin and Company, 1910), pp. 626–32.

[93] Varley, 'Evidence of Mr. Varley', pp. 157–8. Cf. his friend Lord Lindsay, who in "mesmeric experiments" claimed to be able to communicate his will from England to Italy: Lord Lindsay to A. P. Sinnett, [1881], f. 82, Add. 45289B, Mahatma Papers, Volume VII, British Library.

[94] Desmond Fitzgerald's wife and daughter were also mediums: 'E. G.', 'A Test Séance with Mr. Williams', *Medium and Daybreak*, vol. 3 (1872), p. 318. Crookes claimed that his wife once exhibited the power to write under the control of Faraday's spirit: Crookes quoted in minutes for 5 July 1911, Minute Book Volume 7, Add. MS 52264, GC-BL.

[95] Cromwell F. Varley, 'The Reality of Spiritual Phenomena', *Spiritualist*, vol. 9 (1876), pp. 265–6, p. 256.

that could only be explained on the almost impossible supposition of gross fraud and collusion on the part of many ladies and gentlemen present" but also "occurrences which appear to be beyond the domain of any known physical force".[96] The "occurrences" included the movement of untouched tables, intelligent rapping noises that responded to vocalised questions, and nebulous lights that floated in the air. During his mesmeric investigations with Wilson, Barrett observed that one young girl exhibited many of the common phenomena of the mesmeric trance (notably the community of sensation between subject and operator) but what really impressed him was her clairvoyance: she appeared to correctly describe playing cards hidden inside books and the interior of a London scientific instrument maker's shop unknown to her, even though Barrett was satisfied that he was not giving away clues via unconscious facial or bodily gestures.[97] Lodge was equally satisfied with the conditions under which he tested the drapery assistants: the girls exhibited astonishing accuracy in their reproduction of simple images in his mind and of other "agents", despite his "scientific belief" that "no collusion or trickery was possible under the varied circumstances of the experiments".[98]

Having had their first experiences of psychical phenomena, some physical–psychical scientists identified the parallels with historic evidence of ghost-seeing and supernaturalism.[99] Many were quick to identify potentially exciting sources of scientific progress. Thus, Crookes enthused to Tyndall in 1869 about glimpsing "something new and worthy of the notice of the man of science", and these included "a power in some way connected with gravitation" and other unknown forces that, as he put it a few years later, were needed to "do the work of the universe".[100] Barrett interpreted his mesmeric experiences as evidencing a "wonderful exaltation of the perceptive powers" transcending the widely accepted psycho-physiological explanations offered by Carpenter and others.[101] Rayleigh was no less hopeful of ground-breaking results. During his early spiritualistic experiences, he opined to

---

[96] Crookes to Tyndall, 22 December 1869, in Medhurst, *Crookes and the Spirit World*, p. 233. As an unofficial participant, Crookes's name was omitted from the Committee's published *Report*: see *Report on Spiritualism*, pp. 373–95. See also William Crookes to Frederic W. H. Myers, 16 May 1890, SPR/Research/Crookes/1, Society for Psychical Research Archive, Cambridge University Library.

[97] William F. Barrett, 'On Some Phenomena Associated with Abnormal Conditions of Mind', *Spiritualist*, vol. 9 (1876), pp. 85–8.

[98] Lodge, 'Account of Some Experiments', pp. 192–3.

[99] William F. Barrett, 'The Phenomena of Spiritualism', *Nonconformist*, vol. 36 (1875), pp. 1017–20; William Crookes to 'J. H. D.', 10 May 1871 quoted in [Anon.], 'Mr. William Crookes, F.R.S. on Disembodied Spirits', *Spiritualist*, vol. 1 (1869–71), p. 161; Cromwell F. Varley, 'Spiritualism', *Human Nature*, vol. 3 (1869), pp. 367–71, p. 368.

[100] Crookes to Tyndall, 22 December 1869, p. 233; Crookes, *Psychic Force*, p. 5.

[101] Barrett, 'On Some Phenomena', p. 86.

Henry Sidgwick that a "decision of the existence of mind independent of ordinary matter must be far more important than any scientific discovery could be, or rather would be the most important scientific discovery" if more people became interested in spiritualism.[102]

Lodge was at least as interested as Crookes in the specific implications for physics. As he suggested in his address to the British Association meeting of 1891, the apparent capacity of untouched material objects to move by some unknown power (telekinesis) was a puzzle that the "orthodox scheme of physics" could not yet explain and was therefore a "line of possible advance" for this "King of the Sciences".[103] It was the supposed possibility of extending the boundaries of the physical sciences, in theoretical, conceptual and experimental directions, that would sustain the interest of many other physical–psychical scientists in psychical investigation more than any other single factor.

Many physical–psychical scientists were certainly motivated by a strong sense of the duty of scientific practitioners to intervene on questions of social importance. The period in which the older physical–psychical scientists expressed this most vehemently was, not coincidentally, the same period when, as Frank Turner has shown, champions of the professionalisation of the sciences were making their most forceful arguments for the wider public utility of the sciences, whether intellectual, industrial, social, cultural or moral.[104] Carpenter, Faraday and others had already tried to show that psychical phenomena, and especially the physical phenomena of spiritualism, were among the issues where men of science could demonstrate their public utility, even if many commentators denied that their interventions had been satisfactory. Crookes's acknowledgement of his public utility is clear from his manifesto for the scientific investigation of spiritualism, published in his own *Quarterly Journal of Science* in 1870. He regarded it as the

duty of scientific men who have learnt exact modes of working, to examine phenomena which attract the attention of the public, in order to confirm their genuineness, or to explain, if possible, the delusions of the honest and to expose the tricks of deceivers.[105]

---

[102] Lord Rayleigh to Henry Sidgwick, 7 June 1874, cited in Rayleigh, *Third Baron Rayleigh*, pp. 66–7, p. 67.

[103] Oliver Lodge, 'Address', in *Report of the Sixty-First Meeting of the British Association for the Advancement of Science Held in Cardiff in August 1891* (London: John Murray, 1892), pp. 547–57, pp. 554–5.

[104] Turner, *Contesting Cultural Authority*, pp. 203–5.

[105] William Crookes, 'Spiritualism Viewed by the Light of Modern Science', *Quarterly Journal of Science*, vol. 7 (1870), pp. 316–21, p. 317.

Having spent much of the 1860s championing the need for trained scientific expertise rather than mere "popular science" to solve such pressing problems as the cattle plague and water pollution, Crookes was on familiar territory when he contrasted the caution, exactitude and robust knowledge of "modern scientific men" with the methodological sloppiness and impoverished understanding of the "pseudo-scientific spiritualist".[106]

Crookes was certainly not alone in arguing, however, that even trained scientists, notably physiologists, were failing in their public duty because their interventions had clearly not been effective. In 1877, Barrett believed his intervention on spiritualism was a way of challenging the "flimsy explanations, varnished with half-truths, that pass muster at the hands of those psychologists who arrogate to themselves the sole right of instructing the public mind on this subject".[107] Barrett's sense of duty to the "public mind" was strongly rooted in his Christian faith. Like many Victorians, he feared that spiritualism posed a genuine mental and moral threat to the ignorant, credulous and uncultured. As this Congregationalist minister's son warned in a leading non-conformist periodical in 1875, these were the individuals who suffered "mental derangement" after prying into secrets beyond the grave and had lost their moral compass by turning from Christ to a "new religion" based partly on the "lying and contradictory messages given through mediums" and whose best teachings comprised little more than "maxims about the progress of mankind".[108]

By 1875, Barrett was much more optimistic about spiritualism than he had been a few years earlier because of recent positive experiences of spirit-rapping and levitation taking place through the mediumship of a young Irish girl, and in broad daylight and other conditions that he believed precluded fraud and self-deception.[109] As far as he was concerned, scientific enquirers into spiritualism who were also devout Christians (such as himself) were better able to distinguish the facts upon which religious conclusions might be reliably based and had a

---

[106] Crookes, 'Spiritualism Viewed', pp. 318–19. On Crookes's earlier campaigns see Brock, *Crookes*, pp. 53–60, 83–103, 267–97.

[107] William F. Barrett, 'The Demons of Derrygonnelly', *Dublin University Magazine*, vol. 90 (1877), pp. 692–705, p. 700n.

[108] Barrett, 'Phenomena of Spiritualism', p. 1020. Balfour Stewart, who read this article, echoed Barrett's argument against spiritualism as a religion: Balfour Stewart to William F. Barrett, 2 May 1876, WFB-RS. Barrett later argued that the physical phenomena of spiritualism needed to be studied with great caution given that they were caused by an "intermediate race" of spiritual beings who, unlike higher spiritual agencies, were potentially "troublesome & dangerous": William F. Barrett to Oliver Lodge, 18 October 1890, SPR.MS 35/60, OJL-SPR.

[109] See Barrett's more sceptical position in [William F. Barrett], 'Spiritualism and Science', *Nonconformist*, vol. 34 (1873), pp. 445–6.

public duty to do so because it would prevent "thousands of people" being "deluded by the matter".[110] By 1875, he also was prepared to proclaim that spiritualism offered "objective proof" of the Pauline doctrine of the spiritual body, which for him constituted a welcome weapon against the "meshes of materialism" and an "entrance to" but not an alternative to Christianity.[111] By the 1890s, when the "dominant school of scientific thought" appeared to be "essentially, if not *grossly*, materialistic", psychical phenomena presented a welcome "handmaid to faith" as well as a "perfect gold mine for scientific research".[112] But, useful as spiritualism was in aiding the Christian idea of the soul's *survival*, Barrett insisted, like most devout Christians, that the soul's *immortality* was a matter of pure faith rather than of natural enquiry.[113]

Barrett recognised a close parallel between the Christian duty that he sought to fulfil by cleaning up spiritualism and the religious objectives of the *Unseen Universe*, a work co-authored by another strongly Christian-minded physicist with whom he often discussed spiritualism: Balfour Stewart. As we shall see in Chapter 3, one of the arguments of the book was that Christian miracles were not incompatible with scientific theories of the physical universe because the latter suggested the possibility of energy flowing from an unseen universe to the visible one with which it was intimately connected. Barrett insisted that the idea of spirit manifestations was only carrying this argument a "step further", but Stewart and his co-author Tait were not so sure.[114] On the one hand, they were convinced that spirit manifestations were subjective impressions of seance-goers in a state of "mental excitement"; on the other hand, they believed that the possible cause of these impressions – the power of one person to exert a direct mental influence upon another at a distance – was a "valuable inquiry", evidently because this was congruent with the possibility of an invisible realm (in this case mental) acting on a visible one (the brain) at a distance.[115] The strong interest that Stewart showed

---

[110] Barrett, 'On Some Phenomena', p. 88.
[111] Barrett, 'Phenomena of Spiritualism', p. 1020.
[112] William F. Barrett, 'Science and Spiritualism', *Light*, vol. 14 (1894), pp. 539–40, 559–61, 571–2, 583–5, 595–7, on pp. 539, 559 and 571.
[113] William F. Barrett, 'Address by the President', *PSPR*, vol. 18 (1903–4), pp. 323–50, p. 350.
[114] Barrett, 'Phenomena of Spiritualism', p. 1020n.
[115] [Stewart and Tait], *Unseen Universe*, pp. 42–3. See also Balfour Stewart to William F. Barrett, 26 December 1881, SPR.MS 3/A2/19, WFB-SPR. In 1871, Tait had publicly compared spiritualism to materialism as opposing examples of scientific "ignorance and incapacity", with spiritualists denying that even the phenomena of "dead matter" were the subject of physical enquiry and materialists reducing everything to "physical manifestations". Tait's belief that spiritualism was "harmless folly" but that materialism was "pernicious nonsense" suggests his greater fear of materialism – a fear that would

in evidence for thought-reading, both before and after he joined the SPR, was undoubtedly driven by some of the religious goals underpinning his attempt to render Christian miracles physically intelligible.

By the early 1900s, Lodge was proving to be the heir of Barrett and Stewart in arguing that psychical research and ether physics gave scientific intelligibility to the idea of a realm beyond gross matter and of the capacity of the Divine Mind to guide humanity and the cosmos. Yet Lodge's *initial* interests in psychical research were not necessarily religious. Religion certainly occupied an important place in his youth: both his grandfathers were Anglican clergymen, he owed some of his early education to a devoutly Christian aunt, and he was profoundly impressed by what he later called the "mental and spiritual reality" behind the mundane world as evoked in Tyndall's physics lectures and the Reverend James Moorhouse's sermons, both of which he attended as a science student in London.[116] If he did consider his investigations of thought-reading and trance mediumship another way of evincing this "reality" he was only prepared to discuss it publicly from the early 1890s, by which time personal "tuition" from his friend Myers had significantly challenged the materialistic worldview to which he believed leading scientific naturalists had confined him as a student.[117]

Lodge is not the only physical–psychical scientist whose turn to psychical enquiries is difficult to link directly or exclusively to religious motivations. We know, for example, that Rayleigh was a devout Christian, but can only assume that this contributed to his interests in psychical research. In 1911, he told one correspondent that he had "never thought the materialist view possible" and that his strong Christian faith had encouraged him to look to a "power beyond what we see, and to a life in which we may at least hope to take part".[118] Although Rayleigh was never convinced that psychical research had produced conclusive scientific evidence of mind independent of the living brain (via telepathy or survival), the relevance of these areas of psychical investigation to his religious questions is undoubtedly one reason why he never questioned the importance of the enterprise. Other physical–psychical scientists were at least as devout: John Couch Adams was raised as a strict Wesleyan Methodist, Ramsay was a member of the Free Church of Scotland, Stratton was a prominent Unitarian, and J. J. Thomson practised private

---

prompt him to co-author the *Unseen Universe*. See Peter G. Tait, 'Address', *Report of the Forty-First Meeting of the British Association for the Advancement of Science; Held in Edinburgh in August 1871* (London: John Murray, 1872), pp. 1–8, p. 7.

[116] Oliver Lodge, *Past Years: An Autobiography* (London: Hodder and Stoughton, 1931), p. 78.

[117] Lodge, *Past Years*, p. 345.   [118] Rayleigh, *Third Baron Rayleigh*, p. 361.

prayer and was a "regular communicant" of the Church of England.[119] While is it not clear how much and in what way prior religious belief contributed to their psychical interests, they all seemed to have shared a theistic view of the cosmos and would have welcomed any new scientific evidence challenging philosophical materialism.

Crookes never expressed his Christian faith as strongly as Barrett, Lodge and Stewart, but in an early public statement about spiritualism he argued that while spiritualism could be "approached from the sentimental and theological side", he felt he was better suited to studying its "scientific aspect".[120] In many ways Crookes echoed the distinction that Tyndall made in 1867 between religious or emotional and scientific or intellectual approaches to the "credibility of physical facts" associated with miracles: questions about spiritualism that related to its phenomenal or physical aspects – notably the relationship between spirit and matter, the "powers" of spirit when "united" with matter and when free, and the capacity of "intelligent spiritual entities" to communicate with the living – needed to be undertaken by a "man of science" driven by "hard intellect alone" rather than "romantic and superstitious ideas".[121] When, in 1871, Crookes published these words, this ambitious chemist had a significant personal reason to ensure that "sentimental" motivations for studying spiritualism did not affect the desired image of a "hard" scientific enquirer he wanted to project: this was the tragic death of a younger brother in 1867, which surely contributed to his interest in the subject.[122]

Crookes would not be the only physical–psychical scientist for whom bereavement played a role in encouraging psychical enquiries: in 1916, Lodge attended spiritualist seances after receiving news that his son Raymond had been killed in action on the Western Front. Raymond Lodge's death, however, hardly explains Lodge's 'turn' to spiritualism, since it was seances that he attended in 1889 that had privately convinced him of that key spiritualist teaching: the survival of the soul following

---

[119] Chadwick, 'Stratton', esp. 290–1; Harrison, *Voyager in Space and Time*, pp. 12–13; Rayleigh, *Life of J. J. Thomson*, p. 285; Morris W. Travers, *The Life of Sir William Ramsay* (London: Edward Arnold, 1956), pp. 295–6.

[120] Crookes to 'J. H. D.', 10 May 1871 in [Anon.], 'Mr. William Crookes'.

[121] Tyndall, 'Miracles and Special Providences', p. 649; Crookes to 'J. H. D.'.

[122] In December 1870, some eighteen months after first attending seances, Crookes recorded in his diary the hope that he would "continue to receive spiritual communications" from his brother, Philip: Crookes quoted in Fournier d'Albe, *Life of Sir William Crookes*, p. 171. Fournier d'Albe claims that Philip's death (aboard a cable-laying ship) brought William "into close touch" with Varley, who persuaded him to contact Philip by "spiritualistic methods" (on p. 133). This is unlikely given that Crookes does not seem to have known Varley particularly well until 1871. This is implied by William Crookes to Cromwell F. Varley, 10 July 1871 in [Anon.], 'An Experimental Investigation of Spiritual Phenomena', *Spiritualist*, vol. 1 (1869–71), pp. 180–2, p. 182.

bodily death. These were seances given by the American medium Leonora Piper, who was at this time being investigated by the SPR.[123] Piper seemed to be able to commune with the disembodied spirits of some of Lodge's dead relatives whose disclosure of accurate personal family information was, as far as the physicist was concerned, hard to explain by fraudulence, self-deception, telepathy (because the spirits revealed information that he had never known) or any of the "ordinary methods known to Physical Science".[124] Privately convinced that he had communed with personalities in the post-mortem state, Lodge waited until 1902 (when he was much more professionally established) to publicise his belief.[125] But for Lodge, as for Crookes, bereavement was only ever one of many sources of appeal of psychical enquiry.

Crookes and Lodge differed from Myers and Sidgwick, as well as from many other physical–psychical scientists, in not experiencing a crisis of religious faith from which psychical investigation promised a way out. For both Cromwell Varley and William Kingsland, psychical investigation far surpassed Christianity in the rationality of its answers to profound spiritual questions. In 1869, Varley explained that he had received an early religious education from the Christian sect the Sandemanians, but that this had "wholly failed to satisfy [his] anxiety about the future".[126] Disembodied spirits, however, provided him with a better answer to one such anxiety. Having asked spirits "evidently more advanced" than himself why they had not given humanity "some scientific information in advance of any yet possessed by man", he was told that when they "telegraph to mortals" complex ideas, spirits put thoughts directly in the minds of mediums, but that these thoughts became distorted on being turned into words by the "mechanism of the brain and mouth".[127] This explanation was evidently "sound and logical" to the Atlantic cable engineer because it was entirely analogous to the difficulties of communicating complex information via encoded words on electric telegraphic mechanisms.[128]

In 1900, Kingsland attacked the Christian faith in which he had been raised for casting the "so-called *spiritual* world as a region of experience

---

[123] On the Piper investigations see Gauld, *Founders of Psychical Research*, 251–68; Hamilton, *Immortal Longings*, pp. 200–12.

[124] Oliver Lodge, 'Account of Sittings with Mrs Piper', *PSPR*, vol. 6 (1889–90), pp. 443–557, p. 443.

[125] Lodge, *Past Years*, p. 279.

[126] Varley, 'Evidence of Mr. Varley', p. 168. Varley's father, the inventor and landscape painter Cornelius Varley, shared his Sandemanian faith with Faraday. On Sandemanians and Faraday see Cantor, *Michael Faraday*.

[127] Varley, 'Evidence of Mr. Varley', pp. 168–9.

[128] Varley, 'Evidence of Mr. Varley', p. 168.

and action utterly unconnected with the facts of the so-called *natural world*" and making it a "region separate, apart, unknown save by special revelation, and unreachable save through the portals of the grave".[129] Among the "facts" that he believed evinced this convergence of the spiritual and natural worlds were clairvoyance, mesmerism and other hidden powers of the human self that he believed physical science was unable to explore owing to its exclusive focus on matter and force. What especially appealed to this electrical engineer about modern Theosophy was that it was a "higher science" encouraging individuals to train their minds and bodies to apprehend and have a "practical knowledge" of the spiritual world over which established religions wielded questionable authority.[130]

While some physical–psychical scientists had religious motivations for pursuing psychical enquiry, others had little or no such motivations. They joined those who, as Andreas Sommer has shown, had no doubt that psychical enquiry served purely physical and materialistic goals and who had little time for its potential religious significances.[131] The "physical theory" of spirit manifestations that Fournier d'Albe explained in *New Light on Immortality* seems to have been driven more by a desire to explore the implications of electron theory than a concern with the credibility of Christian theology.[132] In 1895, Oliver Heaviside, who at other times mocked Anglican clergymen and denied the Christian doctrine of the immortality of the soul, echoed the attitude of French physiologist Charles Richet when he insisted that the "abnormal phenomena" of spiritualism would ultimately be shown to have a "physical basis" and have no religious significance.[133] The person to whom Heaviside expressed this opinion was Lodge, who would have disagreed with such a purely physicalist goal, and objected to his fellow Maxwellian's related

---

[129] William Kingsland, 'Natural Law in the Spiritual World', *Theosophical Review*, vol. 26 (1900), pp. 441–50, p. 441.

[130] William Kingsland, *The Higher Science* (London: The Theosophical Publication Society, 1889), p. 2.

[131] Andreas Sommer, 'Crossing the Boundaries of Mind and Body: Psychical Research and the Origins of Modern Psychology', unpublished doctoral thesis, University College London, 2013, chapter 2.

[132] Fournier d'Albe, *New Light on Immortality*.

[133] Oliver Heaviside to Oliver Lodge, 11 January 1895, MS Add. 89/50(ii), OJL-UCL. For Heaviside's religious views see Oliver Heaviside, *Electromagnetic Theory Volume II* (London: The 'Electrician' Printing and Publishing Company', 1899), p. 20 and excerpts of his writings in Rollo Appleyard, *Pioneers of Electrical Communication* (London: Macmillan and Co., 1930), pp. 254–7. See also Paul J. Nahin, *Oliver Heaviside: The Life, Work, and Times of an Electrical Genius of the Victorian Age* (Baltimore, MD: Johns Hopkins University Press, 1988), pp. 108 and 252.

belief that life and the soul were inextricably associated with matter and force.

Changing Attitudes to Psychical Investigation

Among the most striking aspects of Tables 2.1 and 2.2 are the different periods for which physical–psychical scientists showed an interest in psychical phenomena. Explaining these trends is limited by the fact that so many physical–psychical scientists left little or no insights into the origins or fate of their interest. Many of those that did provide insights seem to have abandoned psychical research for reasons that were shared by many other enquirers. Some quickly lost interest because of the perceived absurdity of spiritualistic manifestations and rituals, as well as the threat that such enquiries posed to their professional reputation.[134] Others partially or wholly withdrew after disappointing investigations. Dmitrii Mendeleev abandoned spiritualism after the scientific commission he was leading to investigate the subject in 1875 officially declared that the physical effects were due to mediumistic deception.[135] Ten years later, the American astronomer Simon Newcomb was already expressing reservations about psychical research that would eventually persuade him that the whole enterprise had been a waste of time. Addressing the American branch of the SPR, he warned that telepathy, unlike other rare phenomena in the history of the sciences, did not become easier to reproduce or reveal more information about underlying laws and conditions of appearance as investigations proceeded. Phenomena that defied this standard pattern of "scientific progress" were probably due to "accidental or unknown circumstances" rather than "any new law".[136]

Mendeleev's withdrawal from spiritualism was partly driven by serious doubts about the scientific credibility of the methods that spiritualists and some professional colleagues had used when they obtained positive results from tests of mediums. William Ramsay found himself in a similar position. In 1914, he told Fournier d'Albe that he pitied the whole "spiritualistic affair" for being "so overlaid with fraud" and that this was undoubtedly one reason why he had long given up pursuing it.[137]

---

[134] Hans Thirring, 'The Position of Science in Relation to Psychical Research', *British Journal of Psychical Research*, vol. 1 (1927), pp. 165–81.

[135] Gordin, *Well-Ordered Thing*, pp. 87–96.

[136] Simon Newcomb, 'Address of the President', *Proceedings of the American Society for Psychical Research*, vol. 1 (1889–9), pp. 63–86, p. 78. Newcomb's damning verdict on psychical research was in Simon Newcomb, 'Modern Occultism', *Nineteenth Century*, vol. 65 (1909), pp. 126–39.

[137] William Ramsay to E. E. Fournier d'Albe, 13 December 1908, E. E. Fournier d'Albe Papers, Private Collection, Christine Fournier d'Albe.

Moreover, the "affair" had put an unwelcome strain on his professional relationships. He recalled that in 1895, when the SPR had invited him to join Myers, Lodge, Richet and others in tests of the Italian medium Eusapia Palladino, he had declined partly because he did not want to be forced into a situation where he would have to "behave" in ways suggesting his distrust of cherished scientific colleagues.[138] At the time of the tests, however, he doubted his own investigative abilities too. As he explained to Lodge in 1894, he was "so distrustful" of his "own judgement in affairs prestidigitical" that he felt it "imperative to be on one's guard against deception".[139]

Ramsay was not the only physical–psychical scientist whose interest in psychical investigation appears to have declined or died out completely because of misgivings about their own areas of expertise. George F. FitzGerald, a close colleague of Ramsay and Lodge, had been happy to help Barrett and Lodge with occasional psychical investigations in the 1880s and early 1890s, but thereafter seems to have been content merely to speculate on physical theories of psychical effects rather than to investigate them. It was precisely in this latter period that he echoed Carpenter and so many nineteenth-century physiologists, psychologists and physicians in arguing that since psychical phenomena occupied a "borderland" in "close proximity to hysteria, lunacy &c" then its "proper students" were "physicians, not physicists".[140] Similarly, in 1921, years after he had participated in investigations of Palladino, the American physicist Augustus Trowbridge declined an invitation to be involved in a new American Psychical Institute on the grounds that his "own interests and training lie so far away from the subject of psychical research".[141]

Two of the most distinguished British physicists who participated in the SPR's investigations of Palladino well illustrate a tendency among some physical–psychical scientists to remain steady in the level of their commitment to and interpretative positions on psychical investigation. Despite their relative inertness as SPR members, Rayleigh and J. J. Thomson remained optimistic about the value of psychical research until their deaths (in 1919 and 1940 respectively), not least in its potential to illuminate questions that they were still not satisfied had been settled. Thus, in the presidential address that he gave to the SPR only weeks before he died, Rayleigh insisted that while he had difficulties accepting

---

[138] Ramsay to Fournier d'Albe, 13 December 1908.
[139] William Ramsay to Oliver Lodge, 27 July 1894, MS Add. 89/88, OJL-UCL.
[140] FitzGerald quoted in Stead, 'Response to the Appeal', p. 19.
[141] Augustus Trowbridge to Hereward Carrington, 23 June 1921, Folder 50, Box 2, Hereward Carrington Papers (C1159), Manuscripts Division, Department of Rare Books and Special Collections, Princeton University Library.

the idea of telepathy (because it represented a form of communication that seems to have escaped evolutionary development and was so different from speech and writing), "no pains should be spared" to establish its reality.[142] In his 1936 autobiography, Thomson agreed that telepathy's "transcendent importance" justified the need for further research and he shared Rayleigh's humble acceptance that his own negative experience of Palladino's mediumship was not necessarily the final word on the question of whether objects in seances could be moved by some unknown force.[143]

Among the individuals to whom Thomson deferred on the question of Palladino's mediumship was Lodge, who, like Barrett and Crookes but unlike Thomson, displayed interests in psychical research that grew stronger and more complex with time. He, Barrett and Crookes had embarked on psychical investigation with interests in specific questions: he and Barrett had started with the genuineness of thought-reading, while Crookes started with the reality of the physical phenomena of spiritualism. By 1900, all three had, through extensive personal experiences of psychical effects and close friendships with fellow psychical investigators inside and outside the SPR, become interested in a much wider range of phenomena: Barrett and Lodge were much keener on spiritualism than they had been when embarking on psychical investigation, and Crookes had developed an interest in telepathy and the purely psychological aspects of psychical research that he did not have when he started attending seances in the late 1860s. By the early 1900s, all three had become convinced by the evidence for telepathy, telekinesis and the survival of the soul following bodily death, and had therefore come to represent what Eleanor Sidgwick called the "forward" section of the SPR, as opposed to "hard-of-belief" section in which she and many other leading members of the organisation located themselves.[144]

---

[142] Rayleigh, 'Presidential Address', p. 288.

[143] Thomson, *Recollections and Reflections*, p. 158.

[144] Eleanor Sidgwick to William F. Barrett, 23 November 1905, SPR.MS 3/A4/115, WFB-SPR. By this time Sidgwick was more convinced of telepathy than of either telekinesis or survival, although she still believed telepathy needed to be replicated under conditions that ruled out chance coincidence, self-deception and fraudulence: [Eleanor] Sidgwick, 'Presidential Address', *PSPR*, vol. 22 (1908), pp. 1–18, pp. 16–17. By the early 1930s, however, she was more positive about survival: Sidgwick, 'Society for Psychical Research', p. 26. Barrett declared his belief in thought-transference in William F. Barrett, 'Mind-Reading versus Muscle-Reading', *Nature*, vol. 24 (1881), p. 212; in the direct action of mind on matter in William F. Barrett, 'On Some Physical Phenomena, Commonly Called Spiritualistic, Witnessed by the Author', *PSPR*, vol. 4 (1886–7), pp. 25–42, pp. 39–40; and in survival in Barrett, 'Address by the President', p. 350. Crookes had become convinced by telekinesis by the early 1870s: William Crookes, *Researches in the Phenomena of Spiritualism* (London: James Burns, 1874). He declared his belief in survival and telepathy in William Crookes, 'Address by the President',

A combination of old and new preoccupations defined their mature psychical interests. By the turn of the century, Barrett was even more strongly convinced than in earlier decades that psychical research presented an "inexhaustible mine for scientific research" which would be of "inestimable value in destroying a materialism" undermining the foundations of Christianity.[145] But Barrett's interests in psychical research were much broader now. It was something whose results converged with a much wider range of religious and esoteric teachings than he had been prepared to accept before the 1900s (including Swedenborgian and modern Theosophical teachings on the soul).[146] It was also something that lent support to an optimistic view of life on earth, whether by giving credence to new theories that biological evolution was a progressive process partly directed by vital and psychic forces, or by suggesting that telepathy was a genuine faculty that, if allowed to evolve in humans, would accelerate the "sense of sympathy and humanity" in the world and lead to the amelioration of the kind of poverty that he had spent decades tackling in Ireland.[147]

By 1900, Crookes was still interested in the physical phenomena of spiritualism but was very discreet about the results of the handful of investigations that he pursued in private.[148] Compared with the 1870s, the period of his most intense spiritualistic enquiries, he was now much more willing to accept that psychical phenomena required the skills of psychologists and conjurors, as well as physical scientists such as himself, because the physical effects they often involved depended so strongly on

PSPR, vol. 12 (1896–7), pp. 338–55, pp. 339 and 349. Lodge declared his belief in telekinesis in Oliver Lodge, 'Experience of Unusual Phenomena Occurring in the Presence of an Entranced Person (Eusapia Palladino)', *JSPR*, vol. 6 (1893–4), pp. 306–60, pp. 307–8. He announced his conviction of the reality of thought-transference in Oliver Lodge, 'Thought Transference: An Application of Modern Thought to Ancient Superstitions', *Proceedings of the Literary and Philosophical Society of Liverpool*, vol. 46 (1892), pp. 127–45; and in 1902 declared a "personal" conviction of survival, although this could not be justified in a "full and complete manner": Oliver Lodge, 'Address by the President', *PSPR*, vol. 17 (1901–3), pp. 37–57, p. 49.

[145] William F. Barrett, *On the Threshold of a New World of Thought* (London: Kegan Paul, Trench, Trübner and Co., 1908), p. 14.

[146] William F. Barrett, *Swedenborg: The Savant and the Seer* (London: John M. Watkins, 1912); Barrett, *Threshold of a New World of Thought*, pp. 42–3n.

[147] Barrett, *Threshold of a New World of Thought*, p. 99. See also William F. Barrett, 'The Psychic Factor in Evolution', *Quest*, vol. 9 (1917–18), pp. 177–202. For Barrett's charitable and humanitarian work see [Anon.], 'W. F. Barrett, F.R.S.E., M.R.I.A., &c', *Light*, vol. 14 (1894), pp. 439–41.

[148] In 1903, he told the SPR that for "many years" he had been trying unsuccessfully to "find some physical method of testing or measuring supernormal susceptibilities or capacities": William Crookes cited in [Anon.], 'General Meeting', *JSPR*, vol. 11 (1903–4), pp. 152–7, p. 156.

the human mind and on possibly fraudulent human subjects. In 1898, he accepted the more widely used identification of psychical research as "Experimental Psychology" and thus signalled his acceptance that the field of inquiry he had helped to stimulate was more relevant to psychology than to physics.[149] A few years later, he reputedly believed that the "laws" governing the "radiations of Thought" underpinned "all occult phenomena", which made the study of telepathy more important for discerning "psychical truth" than those more physical aspects of psychical research which had interested him for decades.[150] But Crookes, unlike FitzGerald, Trowbridge and others, never lost hope that physics mattered to psychical research, as revealed by his continued interest in physical theories of telepathy and a telegraphic apparatus that allegedly picked up wireless waves from the dead.[151]

Crookes's later writings on psychical research included many defences of his earlier positive evidence of psychical phenomena, but also some unpublished and published pieces exploring psychical research's wider significance.[152] By 1897, he had written (but not published) an 'Essay on Immortality', a topic on which he had rarely touched before, and was much more interested in the way that psychical research underpinned the need for humility in the sciences, especially among those individuals whose overly "terrestrial" or materialistic perspective on the cosmos blinded them to the possibility of an "unseen world" transcending the limits of human faculties.[153] Crookes never seems to have questioned the capacity of spiritualism to provide emotional support, and, following the death of his wife in 1916, the grief-stricken scientist claimed to have gained conclusive evidence of communications from and photographic images of her spirit. Significantly, Crookes was prepared to declare his convictions in spiritualist periodicals and this represented a significant

---

[149] William Crookes, 'Address by Sir William Crookes', in *Report of the Sixty-Eighth Meeting of the British Association for the Advancement of Science Held at Bristol in September 1898* (London: John Murray, 1899), pp. 3–38, p. 32.

[150] Crookes cited in Harold Begbie, *Master Workers* (London: Methuen and Co., 1905), p. 220.

[151] Crookes, 'Address by the President'. In 1915, he tried unsuccessfully to replicate an electrical machine built by a British inventor, David Wilson, for detecting communications from spirits of the dead: Minutes for 6 October 1915, Ghost Club Minutes, Volume 7, Add. 52264, GC-BL.

[152] An example of such defences is Crookes quoted in [Anon.], 'Sir William Crookes on Psychical Phenomena', *Light*, vol. 36 (1916), p. 397. He also expressed psychical interests in his brief memberships of the Theosophical Society and the Hermetic Order of the Golden Dawn: Brock, *William Crookes*, pp. 337–43. Crookes was particularly interested in the claim of theosophists Annie Besant and Charles Leadbeater that they had used clairvoyant means to discern the constituents of atoms.

[153] Fournier d'Albe, *Life of Sir William Crookes*, p. 357; Crookes, 'Address by the President', p. 343.

change in the way he managed his public responses to spiritualism and personal tragedy since the 1870s.[154]

One of the reasons why Crookes may have been more comfortable about revealing the way spiritualism had benefited him personally was Lodge's *Raymond or Life and Death* (1916).[155] This bestselling and controversial account of Lodge's positive evidence for the post-mortem existence of his son Raymond offered consolation to the millions who had lost loved ones in the First World War. It also represented the considerable broadening in Lodge's view of psychical research since he had embarked on the subject in the 1880s. In addition to a fearless acknowledgment of the emotional comforts that spiritualism had brought him personally, Lodge had developed a much stronger interest in the ways in which the results of psychical research constituted paths of advance in different sciences as well as in resolving the "outstanding controversy" between science and religious faith.[156] In myriad books, articles, public lectures and radio broadcasts appearing throughout the early decades of the twentieth century, Lodge tied psychical research to ether physics and biological vitalism, as well as to arguments for the scientific reinterpretation and justification of such fundamental Christian beliefs as the survival of the soul, the efficacy of prayer, and the guidance of humanity and the material cosmos by a benevolent God.

***

This chapter has argued that late-nineteenth- and early-twentieth-century practitioners of the physical sciences, broadly defined, showed an interest in psychical phenomena that was stronger and altogether more complex than historians have claimed. In some ways these physical–psychical scientists did not constitute an especially distinctive portion of the much larger number of individuals who pursued psychical phenomena between the 1850s and 1930s. They echoed more widely shared concerns for decisive verdicts on the reality and provenance of psychical phenomena, and preoccupations with the potential of such phenomena to serve a host of scientific, religious, philosophical and emotional needs. Given the widespread interest in psychical phenomena, we should expect *some*

---

[154] Crookes quoted in [Anon.], 'Sir William Crookes' and Crookes quoted in [Anon.], 'Important Interview with Sir William Crookes', *International Psychic Gazette*, vol. 4 (1917), pp. 61–2, p. 62.

[155] Oliver Lodge, *Raymond or Life and Death* (London: Methuen and Co., 1916). For discussion see Rene Kollar, *Searching for Raymond: Anglicanism, Spiritualism and Bereavement Between the Two World Wars* (Lanham, MD: Lexington Books, 2000).

[156] Oliver Lodge, *Man and the Universe: A Study of the Influence of the Advance in Scientific Knowledge upon Our Understanding of Christianity* (London: Methuen and Co., 1908), pp. 1–24.

practitioners of the physical sciences to follow the trend. Yet we are still left with one unusual trend. The proportion of *physical* scientists in the early SPR and otherwise involved in psychical investigation is more than we might expect on the basis that this enterprise was more concerned with psychological than physical phenomena. Part of the explanation of this is that psychical phenomena seemed to represent *enough* of a physical puzzle to stimulate the interests of many practitioners of the physical sciences. Indeed, what chiefly distinguished these individuals from the large number of other psychical enquirers was their belief in the possibility that psychical phenomena represented a possible extension of the knowledges and practices of the physical sciences.

This chapter has had a distinctly British focus and this reflects the broad trends in Tables 2.1 and 2.2. The focus is partly due to the fact that psychical research, as embodied in the work of the SPR, was a largely British invention and British individuals dominated the enterprise well into the early twentieth century, even when varieties of psychical research were flourishing elsewhere (for example, *métapsychique* in France and *psychische studien* in Germany).[157] The achievements of the SPR in telepathy and other questions ensured that interest in psychical research was particularly strong in Britain, irrespective of whether individuals joined the organisation.

The SPR is, however, only part of the reason for the Anglo-centrism of physical–psychical scientists. It built partly on the achievements of earlier enquirers into mesmerism and spiritualism, among whom the most scientifically distinguished (and strategically useful to the organisation) were Barrett, Crookes and Varley. Their achievements and examples made a strong positive impression on a larger number of younger British physical scientists (for example, Lodge, Rayleigh and J. J. Thomson) who already admired the older individuals for their contributions to the physical sciences. The psychical interests of Lodge, Rayleigh, Thomson and others in turn impressed an even larger number of still younger British physical scientists, many of whom were their students or professional colleagues, and who followed their examples well into the 1920s and '30s. There is evidence to suggest that the creation of such favourable impressions on the controversial question of psychical phenomena owed much to face-to-face interactions, and for this reason physical scientists were more likely to be impressed by the psychical investigations of individuals with whom they could personally acquaint themselves (including

[157] On German *psychische studien* see Sommer, 'Normalizing the Supernormal' and Wolffram, *Stepchildren of Science*; on French *métapsychique* see Brower, *Unruly Spirits* and Lachapelle, *Investigating the Supernatural*.

those outside the physical sciences) than those much further afield.[158] This was surely a significant driver of a snowballing effect of psychical interest among late-Victorian British practitioners of the physical sciences, whether or not it was expressed through SPR membership.

What undoubtedly fuelled this effect was the fact that the scientific individuals most revered by British physical–psychical scientists included William Thomson, Stokes and Maxwell, who gave British physics an anti-materialistic and anti-deterministic edge by arguing that a proper scientific understanding of energy, ether and matter left open the possibility of mind, both human and divine, independent of matter. Although these leaders of British physics would not have approved of it, their interpretations created powerful new possibilities for the convergences between the theories of physics and the results of psychical enquiry. In Chapter 3 we study these convergences in detail.

[158] In 1874, Rayleigh told his mother that after an hour and a half's conversation with Crookes he had "no reason to doubt his trustworthiness" regarding the mediumship of Rosina Showers: Lord Rayleigh to Clara Elizabeth Rayleigh, 3 May 1874 in Rayleigh, *Third Baron Rayleigh*, pp. 65–6, p. 66. Even that arch-sceptic of spiritualism Tyndall revealed that after Varley had visited him in 1868, he was more willing to witness spirit manifestations that the telegraph engineer deemed worthy of his attention: John Tyndall to G. W. Bennett, 22 December 1869, in *Report on Spiritualism*, p. 265.

3   Physical Theories and Psychical Effects

In the preface to his wife's account of her experiences as a spiritualist medium, Augustus De Morgan expressed a key dilemma faced by many early spiritualist enquirers. "I am perfectly convinced", he declared, "that I have both seen, and heard in a manner which should make unbelief impossible, things *called* spiritual which cannot be taken by a rational being to be capable of explanation by imposture, coincidence, and mistake."[1] The problem was that "physical explanations", in which he included "imposture, coincidence, or mistake", were "easy, but miserably insufficient", while the "spiritual hypothesis", which proposed the existence of invisible agencies exerting "will, intellect and physical power", was "sufficient, but ponderously difficult".[2]

The intellectual stature that De Morgan enjoyed by the mid-nineteenth century ensured the currency of this declaration in the emerging literature on spiritualism and, later, psychical research. Its rhetorical value was certainly not lost on Barrett and other physical–psychical scientists seeking a noble scientific precedent to the interpretative problems that they encountered in psychical investigations.[3] In 1870, when publicly defining the scientific 'light' that he proposed to cast on spiritualism, Crookes quoted De Morgan's words to frame a modest refusal to "hazard even the most vague hypothesis" and his dissatisfaction with the "spiritual" theory.[4] Privately, and with Barrett's help, Crookes was keen to develop a purely "physical theory" of the phenomena of spiritualism that involved a force radiated by the body that was capable of causing the movement of

---

[1] 'A. B.' [Augustus De Morgan], 'Preface' in 'C. D.' [Sophia De Morgan], *From Matter to Spirit*, pp. v–xiv, v–vi. The De Morgans' identities had been established shortly after the publication of this work. See, for example, [Anon.], 'From Matter to Spirit', *London Review*, vol. 7 (1863), pp. 547–8.
[2] 'A. B.' [Augustus De Morgan], 'Preface', pp. v–vi.
[3] See, for example, Barrett, *Threshold of a New World of Thought*, p. 4; Crookes, 'Spiritualism Viewed', p. 317.
[4] Crookes, 'Spiritualism Viewed', p. 317.

untouched objects in seances.[5] Although both men accepted that any such theory would be "insufficient" in the way implied by De Morgan's declaration – it could not cope with the obvious intelligent and psychological nature of the phenomena – they maintained that the knowledge established by the physical sciences remained important in establishing the plausibility of the phenomena.

Underpinning Crookes's and Barrett's cautious physical theorising was a much more widely shared and longer-lasting belief that the physical sciences were not irrelevant to the processes of making sense of psychical phenomena. For all their disagreements about the results of psychical investigation, from the 1850s until the 1930s physical–psychical scientists agreed that the theories, concepts and explanations developed within the physical sciences for understanding matter, energy, ether and other aspects of the physical cosmos were, to one degree or another, useful in illuminating psychical questions and in offering interpretations that were more satisfactory than those offered by physiologists, spiritualists and many others engaged in psychical questions. Focussing on the period between the 1870s and the turn of the twentieth century, this chapter analyses the different forms taken by these strategies of illumination and shows that the physical sciences were incredibly fertile sources for psychical thinking. It moves from weaker to stronger claims for the psychical relevance and utility of physical theories, models and concepts. The physical sciences were useful in providing general principles and salutary lessons for developing what many of our protagonists judged to be a more tolerant and scientific way of contemplating psychical phenomena in general, but they also furnished materials for going much further and developing physical analogies and possible explanations.

Removing Scientific Stumbling Blocks

In 1874, John Tyndall delivered the presidential address to the British Association, which met in Belfast that year. For many of its auditors it would rank as one of the most potent of all nineteenth-century statements of materialism.[6] Few passages of the address provoked more alarm than Tyndall's declaration that the

---

[5] William Crookes to William F. Barrett, 15 May 1871, quoted in Fournier d'Albe, *Life of Sir William Crookes*, p. 199. See also Crookes, *Psychic Force*, p. 5; Barrett, 'Phenomena of Spiritualism', p. 1019.

[6] One such auditor was Lodge: Oliver Lodge, *Advancing Science: Being Personal Reminiscences of the British Association in the Nineteenth Century* (London: Ernest Benn, 1931), pp. 35–6.

doctrine of the Conservation of Energy, the ultimate philosophical issues of which are as yet but dimly seen – that doctrine which 'binds nature fast in fate', to an extent not hitherto recognised, exacting from every antecedent its equivalent consequent, from every consequent its equivalent antecedent, and bringing vital as well as physical phenomena under the dominion of that law of causal connection which, so far as the human understanding has yet perceived, asserts itself everywhere.[7]

No less provocative was his later claim that by a "necessity engendered and justified by science" he could cross the "boundary of the experimental evidence" and discern in matter "the promise and potency of all terrestrial life".[8] However much Tyndall insisted that his 'scientific' form of materialism represented a pragmatic scientific methodology rather than a cosmic philosophy, many physicists taking a theistic view of the cosmos interpreted his statements as unwarranted extensions of physical laws and as potent threats to their deep-rooted belief in free will and a universe created and guided by divine will.[9] How Maxwell, Stokes and others responded to Tyndall would shape the way that many physical–psychical scientists would respond to charges that other non-material constituents of the cosmos – psychical effects – were illusory and unworthy of scientific attention.

The most widely read and controversial of all theistic physicists' responses to the materialistic threat of the Belfast Address was undoubtedly Balfour Stewart and Peter Guthrie Tait's *Unseen Universe*.[10] Its main objective was to persuade the "orthodox in religion", who were "aghast at the materialistic statements now-a-days freely made (often professedly in the name of science)", that the "presumed incompatibility of Science and

---

[7] John Tyndall, 'The Belfast Address [1874]', in Tyndall, *Fragments*, vol. 2, pp. 135–201, pp. 180–1.

[8] Tyndall, 'Belfast Address', p. 191.

[9] For critical interpretations of the 'materialism' of this address see Ruth Barton, 'John Tyndall, Pantheist: A Rereading of the Belfast Address', *Osiris*, vol. 3 (1987), pp. 111–34; Bernard Lightman, 'Scientists as Materialists in the Periodical Press: Tyndall's Belfast Address', in Geoffrey Cantor and Sally Shuttleworth (eds.), *Science Serialized: Representations of the Sciences in Nineteenth Century Periodicals* (Cambridge, MA: MIT Press, 2004), pp. 199–237; DeYoung, *Vision of Modern Science*, chapter 3.

[10] The *Unseen Universe* reached its seventeenth edition in 1901. The seventh (1878) edition was the first where Stewart and Tait dropped their anonymity, and coincided with a 'sequel' exploring many of the same arguments in the form of a country house novel: Balfour Stewart and Peter Guthrie Tait, *Paradoxical Philosophy: A Sequel to the Unseen Universe* (London: Macmillan and Co., 1878). For discussion see Graeme Gooday, 'Sunspots, Weather and the *Unseen Universe*: Balfour Stewart's Anti-Materialist Representations of "Energy" in British Periodicals', in Cantor and Shuttleworth, *Science Serialized*, pp. 111–47; P. M. Heimann, 'The *Unseen Universe*: Physics and the Philosophy of Nature in Victorian Britain', *British Journal for the History of Science*, vol. 6 (1972), pp. 73–9; Myers, 'Nineteenth Century Popularisations'.

Religion does not exist".[11] Their argument hinged on a 'principle of continuity' which they interpreted as the means by which humans could confidently reconcile a given state of the universe with its antecedent state. For the authors, it was the essential principle regulating scientific progress and could be used to render apparent breaks of cosmic continuity – including the disappearance of the human soul following bodily death, Biblical miracles and the origin of the visible universe – scientifically intelligible.

Theories of energy, ether and atoms were central to what the book's subtitle modestly announced as "physical speculations". Theories of energy conservation and dissipation suggested that energy no longer available or useful to humanity was not destroyed but dissipated elsewhere. It was not implausible that the abode of this energy was an invisible universe, connected to the visible one via the semi-material luminiferous ether whose absorption of starlight indicated a dissipative capacity. The ether was also the possible domicile of the atomic constituents of matter: William Thomson and others had proposed that atoms could be modelled as vortex ring structures in a perfect fluid, and for many physicists the ether could be regarded as a fluid medium. However, Stewart and Tait argued that, owing to the luminiferous ether's apparent absorptive and frictional properties, its vortices were not permanent and had to originate in vortices in an increasingly subtle, stable and energetic succession of ethers or fluid media, culminating in a truly spiritual, perfectly stable and infinitely energetic invisible universe.

The unseen universe gave continuity to apparent cosmic discontinuities. Events such as the origin of life and the visible universe and the Resurrection of Christ could be accommodated as fluxes of energy from this hidden realm. Moreover, the Christian doctrines of the post-mortem state and immortality of the human soul after bodily death were easier to comprehend: modern physiology suggested that human thoughts (and therefore personality) could be associated with the energies of the molecular constituents of the brain, and it was possible that some of these energies could be "communicated" via the succession of ethers to the unseen universe, where they formed an individual's permanent "spiritual" body that also had the capacity to act on the visible universe.[12]

One of the reasons why the *Unseen Universe* proved so controversial was because it seemed to offer physical proofs of Christian doctrines, which was itself tantamount to the materialism the book was trying to vanquish. In the preface to the second (1875) edition of their work, Stewart and Tait

---

[11] [Stewart and Tait], *Unseen Universe*, p. vii.
[12] [Stewart and Tait], *Unseen Universe*, p. 159.

insisted that far from trying to demonstrate Christian truth from a "mere physical basis", they had merely aimed to show that physical laws helped break down "unfounded objections" to Christian doctrines.[13] Stewart and Tait's modest goal of using physical laws to counter materialism rather than to directly prove the existence of mind and spirit in the cosmos would be adopted, adapted and echoed in numerous arguments made by physical–psychical scientists against the impossibility of psychical effects.

One form of this argument focussed on the limitations of existing physical laws and principles, and thus followed the example of theistic physics, which often warned against extending such laws and principles to questions of life and mind. In a paper presented to the SPR in 1887, for example, Stewart responded to the common charge that telekinesis violated energy conservation by arguing that this physical principle was merely a "sagacious" scientific "assertion" that might be found to break down under certain conditions and domains, including those psychological regions explored by the SPR.[14] While he remained committed to the general principle of the uniformity of nature, he was open to the possibility that this might be embodied in "higher", hitherto undiscovered laws transcending those believed to govern energy transformations.[15]

Two other leading SPR physicists – Crookes and Lodge – drew on a particular line of argument developed by Maxwell and Stokes in the 1870s to highlight the limited scope of physical laws. This involved the metaphor of a pointsman or engine driver.[16] The relationship between the human mind or soul and the body was similar to that between a pointsman or engine driver and a railway locomotive. The laws of energy transformation explained why the locomotive moved but not necessarily why it was shunted or turned at a particular time or place: this required the action of the immaterial will of the pointsman or engine driver. The critical lesson that Maxwell and Stokes wanted to draw from this was that the will, or some other immaterial power, could guide or otherwise act on

---

[13] [Balfour Stewart and Peter Guthrie Tait], 'Preface to the Second Edition', in [Balfour Stewart and Peter Guthrie Tait], *The Unseen Universe; Or Physical Speculations on a Future State* (London: Macmillan and Co., 3rd ed., 1875), pp. x–xi. The Anglican theologian William Josiah Irons criticised the book for its materialism: [William Josiah] Irons, 'On the Scientific Conclusions and Theological Inferences of a Work Entitled "The Unseen Universe, Or Physical Speculations on a Future State"', *Journal of the Transactions of the Victoria Institute*, vol. 11 (1878), pp. 83–139, esp. p. 134.

[14] Balfour Stewart, 'Note on the Above Paper', *PSPR*, vol. 4 (1886–7), pp. 42–4, p. 42. Stewart was responding to Barrett: 'On Some Physical Phenomena'. Barrett later revealed his appreciation of Stewart's remarks: William F. Barrett, 'Dynamic Thought. Part II. The Realm of the Unconscious', *Humanitarian*, vol. 7 (1895), pp. 345–53, p. 351.

[15] Stewart, 'Note on the Above Paper', p. 43.

[16] Stanley, *Huxley's Church*, pp. 222–33; Wilson, *Kelvin and Stokes*, pp. 93–4.

a material system (including the human body) without violating its physical laws.

Crookes's debt to the pointsman metaphor was obvious in the conclusion of the presidential address he gave to the SPR in 1897. As somebody who had relied on Stokes's help in mathematical and physical questions for decades, this debt is not surprising.[17] Crookes told fellow psychical researchers that while physical processes such as striking matches or writing telegrams were governed by the law of conservation of energy, "the vastly more momentous part, which determines the words I use or the material I ignite, is beyond such a law" and such decision-making could only be ascribed to the "mystic forces" of intelligence and free will.[18] Many of his auditors would have appreciated that, having argued for the need to invoke the "guiding and constraining" action of "mystic forces" on known physical processes, Crookes was implicitly challenging the argument that the still-more-mysterious forces investigated in psychical research were impossible simply because they appeared to violate physical laws.[19] By implying that psychical phenomena could perfectly well follow such laws and be coaxed by unknown forces, he was trying to remove "some of those scientific stumbling-blocks" preventing many from involving themselves in psychical research.[20]

Although Crookes's correspondence with Lodge reveals no exchange of ideas about the content of this SPR address, his preoccupation with the capacity of mind to direct or guide otherwise law-like physical processes arguably owed something to the physicist.[21] Only a few years after Crookes's address, Lodge began presenting his own ideas on 'guidance', and these served at least two goals on which Lodge had decided to focus a good deal more of his attention after becoming Principal of Birmingham University: interpreting the results of psychical research and seeking new ways of reconciling science and Christian theology. A key step was tackling what he regarded as scientific, philosophical and religious barriers to understanding how the human mind interacted with the body and how God interacted with the physical universe. In a 1903 paper for the short-

---

[17] Hannah Gay, 'Invisible Resource: William Crookes and his Circle of Support', *British Journal for the History of Science*, vol. 29 (1996), pp. 311–36; Wilson, *Kelvin and Stokes*, pp. 191–208.

[18] Crookes, 'Address by the President', p. 354.

[19] Crookes, 'Address by the President', p. 354.

[20] Crookes, 'Address by the President', p. 354. Another such stumbling block was the "materialistic" idea that spiritual beings could neither exist "without form or matter" nor be "untrammelled by gravitation or space". Crookes's tentative response was to picture such beings as centres of "intellect, will, energy and power" permeating space, much as Faraday represented material bodies as centres of power (see pp. 342–3).

[21] A large amount of the Crookes–Lodge correspondence is in SPR.MS 35/310–366, OJL-SPR.

lived forum of philosophical, theological and scientific debate the
Synthetic Society, he echoed Maxwell and Stokes in insisting that expo-
nents of scientific naturalism pressed physical laws

to what seems their logical and ultimate conclusion, in applying the conservation
of energy without ruth or hesitation, and so excluding, as it has seemed, the
possibility of free-will action, or guidance, of the self-determined action of mind
or living things on matter, altogether. The appearance of control has been considered illusory, and has been replaced by the doctrine of pure mechanism,
enveloping living things as well as inorganic nature.[22]

But Lodge also targeted some of the most vehement critics of scientific
naturalism. These were the philosophers and theologians who, on the
basis of faith, instinct or experience, denied that volition and "non-
mechanical" agencies could be so excluded and had consequently tried
to "undermine the foundations of Physics" by showing that such important laws as the conservation of energy were inexact and "too rapid an
induction" and that "there may be ways of eluding many physical laws
and of avoiding submission to their sovereign sway".[23]

Lodge needed to counter both these positions and, moreover, safeguard physical laws from those who sought to abandon them entirely.
Dynamical laws, he insisted in an earlier contribution to the Synthetic
Society, "do not exclusively cover the realm of Nature even on the
material side", and by this he meant the places where biology and psychology overlapped with the purely physical, where something vital or
mental was "operating consistently and legally no doubt, in no way
upsetting dynamical laws, not in the least infringing the conservation of
energy but contriving interactions and initiating changes which in the
region governed *solely* by dynamics would never have occurred".[24]
Replacing the pointsman with a steamship helmsman, Lodge illustrated
this argument by rhetorically asking Synthetic Society attendees whether
"solely dynamics" could explain the helmsman's actions and the subsequent departure time and direction of the vessel.[25]

Lodge's argument for the possibility of guidance of physical systems by
"non-mechanical" agencies fed into his contributions to the old debate on

[22] Lodge, untitled essay dated 20 February 1903, in *Papers Read Before the Synthetic Society,
1896–1908* ([London]: Spottiswoode and Co., 1909), pp. 385–92, p. 386. The Synthetic
Society held meetings in London from 1896 to 1908: see William C. Lubenow,
'Intimacy, Imagination and the Inner Dialectics of Knowledge Communities: The
Synthetic Society, 1896–1908', in Martin J. Daunton (ed.), *The Organisation of
Knowledge in Victorian Britain* (Oxford University Press, 2005), pp. 357–70.
[23] Lodge, untitled essay, pp. 386–7.
[24] Oliver Lodge, 'Supplement to the Discussion on Mr. Balfour's Paper [1900]', in *Papers
Read Before the Synthetic Society*, pp. 334–40, pp. 336–7.
[25] Lodge, 'Supplement', p. 338.

the efficacy of prayer. By the early 1900s, prayer had taken on an enormous significance for Lodge. The idea that humanity was guided, inspired and progressed by non-corporeal minds seemed to be scientifically more credible in light of the SPR's mounting evidence of the capacity of one mind to directly affect another at a distance.[26] Conclusive scientific proof that prayers influenced the physical world would help resolve the "outstanding controversy" between science and the Christian faith.[27]

Before this distant point was reached, however, Lodge felt that the least he could do was to try to remove a stumbling block in the form of Tyndall's notorious critique of the efficacy of prayer that had fostered much debate since its appearance in 1861.[28] Anticipating much of the determinism that he would propound in the 1870s, Tyndall denied that prayer had any effect in a physical world strictly governed by natural laws. For Lodge, Tyndall's critique showed a theological animus towards prayer which had blinded him to the fact that from the perspective of "pure physics" a prayer for rain no more interfered with energy conservation than the banal case of asking a gardener to fetch a watering can.[29] A proper estimate of the scope of physics would show that prayer, telepathy and communion with the spirits of the dead, as well as uncontroversial requests, supplemented rather than violated cherished physical laws.

The uses to which Stewart, Crookes and Lodge put theistic physics was only one of many ways in which physical–psychical scientists drew psychical implications from critical reflections on physical laws, principles and explanations. Many of these reflections spoke to the growing tendency among late-nineteenth-century physicists to relax claims of the ultimate truth of their theories, models, principles and laws.[30] The tendency was strongly exhibited by the American astrophysicist and SPR member Samuel P. Langley, who in 1901 warned that natural laws were

[26] Oliver Lodge, 'The Reconciliation Between Science and Faith', *Hibbert Journal*, vol. 1 (1902–3), pp. 209–27.

[27] Oliver Lodge, 'The Outstanding Controversy Between Science and Faith', *Hibbert Journal*, vol. 1 (1902–3), pp. 32–61.

[28] On the efficacy of prayer debate see Turner, *Contesting Cultural Authority*, chapter 6.

[29] Lodge, 'Supplement', p. 334.

[30] On the humility of late-nineteenth-century physicists see John L. Heilbron, 'Fin-de-Siècle Physics', in Carl Gustaf Bernhard, Elisabeth Crawford and Per Sörbom (eds.), *Science, Technology and Society in the Time of Alfred Nobel* (Oxford: Pergamon Press, 1982), pp. 51–73; Erwin N. Hiebert, 'The State of Physics at the Turn of the Century', in Mario Bunge and William R. Shea (eds.), *Rutherford and Physics at the Turn of the Century* (New York: Dawson and Science History Publications, 1979), pp. 3–22; Albert E. Moyer, *American Physics in Transition: A History of Conceptual Change in the Late Nineteenth Century* (Los Angeles: Tomash Publishers, 1983).

merely "hypotheses about nature" put together by humans who were susceptible of "fallible judgement" rather than being fixed items in a cosmic "statute book".[31] This fallibility was, according to Langley, painfully clear in the way that progress in one of his areas of research – the science of radiant heat – had been obstructed by a blind adherence to cherished theories.[32] Having spent brief periods in the 1890s studying psychical phenomena, Langley had reasons to conclude from his analysis that a misguided vanity regarding natural laws could prevent an acceptance of genuinely new facts relating to hypnotism and telepathy, however "absurd" they might seem at first.[33]

Among Langley's sources of inspiration for this argument was the same SPR address where Crookes had echoed the pointsman argument. Part of the address discussed the "accessible temper of mind", or healthy sense of scientific ignorance, that Crookes claimed had long driven his interest in psychical research.[34] It was not the first time that Crookes had put scientific humility to such a use. His 1870 manifesto for a new scientific approach to spiritualism had criticised Faraday for insisting that studies of physical principles (including those allegedly operating in table-turning) needed to start with *"clear ideas* of the naturally possible and impossible".[35]

For Crookes, molecular physics, a subject to which he devoted much time from the 1870s onwards, confirmed the perils of adopting a dogmatic attitude towards boundaries of the natural and provided a good "parable" for the "large element of subjectivity" conditioning all knowledge.[36] The parable concerned a homunculus who, owing simply to its microscopic size, understood the physical world very differently from humans. Surface tension, Brownian motion and other molecular forces that "we hardly notice" became so "conspicuous and dominant" for the homunculus that it hardly believed in the existence of a force so conspicuous to us: universal gravitation.[37] Molecular physics provided a salutary lesson to those who took "too terrestrial a view" of the world and who rejected the possibility of an "unseen world" where ideas about

---

[31] Samuel P. Langley, 'The Laws of Nature', in *Annual Report of the Board of Regents of the Smithsonian Institution for the Year Ending June 30, 1901* (Washington, DC: Government Printing Office, 1902), pp. 545–52, pp. 549–50. On Langley see Moyer, *American Physics in Transition*, chapter 10.

[32] Samuel P. Langley, 'The History of a Doctrine', *American Journal of Science*, vol. 37 (3rd Series) (1889), pp. 1–23.

[33] Langley, 'Laws of Nature', p. 551.   [34] Crookes, 'Address by the President', p. 339.

[35] Crookes, 'Spiritualism Viewed', p. 317. He was quoting from Faraday, 'Observations on Mental Education', p. 478, but omitted the italics.

[36] Crookes, 'Address by the President', pp. 344 and 347.

[37] Crookes, 'Address by the President', p. 344.

force, matter and space might differ radically from our own.[38] Although Crookes denied that this "unseen world" was the spiritual or immaterial world supposedly glimpsed in psychical research, his auditors would have found it difficult not to follow him in regarding humility about "terrestrial" laws to be a healthy starting point for contemplating psychical phenomena.

The humility that Crookes judged to be productive in both physical and psychical investigations was closely bound up with his enthusiasm for 'residual' phenomena. The subject of analysis in major Victorian treatises on scientific methodology, residual phenomena embraced those unexpected, anomalous and obscure effects for which there was strong empirical evidence, but which could not be accounted for by otherwise robust theories.[39] The strength of evidence justified further research, which had the potential to demonstrate the existence of hitherto unsuspected agents, causes and laws, and thereby fuel scientific progress.

For many physical–psychical scientists, the potential rewards of the patient study of residuals were not limited to the physical sciences. Crookes believed that his painstaking investigation of unusual and obscure mechanical effects of heat had led to his major discovery of the 'radiant' properties of molecules in rarefied gases, and that a similar faith in the pursuit of residuals would yield important results in studying the physical phenomena of spiritualism.[40] His friend, the Manchester sanitation chemist Robert Angus Smith, agreed. In 1876, he told an appreciative Barrett that he had not been "able to find book which contains all the laws of nature needed to sustain the world", and used the example of the discovery of Neptune from pursuing small "perturbations" in Uranus's orbit to contemplate the vastly more important discovery of mind independent of matter that would follow the investigation of the residual "indications" of apparitions.[41] Fellow Mancunian scientist Balfour

---

[38] Crookes, 'Address by the President', p. 343. Lodge and Ramsay agreed, and in 1894 they attacked SPR member Thomas Barkworth for dismissing evidence of matter passing through matter in spiritualistic seances because it was scientifically impossible. Such an occurrence was improbable but not impossible on the basis of newly discovered properties of matter: Oliver Lodge, 'A Reply to Mr. Barkworth', *JSPR*, vol. 6 (1893–4), pp. 215–17; William Ramsay, 'Mr. Barkworth and "The Unthinkable"', *JSPR*, vol. 6 (1893–4), pp. 254–5.

[39] The methodological virtues of residual effects were discussed in John F. W. Herschel, *A Preliminary Discourse on the Study of Natural Philosophy* (London: Longman, Rees, Orme, Brown and Green, 1830), pp. 156–8; William Stanley Jevons, *The Principles of Science: A Treatise of Logic and Scientific Method*, 2 vols. (London: Macmillan and Co., 1874), vol. 2, pp. 212–16.

[40] William Crookes, 'Another Lesson from the Radiometer', *Nineteenth Century*, vol. 1 (1877), pp. 879–87, esp. p. 886.

[41] Robert Angus Smith to William F. Barrett, 18 October 1876, quoted in Barrett, *Threshold of a New World of Thought*, p. 66.

Stewart would also have agreed. For him, the pursuit of residuals was a skill that he sought to cultivate in his students of experimental physics and was the driver of two important scientific projects that sought to elucidate the obscure causal connections underpinning strongly correlated events: his long-term investigation of close correlation of sunspot cycles and the earth's magnetism and weather, and the SPR's investigation of the close correlation of perceptions of apparitions and reports of death of the individuals represented in the apparitions.[42]

Some of the atmospheric phenomena to which Stewart and his colleagues devoted so much attention were, like the apparitions that fascinated him and other SPR members, capricious, startling, difficult to reconcile with existing knowledge, and often ascribed to the imagination. One such atmospheric phenomenon – ball lightning – inspired another salutary lesson that physical–psychical scientists believed had implications beyond the physical sciences. What often annoyed physical–psychical scientists was that scientific attitudes towards ball lightning could be so radically different from those towards phenomena that, as far as they were concerned, were no less difficult to control and explain. In 1871, Cromwell Varley drew upon his recent experimental studies of ball lightning in a heated reply to the Scottish anatomist Allen Thomson, who had attacked his and Crookes's forays into spiritualism. Thomson's remarks seemed to illustrate the double standards held by the scientific establishment: the Royal Society had not doubted, and recently published his researches on the "very unusual phenomenon" of ball lightning, but it had gone "mad" and rejected Crookes's investigations into a 'psychic force' radiated by the body, a phenomenon that Varley considered "not more startling" than ball lightning.[43]

The Third and Fourth Barons Rayleigh put ball lightning to similar rhetorical use decades later. In their presidential addresses to the SPR (in 1919 and 1939 respectively), they noted that this atmospheric phenomenon remained a puzzle, partly because it seemed to conflict with existing

---

[42] Gooday, 'Sunspots, Weather and the *Unseen Universe*'. Stewart emphasised the importance of residual phenomena in Balfour Stewart, *The Recent Developments of Cosmical Physics: A Lecture* (London: Macmillan and Co., 1870), p. 7. He praised the SPR's work on phantasms in Balfour Stewart, 'President's Address', *PSPR*, vol. 3 (1885), pp. 64–8.

[43] Cromwell F. Varley to Allen Thomson, 19 August 1871, cited in [Anon.], 'Mr. Varley and Professor Allen Thomson', *Spiritualist*, vol. 1 (1869–71), p. 194. Varley was reacting to Allen Thomson, 'Address to the Biology Section', in *Report of the Forty-First Meeting of the British Association for the Advancement of Science; Held at Edinburgh in August 1871* (London: John Murray, 1872), pp. 114–22, esp. p. 121. See also Cromwell F. Varley, 'Some Experiments on the Discharge of Electricity Through Rarefied Media and the Atmosphere', *Proceedings of the Royal Society of London*, vol. 19 (1870–1), pp. 236–42. Ball lightning is the subject of the final section of this paper, most of which covers electrical discharge in rarefied gases (see later in this chapter).

knowledge and had defied attempts at laboratory reproduction and control. Their 'psychical' lesson to scientific peers was a familiar one: these characteristics alone were not legitimate grounds for "rejecting without examination", or dismissing the very possibility of an unusual phenomenon, physical or psychical.[44] Here, an appreciation of the uncertainties surrounding much in the physical sciences was again invoked as an argument against an intolerant attitude towards psychical investigation.

The frustration that Varley and others felt towards men of science and their hostility to psychical investigation was partly driven by conceptions of nineteenth-century progress of the sciences, and in particular the physical sciences. The capacity of the physical sciences to overturn old theories and explanations, to extend the boundaries of knowledge and to apply new theories, explanations and knowledge in spectacular fashion made them a convenient icon of progress. "Human knowledge", Varley opined in a spiritualistic defence of 1869, "has progressed during the last fifty years to such an extent that he seems ridiculous who attempts to indicate any boundary beyond which man's intellect will never be able to pass."[45] As one of the leading scientific engineers of the period, it was unsurprising that Varley picked out the development of photography, rail travel and long-distance telegraphy as examples of such boundaries being challenged and as warnings to those who believed that knowledge could not progress in other directions.

For that pioneer of spectro-chemical analysis, Crookes, it was the development of the spectroscope that furnished a more powerful example of how much progress science had made and could still make. Having accepted this instrument as a revelatory new tool for probing the "ulterior state of things", "modern scientific philosophy" was altogether poorer if it then "scouted" other investigations into hidden forces and powers.[46] As the century wore on, physical–psychical scientists simply adapted such rhetoric in line with new developments. Thus, in his 1911 popular book on psychical research, Barrett turned to the now "commonplace" of communication by submarine telegraph and telephone, as well as the relatively new developments of wireless telegraphy and high-speed photography, to illustrate the progressive character of the boundaries between the possible and impossible.[47] For Varley, Crookes and Barrett, challenging dogmatic denials of psychical phenomena could benefit as much from reflecting on the progress of the physical sciences as from considering the limitations of physical laws.

[44] Rayleigh, 'Presidential Address', p. 285. Rayleigh, 'Presidential Address: The Problem of Physical Phenomena in Connection with Psychical Research', p. 7.
[45] Varley, 'Spiritualism', p. 368.   [46] Crookes, 'Some Further Experiments', p. 483.
[47] Barrett, Psychical Research, pp. 11–12.

Challenging Materiality

When, in 1908, Edward Fournier d'Albe argued that physicists had a right to weigh in on psychological and theological questions about the "relations between mind and matter" because of their "extensive acquaintance" with the properties of matter, he was only making explicit what had been implicit in much of what physical–psychical scientists had been claiming for decades.[48] They had long been using their own and others' researches into electricity, ether and matter in forms of arguments against the impossibility of psychical effects that were more specific than those analysed in the previous section.

A particularly fertile area of research was the startling phenomena associated with the discharge of electricity through gases, which preoccupied and puzzled many British, German and French natural philosophers and electricians from the 1840s onwards.[49] When an electric current from a voltaic cell or other source was connected across the positive and negative terminals of a glass tube from which most of the air had been removed, the tube exhibited a dazzling variety of visual effects that were best seen in rooms as dark as those in which spiritualist seances were typically held: these effects included coloured luminosity around and between the terminals, the fluorescence of the glass and strange dark bands crossing the main glow. The visual effects depended on a host of factors, from the strength of current and pressure of gas to the shape of the tube and source of electrical energy, and stimulated much debate about the relationships between electricity, molecular matter and light. In a Royal Society paper that he published on the subject in 1870, Varley spoke for many fellow investigators when he described the "nature of the action inside the tube" as "involved in considerable mystery".[50]

Only a few days after he had submitted the paper, Varley explained to a leading British spiritualist that spiritualism "must be made a science and not a superstition" and that he aimed to do so via a lecture on "those portions of scientific research with material or physical questions which indicate the existence of spiritual bodies, or rather show that the limits to matter as indicated by the eye and touch are not the only limits".[51] The paper contained exactly the kind of research he needed for this: it showed that the properties of matter could be exhibited by phenomena that

[48] Fournier d'Albe, *New Light on Immortality*, p. vii.
[49] Darrigol, *Electrodynamics*, chapter 7; Falk Müller, *Gasentladungsforschung im 19. Jahrhundert* (Berlin: Verlag für Geschichte der Naturwissenschaften und der Technik, 2004).
[50] Varley, 'Some Experiments', p. 236.
[51] Cromwell F. Varley to Benjamin Coleman, 6 October 1870, cited in [William Stainton Moses], 'Notes by the Way', *Light*, vol. 8 (1888), pp. 361–2, p. 362.

appeared not to be material and that this had the potential to challenge those who denied that "spiritual bodies" could cause material effects. Most of the paper focussed on the nature of the luminous 'arch' produced when a current from a Daniell form of electric battery was discharged between ring-shaped electrodes inside a discharge tube designed by the leading German glassblower and scientific instrument maker Heinrich Geissler. One approach to the "mystery" of the arch was contrasting naked-eye and photographic perspectives on it. After removing the resistance controlling the current strength, Varley reported seeing the negative electrode surrounded by a "bright blue envelope", but that a wet collodion plate exposed for ten seconds detected a "white flattened hour-glass" shape apparently detached from the same electrode.[52] Here and elsewhere the eye and camera did not "tell the same tale" and the capacity of the camera to capture "light too feeble for the eye" had the potential for "other applications".[53] One such application was photographing manifestations of Reichenbach's 'od', which Varley attempted several years later, but in 1870 he also believed that photography was also useful in showing that the luminous discharge had structures and material "limits" that the eye could not see.

Another approach to the mystery of the arch would have served Varley's spiritualistic goals more directly. This focussed on its apparent solidity. Exploiting the well-known susceptibility of electrical discharge to magnetism, he used an electromagnet to deflect the luminosity onto a delicate talc indicator inside the tube, which was subsequently repelled. The physical conclusion that Varley drew from this was that the luminosity comprised "*attenuated particles of matter*" projected from the tube's negative pole.[54] But it also suggested that something as fluid to human touch as a "diffused cloud" of luminosity could 'feel' solid.[55] Varley strategically avoided discussing the spiritualistic implications of this effect in his Royal Society paper, but the lessons he drew from another electrical investigation made at this time strongly suggest that he recognised such implications. At a soirée of the Institution of Civil Engineers in 1871, he and Lord Lindsay showed how a mercury-filled India rubber tube hung between the poles of a powerful electromagnet was deflected upwards into an arch. Varley insinuated that this made the action of spiritual bodies on matter less incredible when he explained to the *Spiritualist* newspaper that "a heavy material object (mercury) is supported upon an imponderable electricity which, in its turn, is supported by an imponderable magnetism"[56] (Figure 3.1).

---

[52] Varley, 'Some Experiments', p. 238.   [53] Varley, 'Some Experiments', p. 238.
[54] Varley, 'Some Experiments', p. 239.   [55] Varley, 'Some Experiments', p. 239.
[56] Varley cited in [Anon.], 'The Force of Gravitation', *Spiritualist*, vol. 1 (1869–71), p. 197.

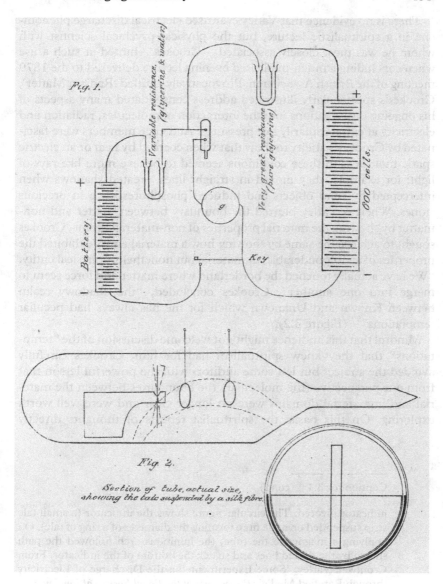

Fig. 1.

Battery.

Variable resistance.
(glycerine & water)

Very great resistance
(pure glycerine)

600 cells.

Key

g

Fig. 2.

Section of tube, actual size,
showing the talc suspended by a silk fibre.

3.1 Cromwell Varley's electrical discharge apparatus. 'Fig. 1' shows the whole electrical circuit containing large batteries, a switch, a variable resistance and a large glass tube containing trace amounts of gas which glowed when electricity was discharged through it. 'Fig. 2' shows a detail of the tube, including the two ring-shaped electrodes between which the luminous arch appeared and an L-shaped bracket on which a delicate

There is no evidence that Varley ever used electrical discharge phenomena in a spiritualistic lecture, but the physical–psychical scientist with whom he was most closely associated – Crookes – hinted at such a use when concluding a much-publicised evening lecture delivered to the 1879 meeting of the British Association. Provocatively entitled 'Radiant Matter', Crookes's spectacularly illustrated address consolidated many aspects of his ongoing investigations into the interaction of molecules, radiation and electricity at extraordinarily low pressures. Audience members were fascinated by Crookes's ability to show that when coerced by heat or an electric spark, molecules in these conditions seemed to behave more like rays of light: for example, they moved in straight lines, created shadows when intercepted by solid objects and induced phosphorescence in precious stones. Whereas Varley blurred the boundary between matter and non-matter by showing the material properties of non-material agents, Crookes sought to achieve the same by showing how a material entity exhibited the properties of an imponderable, immaterial but nonetheless physical entity. "We have actually touched the borderland where matter and force seem to merge into one another", Crookes concluded, "the shadowy realm between Known and Unknown which for me has always had peculiar temptations"[57] (Figure 3.2).

Mindful that this audience might not welcome discussion of the "temptations" that they knew spiritualism had for him, Crookes carefully avoided the subject but left some auditors with the powerful lesson that from one perspective (the molecular) the boundaries between the material and immaterial domains were no longer clear and were well worth exploring. On this basis, the spiritualist reports of thoughts directly

---

Caption for 3.1   (cont.)

indicator pivoted. The circular figure shows the indicator (a small talc strip suspended on a silk fibre forming the diameter of a ring of talc). On applying a magnet to the tube, the luminous arch followed the path shown by the dotted lines and struck the bottom of the indicator. From Cromwell Varley, 'Some Experiments on the Discharge of Electricity through Rarefied Media', *Proceedings of the Royal Society of London*, vol. 19 (1870–1), pp. 236–46. Reproduced by permission of the Syndics of Cambridge University Library.

[57] William Crookes, 'On Radiant Matter', *Nature*, vol. 20 (1879), pp. 419–23, 436–40. Citation on pp. 439–40. Examples of the publicity include [Anon.], 'Meeting of the British Association at Sheffield', *Graphic*, 30 August 1879, p. 1.

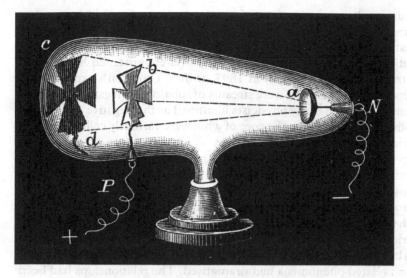

3.2 One of Crookes's 'radiant matter' instruments. This comprises a glass tube containing trace amounts of gas, a disc-shaped negative pole (a) and a Maltese-cross-shaped positive pole (b). When applying a powerful electric current, gas molecules near the negative pole are electrified, projected towards the cross, and many stimulate the phosphorescent coating near the cross. The shadow (d) caused by the cross intercepting the molecular torrent suggested that material particles sometimes behaved like light. From William Crookes, 'On Radiant Matter', *Nature*, vol. 20 (1879), pp. 419–23, 436–40. Reproduced by permission of Universal Images Group/Getty Images.

moving objects at a distance (and thereby exhibiting a material property) or of disembodied spirits materialising out of thin air would have been harder to dismiss as scientifically impossible.[58]

Varley's and Crookes's experiments on the discharge of electricity through rarefied gases would fuel a long-running controversy over the nature of what German physicists had christened 'cathode rays' in the 1860s.[59] These were the mysterious semi-visible and electrically charged emanations that streamed from negative electrodes in Geissler tubes and similar glass vessels used to study electrical discharge through gases. While German physicists tended to interpret cathode rays as forms of

[58] Spiritualist attendees reputedly appreciated Crookes's conclusion: [Anon.], 'The British Association', *Spiritualist*, vol. 15 (1879), pp. 109–14, p. 111.

[59] On this controversy see Darrigol, *Electrodynamics*, chapter 7; Isobel Falconer, 'Corpuscles, Electrons and Cathode Rays: J. J. Thomson and the Discovery of the Electron', *British Journal for the History of Science*, vol. 20 (1987), pp. 241–76.

vibration in the ether of space, Crookes and many other British scientists saw them as torrents of tiny material particles. Later investigations into the nature of cathode rays led, in late 1895, to Wilhelm Conrad Röntgen's sensational discovery of a new form of light (later christened X-rays) that penetrated solid objects, and, in 1897, to J. J. Thomson's experimental evidence that cathode rays were streams of subatomic, negatively charged 'corpuscles', which many physicists identified as the negative form of the hypothetical 'electrons' at the core of what were variously called electron, electric and etherial theories of matter.

Developed in the 1890s by Joseph Larmor in Britain, Hendrik Antoon Lorentz in Holland and others, these theories sought to reduce matter to microscopic, electrically charged electrons and the universe as a whole to electrons and a space-filling ether that carried the electromagnetic field by which the electrons interacted with each other.[60] By building matter from electrical entities, these theories aimed to illuminate the puzzling relationships between matter, electricity and ether, which gas discharge, cathode-ray and related phenomena had dramatised. The relationships had been sidestepped in and proven difficult to solve via the powerful electromagnetic field theory of Maxwell on which architects of electron theories drew.[61]

For the present discussion, the most significant feature of all electron theories was their challenge to conventional ideas about mass. In the 1880s, J. J. Thomson and Oliver Heaviside had shown that moving electrically charged particles seemed to acquire additional inertial mass purely as a result of their interaction with the electromagnetic field. Electron theories proposed that *all* the inertial mass of an electron, and the material bodies that they constituted, was electromagnetic in origin. By the early 1900s, experimental evidence of the existence of negative electrons and the electromagnetic nature of their mass had streamed from many quarters, but the hypothesis that *all* atoms were composed of positive and negative electrons and that therefore *all* mass was fundamentally electromagnetic remained exciting but unproven.[62]

Some physical–psychical scientists were among the earliest popularisers of electron theory and did not hesitate to hint at its possible 'psychical' consequences. It was precisely because electron theory was challenging traditional ideas about matter that one such populariser – Fournier d'Albe – could make the argument with which we began this

---

[60] On the electron, electric and etherial theories of matter see Darrigol, *Electrodynamics*, chapter 8; Hunt, *Maxwellians*, chapter 9; Warwick, *Masters of Theory*, chapter 7.

[61] For the problems of Maxwellian electrodynamics see Buchwald, *Maxwell to Microphysics*.

[62] Oliver Lodge, *Electrons or the Nature and Properties of Negative Electricity* (London: George Bell, 1906), pp. vii–viii.

section. Another populariser was Lodge, who, in a publication of 1901, explained that "[t]he really fundamental dynamics, we are now beginning to see, must have an ethereal and not a material basis" and that the most fundamental property of matter, inertia, "may be susceptible of electric or ethereal explanation".[63] By regarding matter as fundamentally electrical, and electricity as some kind of strain or distortion in the ether of space, Lodge was claiming for matter a quasi-material or possibly immaterial basis. As a result, physical science stood a better chance of embracing those questions of life and mind that it had always sidestepped and of inverting Tyndall's materialism by elevating "matter and all existence to the level of mind and spirit".[64] Although Lodge's publication did not explicitly mention the psychical research with which he was now so closely associated, it was clear that he saw the new dynamics as a way of making it easier to understand how, in many classes of psychical effect, mind could have a direct influence on untouched material bodies: since matter seemed to be fundamentally electrical, etherial, immaterial or even mental then it was no longer so difficult to contemplate telekinesis.

In their interpretations of electron theories Fournier d'Albe and Lodge highlighted the extent to which the theories helped explain another sensational development in the physical sciences during the 1890s: radioactivity.[65] By 1903, the discovery by Henri Becquerel, Marie and Pierre Curie and others that uranium salts and other substances emitted unexpectedly energetic forms of radiation had prompted Ernest Rutherford and Frederick Soddy to draw the startling conclusion that radioactive emanations were the result of the disintegration of the chemical atom. Physical-psychical scientists with interests in modern Theosophy, such as Kingsland and Tunzelmann, linked evidence of the electronic and etherial nature of these emanations to ancient Eastern philosophical teachings on the subtler planes of physical reality from which gross matter developed.[66]

Most physical–psychical scientists alive in 1903 would have agreed with Fournier d'Albe's more modest view that radioactivity had "abolished the dogma of the indestructibility of the atom" and accordingly

[63] Oliver Lodge, 'Scope and Tendencies of Physics', in *The 19th Century: A Review of Progress* (London: G. P. Putnam's Sons, 1901), pp. 348–57, p. 352.

[64] Lodge, 'Scope and Tendencies of Physics', p. 355.

[65] E. E. Fournier d'Albe, *The Electron Theory: A Popular Introduction to the New Theory of Electricity and Magnetism* (London: Longmans, Green, and Co., 1906), pp. 264–79; Lodge, *Electrons*, pp. 163–7.

[66] William Kingsland, *Scientific Idealism or Matter and Force and Their Relation to Life and Consciousness* (London: London Press Co., 1909), esp. pp. 69–88; William Kingsland, *The Physics of the Secret Doctrine* (London: Theosophical Publishing Society, 1910), pp. 92–108; Tunzelmann, *Treatise on Electrical Theory*, pp. 503–5.

prepared the mind for the contemplation of psychical effects.[67] A good example is Langley, who, in the 1890s, was instilling in the American historian Henry Adams his "scientific passion for doubt" and his aversion to scientific dogma.[68] The astrophysicist's recommended reading for his mature student suggested that radioactivity and the notorious critiques of mechanics by Karl Pearson and J. B. Stallo were of a piece with Crookes's SPR address of 1897: they all showed the "overthrow of nineteenth-century dogma" and the need for humility regarding our knowledge of physical laws.[69]

Lodge's 1901 reflections on the electron theory of matter also illustrate the tendency among physical–psychical scientists to mobilise late-nineteenth-century conceptions of the ether of space to challenge older scientific dogmas and create a space of psychical possibilities. In many ways, this use of the ether was another species of the humility that we explored earlier: the capacity of science to humbly accept the need for an ether that had physical properties radically different from those of gross matter suggested that science would also embrace the possibility of other entities and causes with extraordinary properties.

By the late nineteenth century, most physical scientists accepted a long-established argument that some kind of continuous physical medium was required to explain how light waves travelled across 'empty' space at finite speed.[70] The ether concept had its roots in classical antiquity, when an invisible and subtle substance was believed by many to occupy the 'empty' space between the material bodies of the cosmos. The concept was developed during the seventeenth and eighteenth centuries, when a bewildering number of ethers were invented to explain the different physical forces of nature. The most important ether surviving into the nineteenth century was the luminiferous variety, which was invented in the wake of the wave theory of light, which required the existence of a medium within which the vibrations were propagated. The luminiferous ether proved immensely puzzling. Its very existence was inferred rather than being directly detected, it needed to be perfectly invisible and transparent, and it required other, seemingly contradictory, physical properties: for example, it needed to be fluid enough to allow material objects to pass through it unhindered, but rigid enough to support the

---

[67] Fournier d'Albe, *Electron Theory*, p. 264.

[68] Henry Adams, *The Education of Henry Adams* (Boston: Houghton Mifflin Co., 1918), p. 377.

[69] Adams, *Education of Henry Adams*, p. 450.

[70] On the ether question see Geoffrey Cantor and M. J. S. Hodge (eds.), *Conceptions of Ether: Studies in the History of Ether Theories 1740–1900* (Cambridge University Press, 1981); Navarro, *Ether and Modernity*.

transverse forms of vibration constituting light. James Clerk Maxwell's electromagnetic theory of light of 1865 compounded the ether 'problem' by arguing that the luminiferous ether was also the seat of all electrical and magnetic activity, with light being a form of vibration in this electromagnetic medium.

One of the most pressing of all ether problems was its relationship to gross or 'ordinary' matter. The elaborate interferometric experiments of Albert A. Michelson and Edward W. Morley of the 1880s produced inconclusive evidence for Augustin-Jean Fresnel's theory that the earth moved relative to a stationary ether and it suggested that the ether was dragged by matter passing through it; Lodge's equally elaborate experiments of the 1890s, however, appeared to contradict this by evidencing no significant dragging of the ether by matter.[71] While most physicists accepted that there had to be some of kind of mechanism that gave the ether its extraordinary physical properties, they struggled to produce satisfactory models of it.[72] Late-nineteenth-century theories that electricity and magnetism, as well as the constituents of gross matter, might be the result of motions or structures within the ether put a huge strain on older ideas that the ether was a form of elastic solid or fluid, and prompted much of the work leading to electron theories of matter and mounting challenges to the very idea of the ether being mechanical.

Despite, or rather because of, the ether's puzzling nature, it well served physical–psychical scientists' more specific strategies of removing scientific stumbling blocks to the contemplation of psychical phenomena. One strategy was to make such phenomena seem no more far-fetched than the ether. In 1877, for example, Crookes ridiculed the idea that his claims about puzzling new forces associated with the bodies of mediums and with heat were implausible because they defied the "'common sense of educated mankind'".[73] The proposal was preposterous when considering that the same court of appeal would deem "utterly incredible" the idea of empty space being filled by "*something* indefinitely more elastic and immeasurably more solid than tempered steel, a medium in which suns and worlds move without resistance".[74] William Ramsay would have agreed with the spirit, if not the letter, of Crookes's argument. He insisted that training in chemistry and physics was better than common sense in

---

[71] Bruce J. Hunt, 'Experimenting on the Ether: Oliver J. Lodge and the Great Whirling Machine', *Historical Studies in the Physical Sciences*, vol. 16 (1986), pp. 111–34; Swenson, *Ethereal Aether*.

[72] Darrigol, *Electrodynamics*, chapters 5 and 8; Hunt, *Maxwellians*, chapter 4.

[73] Crookes, 'Another Lesson', p. 887. Crookes was quoting from one of Carpenter's attacks on his spiritualistic investigations.

[74] Crookes, 'Another Lesson', p. 887.

judging spiritualists' claims insofar as the idea that solid objects could pass through others was not impossible in light of the "view" (taken by Ramsay and many other physical scientists) that matter was "merely one of the factors of energy" in the form of the vortical motion of a rarefied etherial medium.[75]

A related strategy was to turn *specific* physical properties of the ether into arguments for the plausibility, as opposed to the reality, of some of the effects evidenced in psychical research and spiritualism. The argument owed something to theistic physicists who argued that such etherial properties as universality and perfect continuity made it easier to contemplate divine omniscience and omnipotence.[76] Stokes gave a particularly telling, but non-psychical, example in his 1893 series of Gifford Lectures on natural theology. The fact that the ether was found to have properties so extraordinary and so different from what physicists anticipated made it an example of the "unreasonable-ness" of "*à priori* rejection of all that transcends the ordinary course of nature", including ostensibly "supernatural" events.[77] It also seemed to have one property – the ability to transmit longitudinal vibrations of potentially limitless velocity – that led to the "contem-plation of the possibility of the communication of intelligence from one part of the universe to another", which, as Stokes pointed out, included divine will.[78]

Only a year after Stokes's address, somebody with whom he often discussed physics and Christian theology – Barrett – applied similar reasoning in an address to London spiritualists. He asked whether, given that it was now

scientific heresy to disbelieve in an imperceptible, imponderable, infinitely rare and yet infinitely elastic all-pervading kind of matter, the so called luminiferous ether, which is both interstellar and interatomic (a material medium of a wholly different order of matter to anything known to our senses, and the very existence of which is known inferentially), is it so very incredible a thing to suppose that in this ether (or in some other unseen material medium) life has originated.

Since the "Divine law" of evolution had created life from "grosser mat-ter", there was "nothing extravagant" in the ideas that it had fashioned life from the "finer and more plastic matter of the ether", and that "myriads of

---

[75] William Ramsay, 'Mr. Barkworth and "The Unthinkable"', p. 254.

[76] See Geoffrey Cantor, 'The Theological Significance of Ethers', in Cantor and Hodge, *Conceptions of Ether*, pp. 135–55; Schaffer, 'Where Experiments End', pp. 263–7.

[77] George G. Stokes, *Natural Theology: The Gifford Lectures Delivered Before the University of Edinburgh in 1893* (London: Adam and Charles Black, 1893), p. 24.

[78] Stokes, *Natural Theology*, p. 33.

living creatures" existed in this invisible etherial realm which occasionally communicated with us.[79]

If Lodge ever read this passage in the early 1900s he would have demurred to its conceptions of the ether but agreed with its overall ambition of removing the incredibility of the idea of intelligences in domains beyond the visible or material. By this period, experimental and theoretical studies of electrons had persuaded him that if electrons were structures in the ether then the ether needed to have colossal density and energy. Yet in a lecture of 1908, he effectively updated Barrett's argument by contending that since gross matter was known to have psychical significance (the brain), then it was difficult to believe that the ether, "with all its massiveness and energy", had none.[80] In this lecture, as in so many of the scientific papers and addresses analysed in this section, physical–psychical scientists only came to the brink of harder physical theorising about psychical phenomena in which, as we shall see later, we know they indulged elsewhere.

Dim Analogies

In 1896, one of Lodge's closest scientific colleagues and friends, George F. FitzGerald, observed:

May we not hope that studies of physiological actions, of chemical constitution and change, of vortex motion, of the laws of matter and ether, may some day enable us to discover the motions in our brains underlying sound and light, and smell and touch, and pain and pleasure, hate and love. And may we not hope, then, to be able to form some dim analogies by which we may divine what underlies the much more complex motions of organic nature as a whole, and have a scientific basis for investigating what underlies the whole sequence of organic evolution.[81]

This appeared towards the end of a memorial lecture to the Chemical Society of London on Hermann von Helmholtz, the German physicist and physiologist who had died two years earlier and who was one of the architects of the vortex theory of matter to which FitzGerald alluded. As Bruce Hunt has argued, FitzGerald's lecture expressed the hopes of a 'radical mechanist' school of physics in late-nineteenth-century

---

[79] Barrett, 'Science and Spiritualism', p. 572. In 1880, Barrett and Stokes had discussed the analogy between thermodynamic and spiritual dissipation: George G. Stokes to William F. Barrett, 11 September 1880, f. 127, WFB-RS.

[80] Lodge, 'Ether of Space', p. 543.

[81] George F. FitzGerald, 'Helmholtz Memorial Lecture', in *Memorial Lectures Delivered Before the Chemical Society, 1893–1900* (London: Gurney and Jackson, 1901), pp. 885–912, p. 912.

Britain, a school that included Lodge, J. J. Thomson and others who sought to reduce matter to mechanical motions in an ether and which was preoccupied with the idea that motion was a fundamental property of all things.[82]

FitzGerald's observation also highlights the possibilities and problems of analogies used in late-nineteenth-century physics. On the one hand, analogies drawn from mechanics, hydrodynamics and other established fields had proven especially useful in the investigation of the more unfamiliar phenomena of electricity, magnetism and heat.[83] On the other hand, as FitzGerald's strategic use of the word 'dim' suggests, analogies needed to be deployed carefully, since they often implied a *likeness* between the established and unfamiliar that was not intended. Oliver Lodge was acutely aware of this in the wake of the publication of his *Modern Views of Electricity* (1881), which attempted to make intelligible the new Maxwellian conceptions of electricity, magnetism and ether via a host of literal and imaginary mechanical models. While these models reflected a more widely shared belief in the ultimately mechanical nature of electricity, magnetism and ether, they were often criticised for implying close but unproven mechanical analogies.[84]

Despite these problems with mechanical analogies, analogies per se pervaded nineteenth-century sciences, and their frequent use by physical–psychical scientists in the context of psychical questions was entirely consistent with their enthusiasm to explore them, as FitzGerald hinted, in domains bordering on and beyond the physical. The principle purpose of analogies drawn by physical–psychical scientists from the physical sciences was, like other analogies, to render psychical phenomena more plausible, and represented a stronger use of physical theories than challenging materialism or the impossibility of psychical effects. In this sense, the ambitions of physical–psychical scientists overlapped with those of mesmerists, spiritualists, modern Theosophists and proponents of other forms of occultism, who frequently drew analogies between 'mineral' and 'animal' magnetism and between material and 'spiritual' electricity to emphasise the scientific character of their enterprises.

Many of these analogies operated at a very general level. Those drawn by late-eighteenth- and early-nineteenth-century mesmerists between animal magnetism and physical forces were still being deployed in the

---

[82] Hunt, *Maxwellians*, pp. 96–104. See also Martin J. Klein, 'Mechanical Explanation at the End of the Nineteenth Century', *Centaurus*, vol. 17 (1972), pp. 58–82.

[83] For discussion of analogies in British physics see P. M. Harman, *The Natural Philosophy of James Clerk Maxwell* (Cambridge University Press, 1998), chapter 4; Smith and Wise, *Energy and Empire*, esp. chapters 6–8.

[84] Hunt, *Maxwellians*, pp. 91–5.

late 1800s. In 1874, the physics populariser and mesmerist John T. Sprague asked whether,

from every point there proceeds something of the nature of spheres or circles of undulations analogous to those of light or magnetism, and extending to unknown distances, and that man and animals may in some cases possess the power of observing or being actuated by these undulations in some way different from the functions of the ordinary senses?[85]

The analogy that Crookes made in 1862 between the indestructible energy of a candle flame and the immortal soul also resurfaced many times in the popular and semi-popular writings of physical–psychical scientists. The *Unseen Universe* made much of "analogies in nature" which seemed to "throw light upon the doctrines of Christianity", and particularly the conservation of energy and the Christian idea of the future state and immortality of the soul.[86] The evidence of the survival of the soul amassed by the SPR and new theories of the ether made this analogy particularly compelling to Lodge. In the early 1900s, the very period when he started signalling his debt to the *Unseen Universe*, he asserted that "no really existing thing perishes", and argued that just as the under-lying realities of the physical world – energy and ether – have an "attribute of immortality", so life, mind and other qualities associated with the human self were never annihilated.[87]

One of the obscurer chapters in the *Unseen Universe* explored analogies between living beings and inanimate machines, partly to understand the capacity of life to produce complex energy transformations. One of these analogies was between a living being and a rifle trigger or some other "delicately constructed machine".[88] Its function was to illustrate something that excited Maxwell in his quest for an argument against the impossibility of free will in a system operating according to physical laws: it showed the capacity of unstable systems (both animate and inanimate) to produce significant consequences from impetuses that were incalculably small and therefore beyond the scope of a strictly mate-rialistic view of the cosmos.

Barrett's enthusiasm for the *Unseen Universe* may have owed something to the fact that by the early 1870s he too was exploring an analogy between the instabilities in living and non-living systems – specifically, between mesmerised human beings and the 'sensitive' flame. First

---

[85] 'Sigma' [John T. Sprague], 'Perceptive Powers of Men and Other Animals', *English Mechanic and World of Science*, vol. 20 (1874), p. 223.

[86] [Stewart and Tait], 'Preface to the Second Edition', p. xi.

[87] Lodge cited in [Anon.], 'Memorial to Mr. Myers at Cheltenham College', *JSPR*, vol. 13 (1907–8), pp. 148–52, pp. 149–50.

[88] [Stewart and Tait], *Unseen Universe*, p. 145.

observed in the United States in the late 1850s, and explored by Barrett and Tyndall the following decade, sensitive flames were spectacular additions to Tyndall's popular Royal Institution lectures covering the invisible vibrations of heat, light and sound. When the flame's gas supply was adjusted to near-flaring point, it became very unstable and extraordinarily sensitive to sounds containing frequencies matching those in the hissing sounds caused by friction between the rapidly moving gas and the burner nozzle. Sounds containing particular frequencies instantly caused "long, straight, and smoky" flames to become "short, forked, and brilliant" and vice versa, and flames could be made to jump up and down in exact synchrony with the blowing of a whistle, the clapping of hands, the singing of notes, and even high-pitched sounds that were too faint to hear[89] (Figure 3.3).

Despite uncertainties about their precise physical cause, sensitive flames were interpreted by Barrett, Tyndall and others as examples of the more general phenomenon of sympathetic vibration or resonance, in which the vibrations of one system could be excited by subjecting it to oscillations of the appropriate frequency in another vibrating system.[90] For many physicists, including Rayleigh and Eleanor Sidgwick, sensitive flames became an important puzzle for developing theoretical and experimental studies of the larger questions of instability in hydrodynamic and other purely physical systems.[91]

For Barrett, sensitive flames fulfilled physical *and* non-physical purposes. This had become apparent to him by 1868, when, in a lecture on sensitive flames and other examples of sympathetic vibration, he concluded that whether by "disease or nervous derangement" the human body could be put into a "state of unstable equilibrium" and this made it sensitive to "the slightest stimuli, if of the proper kind".[92] Barrett gave homeopathy as an example of this, but could have given another. This derived from his recent experiences of several Irish girls who, in the mesmeric state, displayed exalted sensory powers far surpassing anything that physiological and psychological theories could satisfactorily explain. His 1876 account of

[89] John Tyndall, *Sound. A Course of Eight Lectures Delivered at the Royal Institution of Great Britain* (London: Longmans, Green, and Co., 1867), p. 236.
[90] William F. Barrett, 'Note on "Sensitive Flames"', *Philosophical Magazine*, vol. 33 (4th Series) (1867), pp. 216–22; Tyndall, *Sound*, pp. 217–54.
[91] Lord Rayleigh, *The Theory of Sound*, 2 vols. (London: Macmillan, 2nd ed., 1894), vol. 2, pp. 376–414. For Sidgwick's collaboration with Rayleigh on this puzzle see Lord Rayleigh, 'Further Observations upon Liquid Jets, in Continuation of Those Recorded in the Royal Society's 'Proceedings' for March and May, 1879', *Proceedings of the Royal Society of London*, vol. 34 (1882–3), pp. 130–45.
[92] William F. Barrett, 'On Musical and Sensitive Flames', *Chemical News*, vol. 17 (1868), pp. 220–2, p. 221.

Fig. 1.

Fig. 2.

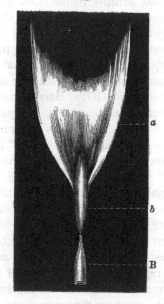

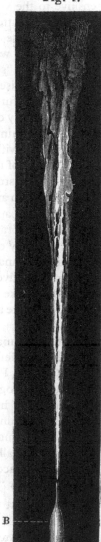

Fig. 3.

3.3 Barrett's representation of the sensitive flame. When subjected to high-frequency sounds, a tall and loose flame depicted in 'Fig. 1' becomes the short and divergent flame in 'Fig. 2'. 'Fig. 3' shows the tapering burner that Barrett invented to increase the sensitivity. From William F. Barrett, 'Note on "Sensitive Flames"', *Philosophical Magazine*, vol. 33 (4th Series) (1867), pp. 216–22. Reproduced by permission of the Syndics of Cambridge University Library.

these experiences strongly suggests that he perceived analogies between the girls and sensitive flames. One girl became so sensitive to the voice and gestures of a mesmeric operator that it was "impossible for the latter to call the girl by her name, however faintly and inaudibly to those around, without at once eliciting a prompt response".[93] Her behaviour was comparable to that kind of flame that he was already likening to a "sensitive, nervous person uneasily starting and twitching at every little noise".[94]

By the mid-1870s, Barrett had pushed the analogy even further. In another popular lecture on sympathetic vibrations, he reputedly declared that, like sensitive flames and other resonant systems, one mind could respond to a "slight disturbance" from another if in synchrony with it, and that this would illuminate the evidence for the direct "action of mind on mind" he had been collecting.[95] By 1882, however, his studies of thought-reading had prompted him to question this analogy. In an article for the SPR, he warned that "speculations" about the brain's capacity to radiate nervous energy, which would excite sympathetic vibrations in distant brains in a state of unstable equilibrium, were "merely of use in suggesting lines of experiment" and doubtful as ultimate explanations.[96] This represented a relaxation of Barrett's analogy from sensitive flames *and* from electrical induction, which he had earlier used to make sense of the way energy in the nerve fibres of one person seemed to excite a similar state of activity in the nerve fibres of another.[97]

Barrett's doubts about the "insufficiency of any physical analogy or materialistic explanation" of thought-reading, thought-transference or telepathy were expressed much more emphatically by Gurney, Podmore and Myers in *Phantasms of the Living*.[98] "Let us use every analogy which helps us", they advised, "but let us recognise that nothing has been discovered which shows that thought-transference has anything to do with ether or with vibrations."[99] Apart from being rather too "materialistic" for understanding an effect representing a welcome challenge to materialism, etherial and other physical vibrations were more acceptable as a "metaphor" than an analogy for several reasons.[100] No "definite

[93] Barrett, 'On Some Phenomena', p. 85.
[94] Barrett, 'Note on "Sensitive Flames"', p. 219. For further discussion see Noakes, 'The Bridge' pp. 431–8.
[95] [Anon.], 'Sympathetic Vibrations', *Times*, 29 December 1876, p. 3.
[96] William F. Barrett, 'Appendix to the Report on Thought-Reading', *PSPR*, vol. 1 (1882–3), pp. 47–64, p. 62.
[97] Barrett, 'On Some Phenomena', p. 87.
[98] Barrett, 'Appendix to Report on Thought-Reading', p. 62.
[99] Edmund Gurney, Frederic W. H. Myers and Frank Podmore, *Phantasms of the Living*, 2 vols. (London: Rooms of the Society for Psychical Research, 1886), vol. 2, p. 315.
[100] Gurney, Myers and Podmore, *Phantasms of the Living*, vol. 1, p. 112.

experiment" had been performed to show that telepathy displayed the properties of reflection and interference; the intensity of telepathic impressions, unlike those of physical vibrations, did not appear to diminish with distance; and sometimes the impression in the mind of the person receiving it was only symbolically related to that in the mind of the person transmitting it, as in cases where one person perceived the form of a distant dying person rather than what was in the mind of the latter individual.[101]

While many physical–psychical scientists followed Barrett in accepting that strict analogies between telepathy and physical vibrations were not supported by experimental evidence, they did not think that this was a reason to abandon altogether the idea that analogies were useful in rendering psychical effects more intelligible. It is not surprising that such analogies were more likely to be upheld outside the SPR's publications. In 1891 articles for the evangelical monthly *Good Words*, even Barrett maintained that the general "law" of "sympathetic vibrations" applied to sensitive flames and telepathy and furnished "spiritual analogies" with the Christian idea that the "pure heart responds only to what is beautiful and true".[102] Within 20 years he had come to accept that this kind of general analogy reflected a deeper "correspondence" between the physical and spiritual domains taught in Swedenborgianism and Modern Spiritualism.[103] Barrett's analogies may well have impressed the American physicist Amos E. Dolbear long after the warnings of *Phantasms of the Living*. In an 1899 popular book on physics, he explained that while sympathetic vibration originally applied to physical phenomena it was "discovered to be a mode of action quite analogous to mental phenomena between individuals in which similar mental states are induced".[104]

Lodge had long been aware of the limitations of strict analogies between physical and psychical systems but believed they served useful functions. In his first paper for the SPR, he offered what, for this audience, he strategically called a "rough and crude analogy" for thought-transference, the latter term implying for him too many questionable ideas about the location and motion of the mind.[105] Lodge was too keen a user of physical analogies to deny their possible use in illuminating

[101] Gurney, Myers and Podmore, *Phantasms of the Living*, vol. 2, p. 315.

[102] William F. Barrett, 'Sympathetic Vibration', *Good Words*, vol. 32 (1891), pp. 41–6, p. 46; William F. Barrett, 'Psychical Research', *Good Words*, vol. 32 (1891), pp. 467–71.

[103] William F. Barrett, 'Discrete Degrees', *The New Church Magazine*, vol. 33 (1914), pp. 415–25, p. 418.

[104] Amos E. Dolbear, *Matter, Ether and Motion: The Factors and Relations of Physical Science* (London: Society for the Promotion of Christian Knowledge, 1899), p. 312.

[105] Lodge, 'Account of Some Experiments', p. 191.

the apparent discarnate nature of thought. The source of his analogy was predictable enough. The Maxwellian conception of electricity that he was now busy promulgating in Victorian culture located most of the electrical energy in the space around charges and conductors, and this made it easier to conceive of consciousness existing "like a faint echo in space, or in other brains" without contradicting the widely accepted fact that the brain was the material mediator or organ of consciousness.[106] The emphasis that Maxwellian electrical theory placed on the space around material bodies seemed entirely analogous to the emphasis that spiritualists and psychical investigators placed on the importance of the psychical and spiritual 'space' around human bodies.

Similar to Barrett's changing attitude towards sensitive flames, Lodge clearly considered his Maxwellian analogy particularly "rough and crude" in light of the mounting evidence for telepathy's dissimilarity to physical vibrations. Yet even after 1903, when he publicly rejected assumptions that telepathy had to be etherial or physical and defended its possible "psychical" nature, he was still prepared, mostly in non-SPR publications, to describe telepathy as a "syntonic" process, which drew implicitly on an analogy with the etherial or wireless telegraphic work that he had been pursuing since the mid-1890s.[107]

The utility of Lodge's etherial analogy was certainly not lost on his former laboratory assistant, the wireless and cable telegrapher Benjamin Davies. In a 1909 lecture on psychical research, this devout Christian found in the capacity of the immaterial ether to connect the material bodies of the cosmos an "analogy in the physical world" for the "yet unknown means of communication between human beings, quite independent of the usual purely material channels".[108] Davies was one of many physical–psychical scientists who, well into the early 1900s, displayed an enormous amount of confidence in the analogies between physical and psychical domains and indeed believed that late-nineteenth-century developments in physics made it possible to propose closer analogies, tentative models and explanations of psychical effects.

[106] Lodge, 'Account of Some Experiments', p. 191.
[107] Oliver Lodge, 'Presidential Address', PSPR, vol. 18 (1903–4), pp. 1–21, p. 19; Oliver Lodge, The Survival of Man: A Study in Unrecognised Human Faculty (London: Methuen and Co., 2nd ed., 1909), pp. 61 and 90. 'Syntonic' was a term used by Lodge to describe the capacity of one oscillating electrical circuit to resonate with or be tuned to another circuit.
[108] Benjamin Davies, Notes for a lecture dated 29 November 1909, f. 13, NLW 21/1, BD-NLW.

Maxwellian Psychics

In an 1882 lecture on the ether, Lodge echoed his hero Maxwell in privileging a medium of physical communication over direct action-at -a-distance in explaining how material bodies interacted across 'empty' space. The "mass of experience" of humanity made action-at-a-distance "absolutely unthinkable", whereas the idea of a "connecting medium" made the interaction "simple and intelligible".[109] But Lodge's conception of the scope of such an explanation was much more ambitious than Maxwell's had ever been. This was clear from his analysis of the banal example of somebody communicating to a dog. The dog could be summoned by direct mechanical contact (prodding or throwing a stone), light (beckoning), sound (whistling), or by electrical and magnetic attraction. Perhaps surprisingly for some auditors, Lodge also accepted "mesmerism" as a possible form of communication, but in this and all other cases he urged looking for the "medium which conveys these impressions".[110] Lodge thereby challenged a long-held view that mesmerism, clairvoyance and related obscure psychological powers operated via direct action at a distance.[111]

Nine years later, in his address to fellow physicists at a British Association meeting, Lodge was even bolder in the psychical functions he contemplated for the ether. Just as the ether had been invented partly to explain how physical interactions across 'empty' space did not violate energy conservation, so, according to Lodge, it had the potential to explain how, contrary to much scientific criticism, untouched objects could move in spiritualist seances without upsetting the same physical principle.[112] It also had the potential to avoid the troubling idea of mind directly pushing matter, which, following Newton's third law of motion, required a mechanical force to react on the mind.[113] Before denying the possibility of telekinesis on energetic and mechanical grounds, it was important to consider a "novel mode of communicating energy", which

---

[109] Oliver Lodge, 'The Ether and Its Functions', *Nature*, vol. 27 (1882–3), pp. 304–6, 328–30, p. 304. Cf. Maxwell, 'On Action at a Distance'.

[110] Lodge, 'Ether and Its Functions', p. 305.

[111] This view is expressed in J. C. Colquhoun, *An History of Magic, Witchcraft and Animal Magnetism*, 2 vols. (London: Longman, Brown, Green and Longmans, 1851), vol. 1, p. 190; Gustavus G. Zerffi, *Spiritualism and Animal Magnetism: A Treatise on Dreams, Second Sight, Somnambulism, Magnetic Sleep, Spiritual Manifestations, Hallucinations and Spectral Visions* (London: Robert Hardwicke, 1871), pp. 67–71.

[112] On the ether and energy conservation see M. Norton Wise, 'German Concepts of Force, Energy and the Electromagnetic Ether: 1845–1880', in Cantor and Hodge, *Conceptions of Ether*, pp. 269–307.

[113] This issue was raised by Lodge in 'Supplement', p. 339.

might take place via an "immaterial (perhaps an etherial) medium of communication" and this might also be the medium for the transfer of thoughts between individuals.[114]

There are many reasons for Lodge's growing confidence in the ether's possible psychical functions. Between 1882 and 1891, he had added to the mass of experimental evidence for telepathy and contributed significantly to a milestone in Maxwellian physics: this was the experimental generation and detection (in 1888) of electromagnetic waves along wires, an achievement reinforcing Heinrich Hertz's generation and detection of the electromagnetic waves along wires (in 1887) and, more impressively, in free space (also in 1888).[115] For many physicists, this provided a ringing endorsement of Maxwell's electromagnetic theory of light and of the corresponding identity between the luminiferous ether and the space-filling medium that the theory required to store electrical and magnetic energies. In concluding his semi-popular book on Maxwellian "views of electricity", Lodge boasted that the "whole domain of Optics is now annexed to Electricity, which has thus become an imperial science".[116] The extension of electricity's empire into optics invigorated debate on the mechanisms by which the ether might give rise to electrical, magnetic, gravitational and other attributes of the physical world, but it also stimulated physical–psychical scientists besides Lodge into speculating on the etherial means by which psychical phenomena might be annexed to physics. With evidence that the ether had more functions than previously assumed, Gurney, Myers and Podmore's scepticism about its relationship to telepathy would have seemed less convincing.

Some of this speculation drew on a much older debate about the relationship between electricity, ether and the human body. Historians have underestimated the significance of the human body in nineteenth-century physics and the extent to which the nascent discipline was nourished via interactions with physiology and medicine.[117] For the overlapping mid- and late-nineteenth-century circles of electricians, physicians, physiologists and physicists, the human body was an important site for researching the limits of physical theories and principles relating to heat, electricity, magnetism and the ether. Nothing captures this importance more clearly than the vigorous debates about electrophysiology and medical electricity. By

---

[114] Lodge, 'Address', p. 555. Lodge also argued that since we did not know how our ideas caused us to move our bodies we could not rule out the possibility of us moving objects without mechanical contact.

[115] Buchwald, *Creation of Scientific Effects*; Darrigol, *Electrodynamics*, pp. 234–64; Hunt, *Maxwellians*, chapters 6–7.

[116] Oliver Lodge, *Modern Views of Electricity* (London: Macmillan and Co., 1889), p. 307.

[117] For discussion of this issue see Morus, 'Physics and Medicine'.

the 1870s, there was a strong consensus that the body produced, and that its health could be improved by, electrical currents, but the precise relationship between the body and electricity remained a subject of debate and ongoing investigation.[118] The problem was succinctly expressed by William H. Stone, who, in an address to fellow physicians in 1886, warned that the foundations of physiological electricity were "unsafe" without further research into the enigmatic relationship between electricity and nerve force, and that this question could only be effectively solved by physiologists and medical practitioners adopting the metrological techniques of experimental physics.[119]

From the early 1870s onwards, Stone was rare among physicians, electricians and physicists in regarding the relationship between the human body and *magnetism* as another source of scientific progress.[120] Given that electric currents were known to flow in the body, it was not impossible that by electromagnetic induction, the body was susceptible to external magnetic fields. On this basis, Stone tested his own physiological response to a strong magnetic field and although the trials were inconclusive, he maintained that the susceptibility was plausible and worthy of further investigation.[121] Stone's conclusion was shared by no less a person than William Thomson. In a lecture of 1883, Thomson considered it "marvellous" that Cromwell Varley (who had died only recently) and Lord Lindsay had not experienced any physiological effect when placing their heads within a strong magnetic field.[122]

The research potential of this magnetic 'sensitivity' was not lost on Barrett, who would join Stone and other SPR members in approaching the subject via a critical study of Reichenbach's researches. Neither was the potential lost on Barrett's Dublin colleague FitzGerald. In 1883, he suggested to Barrett that a possible explanation of the magnetic luminosity attributed to od derived from his ongoing theoretical study of how, based on Maxwell's theory, light could be experimentally produced from electromagnetic vibrations: it was possible that when moving close to

[118] Morus, 'Measure of Man'; Morus, *Shocking Bodies*.

[119] William H. Stone, 'Abstracts of the Lumleian Lectures on the Electrical Condition of the Human Body: Man as a Conductor and Electrolyte. Lecture III', *Lancet*, vol. 1 (1886), pp. 863–5, p. 865.

[120] William H. Stone, 'The Persecution of Dr. Stone – Medical Electricity', *English Mechanic and World of Science*, vol. 18 (1873), p. 312.

[121] William H. Stone, 'Reichenbach's Experiments', *English Mechanic and World of Science*, vol. 20 (1874), p. 120; William H. Stone, 'Hysteria and Hystero-Epilepsy', *Saint Thomas's Hospital Reports*, vol. 11 (1880), pp. 85–102, pp. 100–1.

[122] William Thomson, 'The Six Gateways of Knowledge [1883]', in William Thomson, *Popular Lectures and Addresses*, 3 vols. (London: Macmillan and Co., 1889), vol. 1, pp. 253–99, p. 261.

a magnetic field, air molecules were more likely to collide with each other and that the resulting changes in direction induced electrical oscillations in the molecules that generated light.[123] Maxwell's work seemed to lend plausibility to Reichenbach's proposed link between light and magnetism, just as Faraday's had done in the 1840s.

FitzGerald's theoretical argument for the possibility of generating electromagnetic waves provided, along with Poynting's and Heaviside's Maxwellian analysis of energy flow in the electromagnetic field, the crucial interpretative framework for Hertz's and Lodge's evidence for the experimental generation and detection of electromagnetic waves in 1887–8. This theoretical and experimental work gave many physical–psychical scientists new reasons to intervene in an older debate about the spatial boundaries of the forces and energies associated with human vitality and the nervous system. This was a debate in which some physical–psychical scientists and their adversaries had already participated. In 1871, Crookes had dipped his toe into the debate when publishing evidence of 'psychic force', a force first proposed by the barrister and psychological writer Edward Cox to explain how spiritualist mediums were able to move untouched objects in seances. To give Cox's proposal more credibility, Crookes linked it to the work of the British physician Benjamin Ward Richardson. Richardson had recently put forward the theory of a gaseous "atmosphere" associated with the material constituents of the nervous system, and which shrouded the body and mediated sensory experiences.[124]

Four years later, William Benjamin Carpenter opined in a major text-book of physiological psychology that by viewing nerve force as a "form of Physical energy", it was "not altogether incredible that it should exert itself from a distance, so as to bring the Brain of one person into direct dynamical communication with that of another".[125] Carpenter's caution reflected the fact that for many medical practitioners and physiologists, there was simply no evidence for this kind of leakage of nervous activity beyond the body.[126] For at least one physical–psychical scientist,

---

[123] George F. FitzGerald to William F. Barrett, 23 April 1883, in William F. Barrett et al., 'First Report of the "Reichenbach" Committee', *PSPR*, vol. 1 (1882–3), pp. 230–7, pp. 236–7. On FitzGerald's work at this time see Hunt, *Maxwellians*, pp. 33–47.

[124] Crookes, 'Some Further Experiments', p. 477n; Benjamin Ward Richardson, 'A Theory of a Nervous Atmosphere', *Medical Times and Gazette*, vol. 1 (1871), pp. 507–9.

[125] William B. Carpenter, *Principles of Mental Physiology* (London: Henry S. King, 1875), p. 633. On Carpenter's attitude towards brain radiation see Shannon Delorme, 'Physiology or Psychic Powers? William Benjamin Carpenter and the Debate over Spiritualism in Victorian Britain', *Studies in History and Philosophy of Biological and Biomedical Sciences*, vol. 48 (2014), pp. 57–66.

[126] See, for example, [Anon.], 'Psychical Force', *Medical Times and Gazette*, vol. 2 (1876), pp. 545–6.

however, Carpenter's speculation was particularly welcome because it showed that even this notorious critic of thought-reading was prepared to accept that this phenomenon might not be as opposed to physical and physiological laws as some insisted.[127]

The experimental generation of electromagnetic waves gave Crookes an incentive to speculate further on the psychical possibilities of physical forces and energies associated with the body. In a semi-popular article that he contributed to the February 1892 number of the *Fortnightly Review*, Hertzian waves or 'electric rays' occupied a prominent place in a survey of the actual and possible technological applications and intellectual significances of electrical discovery per se.[128] One of the most striking of these 'possibilities' was the reinvigoration of older theories of 'brain waves'. The English journalist James Knowles had put forward one of the most widely read of these theories in 1869. His "crude hypothesis" proposed that thoughts and feelings caused movements in the "constituent particles" of the brain, and that these movements resulted in vibrations in the ether.[129] His speculation depended partly on an increasingly common view among nineteenth-century scientists that thought could be correlated with some kind of "atomic movements" in the brain, a view that Tyndall and Stewart and Tait exploited in their radically different interpretations of the cosmos.[130] Possibly drawing on the principle of sympathetic vibration popularised by Barrett and Tyndall, Knowles also suggested that atomic movements produced etherial vibrations that enabled particular thoughts and feelings to be stimulated in distant individuals possessing brains with a "sympathetic substance".[131] With this hypothesis, Knowles sought to explain the evidence of psychological communion in the phenomena of mesmerism, electrobiology, spiritualism and ghost-seeing.

Crookes's version of the theory merely updated Knowles's. The discovery of the capacity of a "receiver" of electric rays to be open to some wavelengths but closed to others made it "not improbable" that "other sentient beings" had sensitivities to electromagnetic vibrations that differed from those of the human eye.[132] Moreover, within the human brain

---

[127] Barrett, 'On Some Phenomena', p. 87. Barrett also cited the work of the early-nineteenth-century British physician and mesmerist Herbert Mayo, who proposed an 'exoneural action' of the brain to explain mesmeric phenomena: Barrett, 'Phenomena of Spiritualism', p. 1019n.

[128] Crookes, 'Some Possibilities of Electricity'.

[129] J[ames] T[homas] K[nowles], 'Brain-Waves – A Theory', *Spectator*, vol. 42 (1869), pp. 135–7, p. 136.

[130] Knowles, 'Brain-Waves', p. 136. See also Tyndall, 'Scientific Materialism [1868]', p. 87; [Stewart and Tait], *Unseen Universe*, pp. 159–60.

[131] Knowles, 'Brain-Waves', p. 137.

[132] Crookes, 'Some Possibilities of Electricity', p. 176.

there "may lurk an organ capable of transmitting and receiving other electrical rays of wave-lengths hitherto undetected by instrumental means" and that these would render cases of thought-transference "explicable".[133] Many readers of the *Fortnightly Review* may have been disappointed by Crookes's refusal to speculate on the "possibilities" of catching and harnessing such brain waves, but any interest they showed in possibility of an electrical transmission organ would have been further excited months later when, in the May number of the periodical, the British physiologist John G. McKendrick outlined the case for 'human electricity', including new evidence raising the probability that vital processes were due to electrical changes.[134]

Someone who seems to have been aware of the latest findings in electro-physiology *and* was prepared to speculate on catching brain waves was Samuel Tolver Preston, a civil engineer who from the 1870s onwards devoted most of his time to technical writings on the physics of molecules, energy, gravity and the ether. In 1890, two years before Crookes's speculation appeared, Preston wrote to FitzGerald and Barrett with a vague supposition that thought-reading might be a form of "'telephoning' thought" along the infinitely thin vortex filaments that he (and FitzGerald) proposed as the etherial connecting mechanism of the cosmos.[135] FitzGerald's and Barrett's responses are unknown, but we do know that by 1893 Preston had arrived at a more elaborate theory. By this time, he was trying to publish a manuscript entitled 'On the Physics of Thought-Reading', which he also sent to J. J. Thomson and Barrett for their comments.[136] The proposed "mechanism" owed something to Knowles, Crookes and the American electrical engineer Edwin J. Houston, who had recently speculated that thought-reading could be a form of resonance involving electromagnetic waves emitted and absorbed by vibrating constituents of the brain.[137]

---

[133] Crookes, 'Some Possibilities of Electricity', p. 176.

[134] John G. McKendrick, 'Human Electricity', *Fortnightly Review*, vol. 51 (1892), pp. 634–41.

[135] S. Tolver Preston to George F. FitzGerald, 3 September 1890, f. 11/63, George Francis FitzGerald Papers, RDS Library and Archives. Preston's speculation was consistent with his strong commitment to mechanical or molecular theories of interaction between material bodies across 'empty' space. These were vastly preferable to dubious ideas about action-at-a-distance forces that spiritualism seemed to be promulgating: S. Tolver Preston, 'On the Importance of Experiments in Relation to the Mechanical Theory of Gravitation', *Philosophical Magazine*, vol. 11 (5th Series) (1881), pp. 391–3.

[136] S. Tolver Preston to J. J. Thomson, 5 May 1893, P45, JJT-CUL; S. Tolver Preston, 'On the Physics of Thought-Reading [1893]', P45a, JJT-CUL.

[137] Preston, 'Physics of Thought-Reading'; Edwin J. Houston, 'Cerebral Radiation', *Journal of the Franklin Institute*, vol. 133 (1892), pp. 488–97.

Preston agreed with Knowles, Crookes and Houston that there was an impressive body of evidence for thought-reading, and that this was probably a form of resonance involving vibrations exchanged between different brains. Like Crookes and Houston, he was also impressed with the biological and physiological evidence of the electrical activity of the brain, but, as a Maxwellian, he located the probable source of brain waves in the ether pervading the brain rather than brain cells themselves. The motion of brain cell groups caused electrical oscillations, which then caused resonance in the ether contained within the groups, and this generated electromagnetic waves (just like a resonant mass of air was a source of sound waves). As he explained in his manuscript, these waves induced "synchronous resonance" in the ether pervading a second brain, which generated the cellular motions and thoughts corresponding to those etherial vibrations.[138] Preston admitted to Thomson that the theory was not "free from hindrances", but he had taken the trouble to tackle a number of key physical and psychological questions: he proposed that the emitted waves could be between a millimetre and a centimetre in wavelength, which would ensure that they were not deflected by the head, and suggested that different thoughts could be defined by the "infinite" variety of ether waves and the shape of molecular vibrations.[139]

Unsurprisingly for someone who had applied sympathetic vibration to thought-reading, Barrett was reputedly interested enough in Preston's theory to recommend it for publication in the SPR's privately circulated *Journal* as a speculation suggesting new lines of experimental inquiry.[140] Similarly, Thomson was sufficiently intrigued by the theory to suggest that the SPR explore something at which Crookes, as well as Gurney, Myers and Podmore, had only hinted: the possible effect of metallic screens on stopping brain waves.[141] However, other SPR figures were less positive: its president Henry Sidgwick reputedly held that thought-reading was already too capricious an effect to render the metallic screen tests decisive and no such tests seem to have been staged; other leading figures in the organisation were, as we have seen, increasingly doubtful that thought-reading or telepathy were physical phenomenon at all; and, for all his enthusiasm, Barrett only recommended publication of the theory with his "critical remarks" added.[142] Despite Preston's confidence

---

[138] Preston, 'Physics of Thought-Reading'.
[139] S. Tolver Preston to J. J. Thomson, 15 May 1893, P46, JJT-CUL; Preston, 'Physics of Thought-Reading'.
[140] Barrett's reaction is mentioned in S. Tolver Preston to J. J. Thomson, 18 May 1893, P47, JJT-CUL.
[141] S. Tolver Preston to J. J. Thomson, 18 May 1893, P47, JJT-CUL.
[142] S. Tolver Preston to J. J. Thomson, 12 June 1893, P48, JJT-CUL.

in the credibility and importance of the theory, these criticisms may explain why he does not seem to have written or engaged others on the subject again.

Given the currency that Preston's thought-reading theory had among leading physical–psychical scientists, it is possible that it had some effect in sustaining their hopes for a physical theory of telepathy. Such a hope was boosted in the wake of the discovery of X-rays in late 1895. Whilst spiritualists hailed evidence of a new form of 'seeing' occult or hidden aspects of material bodies as lending support to occult or psychical modes of apprehension, physical–psychical scientists engaged in new discussions about telepathy as a form of resonance. In July 1896, Heaviside, who shared some other physical–psychical scientists' belief in a probable "physical basis of all abnormal phenomena", asked fellow Maxwellian Lodge if he had contemplated whether X-rays constituted a "physical example of the possibility of the direct action of one brain upon another at a short distance, even a selective action".[143] Heaviside revealed that the basic idea had occurred to him about 20 years earlier, but Röntgen's discovery provided the plausible physical basis, insofar as X-rays might be generated by electrical oscillations in "brain atoms" (imagined to be storage cells) and that these would "sympathetically excite similar cells in other brains in an imperfect way, & so provoke a vague impression, which the usual action of the brain may then develop to a picture".[144] The idea was certainly familiar to Lodge from Crookes's writings, and he was sufficiently intrigued by Heaviside's idea to share it with Myers, who frequently sought Lodge's advice on physical analogies and models of psychical effects.

In forwarding Heaviside's speculation to Myers, Lodge asked his friend not to pass it on to Crookes, who needed to remain "independent" in the matter.[145] Lodge was probably referring to Crookes's post-Röntgen speculations on brain waves, which, unlike Heaviside and Preston, he *was* willing to publicise and boasted an appropriate platform for doing so: his presidential address to the SPR. X-rays were doubly useful here: their unusual physical properties helped Crookes's general case for scientific humility, and one such property – their astonishingly high frequency – suggested the possibility of other high-frequency vibrations by which intelligence was transmitted between individuals via ganglions in the brain that were more developed in some than in others. Anticipating

[143] Oliver Heaviside to Oliver Lodge, 11 January 1895, MS Add. 89/50(ii), OJL-UCL; Oliver Heaviside to Oliver Lodge, 28 July 1896, MS Add. 89/50(ii), OJL-UCL.
[144] Oliver Heaviside to Oliver Lodge, 26 August 1896, MS Add. 89/50(ii), OJL-UCL.
[145] Oliver Lodge to Frederic W. H. Myers, 25 August 1896, attached to Heaviside to Lodge, 28 July 1896.

objections (mainly from fellow SPR members) that an X-ray theory of brain waves or telepathy would still require the intensity of mental impressions to decline with distance, Crookes suggested that an intense "telepathic chain of brain waves" might get round the difficulty and questioned whether this intensity law was applicable in "these subtile regions".[146] Despite the brevity of this speculation, Crookes's stature in the scientific world ensured that it sparked much debate within scientific, intellectual and occult circles.[147] For some critics, Crookes's position was doubly problematic because it represented a far-fetched theory of an unproven human faculty. Lodge showed his awareness of this wider apprehension about brain-wave theorising when, in 1898, he persuaded Crookes to shorten the discussion of the topic he wanted to include in his presidential address to the British Association.[148]

Lodge's input on Crookes's address reflected his sensitivity to specific scientific constituencies whose hostility towards psychical research had often caused problems for the image that the SPR wanted to project. He knew that Crookes's audience comprised many physiologists, psychologists and medical practitioners who had had enough problems with the evidence for telepathy, let alone bold hypotheses about it. Thus, as a warning to Crookes, Lodge recommended reading a recent paper by Francis Gotch, a physiologist at University College Liverpool with whom he had collaborated on studies of the inhibiting effect of rapidly alternating electric currents on nerve transmission.[149] Gotch argued that the new knowledge of nerve inhibition and fatigue furnished stronger evidence for the physiological basis of the extraordinary means by which hypnotised subjects sensed their immediate physical environment. But he also denied that there was any convincing physiological basis for other psychical effects: no organ for telepathy had been discovered and the idea of a brain being able to induce its state in a distant one was radically "opposed to the whole of physiological science".[150] Even if Crookes

[146] Crookes, 'Address by the President', p. 352.
[147] [Anon.], 'Electric Waves', *Electrical World*, vol. 29 (1897), p. 252; [Anon.], 'Lessons in Humility', *Light*, vol. 17 (1897), p. 78; [Anon.], 'Professor Crookes on Ethereal Bodies', *Spectator*, vol. 78 (1897), pp. 200–1.
[148] William Crookes to Oliver Lodge, 15 August 1898, SPR.MS 35/340, OJL-SPR; Crookes, 'Address by Sir William Crookes', p. 31. Crookes updated his speculation by suggesting that, following recent physiological studies, nerves acted similarly to 'coherers' in wireless telegraphy: they allowed electrical currents to pass when subject to ether waves of the appropriate frequency.
[149] William Crookes to Oliver Lodge, 29 August 1898, SPR.MS 35/342, OJL-SPR; Oliver Lodge, 'The Work of Hertz', *Nature*, vol. 50 (1894), pp. 133–9, pp. 135–6.
[150] Francis Gotch, 'Some Physiological Aspects of Hypnotism', *Science Progress*, vol. 1 (1897), pp. 511–30, p. 529.

read this, he did not abandon the idea that brain waves might exist and be experimented upon.[151]

Given Lodge's interest in contemplating the possible psychical functions of the ether, as well as his involvement in producing evidence for telepathy and the development of etherial or wireless telegraphy, his reservations about Crookes's brain-wave theory might seem odd. Yet from the mid-1890s onwards, Lodge's attitude towards the function of the ether in understanding psychical phenomena become more complex. In 1894, only three years after he had tentatively suggested to the British Association that the ether might function as the vehicle of telepathic transmission, he echoed many leading SPR figures in arguing that if experimental evidence failed to show that telepathy varied "according to some law of distance", then it was likely to be a "purely psychological" process not involving a "physical medium".[152] As somebody more closely acquainted than Crookes (and, for that matter, Houston and Preston) with the nature of telepathy and with the work of such leading interpreters as Myers, Lodge clearly felt in a better position than other physical–psychical scientists to adjudicate on the ether's role in telepathy.

This hardly stopped Lodge from contemplating *other* psychical functions for the ether. From the mid-1890s onwards, he had accepted that the difficulty of describing the ether in terms of the laws of force and motion meant that such laws might not be fundamental, and that more fundamental mechanical laws might relate to this quasi-material or immaterial medium, which might then make it easier to understand the way mind and spirit interacted with matter and to relate the immaterial domains glimpsed in psychical research to physics.[153] By the early 1900s, he had extended this thinking to that most difficult of all psychical questions: materialised spirits. In the earliest form of one of his most notorious theories of psychical phenomena, Lodge proposed that just as animals unconsciously used material from their surroundings to build anatomical features (for example, shells and muscle), so discarnate intelligences might use their "etherial" bodies (by which they were already in touch with the "physical universe") to mobilise "terrestrial particles" that appealed to human senses.[154]

[151] Crookes, 'Address by Sir William Crookes', pp. 31–2.
[152] Oliver Lodge, 'On the Difficulty of Making Crucial Experiments as to the Source of the Extra or Unusual Intelligence Manifested in Trance-Speech, Automatic Writing, and other States of Apparent Mental Inactivity', *PSPR*, vol. 10 (1894), pp. 14–24, pp. 17–18. By 1903, Lodge was arguing that the possibility of telepathy being a purely psychical rather than physical or etherial process meant that it could also be a form of action at a distance, but that physicists had no right to reject psychical processes on this basis: Lodge, 'Presidential Address', pp. 19–20.
[153] Lodge, 'Interstellar Ether'.   [154] Lodge, 'Address by the President', p. 47.

The writings of physical–psychical scientists on telepathy reveal that they were often a good deal bolder in theorising and speculating on psychical phenomena in private than in public. Mindful that such activity in public could threaten the professed aim of the SPR not to be associated with "any particular explanation of the phenomena investigated", physical–psychical scientists closely associated with the organisation often limited such creative scientific thinking to private correspondence.[155] For example, Crookes and Lodge contributed little to the public debate about the psychical interpretations of mathematical theories of hyperspace, the most notorious of which was Zöllner's suggestion that disembodied spirits were four-dimensional beings and that this enabled them to help mediums achieve such startling physical feats as passing matter through matter.[156] In private, however, they were happy to discuss such matters at length.[157]

Of particular interest to Lodge as a subject of mainly private debate was a 'Maxwellian' way of thinking about the telekinetic powers of the spiritualist mediums D. D. Home, William Stainton Moses and Eusapia Palladino. In the mid-1890s, all three were the subject of much debate in the SPR: Myers wrote critical studies of Home and Moses (both of whom were dead by 1892), and Myers, the Sidgwicks, Lodge and other SPR members conducted several investigations of Palladino. One of Myers's preoccupations with Home and Moses was that their invisible spirit 'controls' appeared to violate natural laws: for example, Home's controls caused cool dew to drop from a ceiling and extracted brandy from a brandy-and-water mixture, while Moses's controls could unequalise the temperature of a room and produce scents and imitation gems seemingly out of nowhere. To help Myers interpret these phenomena, Lodge advised him to read about Maxwell's 'sorting demon'.[158]

The 'sorting demon' was William Thomson's alternative name for an imaginary microscopic sentient being that Maxwell had devised in the late 1860s to illustrate the probabilistic nature of one of the great generalisations in nineteenth-century physics: the universal dissipation of heat. The demon was imagined operating a frictionless door between two compartments of a closed box containing a large number of gas

---

[155] [Anon.], 'Objects of the Society', p. 5.

[156] Johann Karl Friedrich Zöllner, *Transcendental Physics* (translated by Charles Carleton Massey) (London: W. H. Harrison, 1880). For discussion see Staubermann, 'Tying the Knot'.

[157] Frederic W. H. Myers to Oliver Lodge, 21 June 1894, SPR.MS 35/1400, OJL-SPR; William Crookes to Oliver Lodge, 22 May 1909, SPR.MS 35/356, OJL-SPR. Here Crookes revealed that for many years he had been "speculating" that the ether of space might be "matter in the 4th dimension" and the domicile of spiritual beings.

[158] Frederic W. H. Myers to Oliver Lodge, 22 May 1894, SPR.MS 35/1397, OJL-SPR.

molecules.[159] By careful operation of the door, the demon could separate the faster from the slower-moving molecules, thus causing one compartment to be hotter than the other. In doing this, the demon would neither have been violating the microscopic laws governing the motion of individual molecules (the kinetic theory of gases) nor expending energy on separating molecules, but the result would contradict the ordinary experience of the tendency in nature towards temperature equilibrium. It was therefore theoretically possible but unlikely for this tendency to be reversed. Similar to Crookes's 'homunculus' (which it doubtless inspired), the demon was designed to show the limited human perspective on the world – in this case, a perspective that saw dissipation as inevitable.

Maxwell's demon was just what Myers was looking for because it showed that it was "theoretically possible" for some kind of intelligence present in seances to manipulate physical processes at the microscopic level and produce results that seemed contrary to ordinary experience or supernatural at the macroscopic level.[160] By simply directing individual molecules, such intelligence could have caused the strange belts of hot and cold air often felt in seances, as well as the effects associated with Home and Moses. Myers also envisioned the possibility that the intelligence could displace matter on the molar as well as the molecular scale and that this would help understand the "coarser phenomena" of telekinesis – phenomena that both he and Lodge had witnessed during seances with Palladino.[161]

What excited Lodge about this use of Maxwell's demon was that FitzGerald suggested something similar entirely independently of Myers.[162] In helping Lodge cope with criticisms of his evidence for Palladino's genuineness, FitzGerald offered a theory of her telekinetic abilities even though he did not think the "time is at all ripe for theories".[163] His tentative theory denied the "theoretical impossibility" of some people being able unconsciously to "so *arrange* the motions of molecules of their hands" that all the irregularly moving air molecules flanking the skin of the hands were given the same "motion of translation", which would create enough of a molecular torrent to cause the kinds of movements observed in seances.[164] Unlike Myers, FitzGerald attributed the sorting power to the unconsciousness of the medium.

---

[159] On Maxwell's demon see Stanley, *Huxley's Church*, pp. 233–40.
[160] Myers, *Human Personality*, vol. 2, p. 531.
[161] Myers, *Human Personality*, vol. 2, p. 535.
[162] Oliver Lodge to George F. FitzGerald, 6 January 1895, f. 2/54, George Francis FitzGerald Papers, RDS Library and Archives.
[163] George F. FitzGerald to Oliver Lodge, 5 January 1895, MS Add. 89/35, OJL-UCL.
[164] FitzGerald to Lodge, 5 January 1895.

Lodge agreed that cases of telekinesis did not need to invoke discarnate intelligences, but was only marginally less reserved than FitzGerald in how he used sorting demons in a psychical context: not surprisingly, FitzGerald never published his theory, and Lodge was only prepared to declare in an SPR presidential address that demons were the "most likely direction for groping after a theory" of telekinesis and other unknown powers of mediums to deviate the usual course of physical nature.[165]

The restraint that Lodge exercised in SPR publications over physical theories of telekinesis and telepathy contrasts with what Fournier d'Albe attempted in *New Light on Immortality*, a book over which the SPR had little control. The final part of a trilogy of popular books on electron theory, radioactivity and cosmology, *New Light on Immortality* drew parallels between the atomic and astronomical orders of things, and argued that the natural laws of entities in one level could be seen as the social laws of entities in an inferior level. What humans interpreted as chemical laws were "*laws of life of the atomic species*", which radioactivity displayed with such 'vital' characteristics as growth and decay.[166] Fournier d'Albe used this to challenge philosophical dualism: it was not impossible that what seemed like a lifeless entity at one level was the aggregate of what were 'life units' at an inferior level. Assuming that life existed in some form at every level, Fournier d'Albe argued that the directive agency that humans associated with life was probably present at the microscopic level – in the form of hypothetical particles within each cell that he christened 'psychomeres'. In aggregate, psychomeres formed a fine invisible mist that was a possible domicile of the soul. Since recent studies of atoms revealed them to be mostly empty space, the human body could also be seen as a mist, but one that it was not impossible for this psychomere aggregate to penetrate and part from. The temporary departure of the aggregate was Fournier d'Albe's tentative theory of the materialised spirit forms seen in seances, their permanent departure from the material being constituting a "Physical Theory of Immortality".[167]

Given the strong relevance of Fournier d'Albe's book to the physical and psychical questions addressed by leading physical–psychical scientists, their relative silence about it might be surprising. But they probably refused to pay attention to a work that exceeded the limits that they believed needed to be placed on speculations about the psychical uses of physics, especially within the auspices of the SPR. Frederick Stratton

---

[165] Lodge to FitzGerald, 6 January 1895. See also Lodge, 'Address by the President', pp. 49–50 and Oliver Lodge, 'On the A Priori Argument Against Physical Phenomena', *PSPR*, vol. 25 (1911), pp. 447–54, p. 448.
[166] Fournier d'Albe, *New Light on Immortality*, p. 90.
[167] Fournier d'Albe, *New Light on Immortality*, p. viii.

spoke for Lodge, Crookes and Barrett when, in reviewing Fournier d'Albe's book for the SPR, he attacked the author for building his theory from ignorance rather than knowledge of the nature of matter and for not carefully distinguishing the "purely speculative parts of a theory and those which are experimentally demonstrable".[168]

Yet Fournier d'Albe's book illustrates the existence, in the early 1900s, of places besides SPR publications where physical–psychical scientists could engage in theories and speculations about psychical effects, and to an extent that was often deemed excessive within the SPR. For some of these physical–psychical scientists, the likes of Barrett, Crookes and Lodge were too cautious and not bold enough in drawing psychical lessons from the rapid developments in the physical sciences. Lodge's scepticism of etherial theories of brain waves struck them as oddly conservative from somebody who also seemed to think that the ether was the link between 'physics and psychics' and who was so closely involved in the invention of ether-based intelligence transmission. He would certainly have frustrated the telegraph engineer and science populariser William Lynd, who, in the wake of Guglielmo Marconi's feats in long-distance wireless telegraphy, told readers of one popular magazine that "etheric waves" could now explain "many of the puzzling phenomena of Hypnotism and Thought-Transference" and that the "day was not far distant when Telepathy will be studied like any other branch of physics".[169]

Doubts and Criticisms

In a posthumous book of 1917, Silvanus Thompson, a physicist who counted Barrett, Crookes and Lodge among his closest friends, warned that the "truly spiritual man" seeking "eternal truth" needed to "avoid the well-intentioned but crude divulgations of the new school of sentimental philosophers who mix physics and psychics for us in facile and graceful oratory". Just because

radium emits mysterious corpuscles, or ether-waves are found to be generated by electric sparks, we are not bound to deduce therefrom the probability that our dead friends can speak to us through a planchette, or that the doctrine of the Trinity can be developed from the triple conservation of matter, of energy, and of ether.[170]

[168] Frederick J. M. Stratton, 'Review', *JSPR*, vol. 14 (1909), pp. 78–9, p. 79.
[169] William Lynd, 'Thought Transference and Wireless Telegraphy', *Surrey Magazine*, vol. 1 (1900), pp. 24–7, p. 27. Lynd's activities popularising telepathy and wireless telegraphy are documented in his diary: William Lynd, Diary for 1899, GB 29 EUL MS.14, University of Exeter Library Special Collections.
[170] Silvanus P. Thompson, *A Not Impossible Religion* (London: Bodley Head, 1917), pp. 32–3.

This was undoubtedly a response to numerous early-twentieth-century attempts to draw out psychical, religious and spiritual significances from the discovery of radioactivity, the development of wireless telegraphy and the invention of ethereal theories of matter. Thompson's chief targets were probably those liberal Christian thinkers, spiritualists and modern Theosophists who readily seized on any sign that the triumphant physical sciences were converging with their own beliefs. One of the best-known Quaker scientists of his day, Thompson's attitude was entirely consistent with his objections to forms of religion and spirituality whose foundations were shallow, vague, hastily assembled and which privileged appeals to the external world over intuition.[171]

If any other physical–psychical scientist alive in 1917 had read Thompson's remarks, however, it would have put them in a dilemma. On the one hand, most of them would have agreed that straightforward deductions of psychical, religious and spiritual lessons from physics were perilous; on the other hand, they would have felt guilty of the charges made by Thompson. They, and those physical–psychical scientists who were now long dead, had spent much time thinking and writing about the way that ideas in the physical sciences changed the "probability" that extraordinary claims made about the human body, mind and spirit were true.

Thompson was hardly the first scientific colleague of leading physical–psychical scientists to express grave doubts about the uses of physics in contemplating the kind of results to which psychical investigation was leading. While Stokes agreed that the extraordinary properties of the ether prepared the mind for weighing evidence for such incredible claims as the survival of the soul following bodily death, neither the ether nor other uses of our "natural faculties" could ever "demonstrate, or even render probable" the post-mortem state because this was purely a matter of religious faith.[172] Similarly, in one of his last works, Maxwell insisted that questions of human personality and its post-mortem existence were beyond the "limits of science" and this is why he thought his close colleagues Stewart and Tait were, in the etherial afterlife of their *Unseen Universe*, "far transcending the limits of physical speculation".[173] Invoking physical theories to explain psychical, spiritual and non-material realms clearly

---

[171] On Thompson's Quakerism see Geoffrey Cantor, *Quakers, Jews and Science: Religious Responses to Modernity and the Sciences in Britain, 1650–1900* (Oxford University Press, 2005), chapter 6.

[172] Stokes, *Natural Theology*, p. 57. See also Stokes quoted in Stead, 'Response to the Appeal', p. 18.

[173] James Clerk Maxwell, 'Ether', in W. D. Niven (ed.), *The Scientific Papers of James Clerk Maxwell* (Cambridge University Press, 1890), vol. 2, pp. 763–75, p. 775.

threatened the kind of materialism that such theorisation was often designed to thwart.

That most widely debated application of physical theories to psychical questions – brain waves – prompted similar doubts. The American *Electrical Engineer* worried that Crookes's theory of 1897 looked like the "scientific imagination run riot".[174] More telling was the response of this periodical's British equivalent, *The Electrician*, to Preston's 'Physics of Thought-Reading' paper which he had offered them. It accepted the possibility that thought waves could become part of physical science if, like other physical phenomena, their wavelength and other properties could be measured. However, neither physical measurement nor physical theories of the kind Preston advanced could cope with the complexity and emotional significance of thought. How, *The Electrician* asked sardonically, was the "coefficient of mendacity" to be measured, and what was the "focal length of a gross exaggeration, or the negative focus of a disparaging remark?"[175]

For some critics of physical–psychical scientists, the problem was that no amount of theorising and modelling could hide the fact that the phenomena being interpreted seemed to violate so many cherished physical laws and principles. Huggins found materialised spirits hard to accept because the idea of essentially human-like entities being created and annihilated instantaneously in seances involved rejecting "all that science and experience have taught us".[176] Similarly, when, in 1894, Lodge asked his former physics mentor George Carey Foster to comment on a paper describing telekinetic effects, Foster asked sternly whether or not the "whole progress of physics is based on the assumption that these things do *not* happen?".[177] Echoing Hume's famous argument against miracles, he was more inclined to believe that a "mistake" had been made than that mechanical laws and energy conservation had been upset.[178]

For other critics, the problem with much psychical speculation was that the physical concepts and theories being invoked were not themselves reliable. A common objection to the *Unseen Universe* was that it based its arguments for the survival of the soul on speculations about, rather than certain knowledge of, the ether's capacities.[179] For some critics, ideas

---

[174] [Anon.], 'Crookes Brain Waves', *Electrical Engineer*, vol. 23 (1897), pp. 220–1, p. 221.

[175] [Anon.], 'Notes', *Electrician*, vol. 30 (1892–3), pp. 587–8, p. 588.

[176] William Huggins to William F. Barrett, 4 October 1908, f. 71, WFB-RS.

[177] George Carey Foster to Oliver Lodge, 25 October 1894, MS Add. 89/38, OJL-UCL.

[178] Foster to Lodge, 25 October 1894.

[179] [Anon.], 'The Possibility of a Future Life', *Quarterly Journal of Science*, vol. 5 (New Series) (1875), pp. 472–86; [Anon.], 'The Unseen Universe', *London Quarterly Review*, October 1875, pp. 49–83, p. 78; William Kingdon Clifford, 'The Unseen Universe', *Fortnightly Review*, vol. 17 (1875), pp. 776–93, esp. pp. 789–90.

about the ether, atoms and molecules were already too perplexing to solve the difficulties of imagining how distant brains could directly interact with each other.[180]

Most scathing were those critics who linked physicists' belief in disembodied spirits to their attachment to questionable aspects of their science. In his notorious 1884 critique of the concepts and theories of 'modern physics', the American lawyer and science writer J. B. Stallo attacked the "pre-scientific survivals" haunting physicists, notably their tendency to banish physical facts to the "regions of the Extra-sensible".[181] This included invoking such dubious concepts as the elasticity of atoms, and mechanical force, and even supernatural entities: "[f]aith in spooks (with due respect be it said for Maxwell's thermodynamical "demons" and for the population of the "Unseen Universe")", he warned, "is unwisdom in physics no less than in pneumatology".[182] Similarly, that more redoubtable critic of mechanics, the Austrian physicist and philosopher Ernst Mach, lamented in 1900 that the "fetishism" typically shown by savages could be found among physicists, insofar as they showed a tendency to invest such "mysterious and impalpable entities" as heat, electricity and magnetism with "wonderful properties".[183] This tendency usually sprang from a "biased intellectual culture and from a lack of philosophical training", which was tragically manifest when a famous chemist, a "noted" physicist and a "renowned" biologist espoused spiritualism.[184]

***

The writings of physical–psychical scientists on brain waves, telekinesis and other psychical phenomena suggest that they were willing and able to theorise about psychical effects a good deal more than we have hitherto assumed. By the end of the nineteenth century, most of them were acutely aware, not least from critics near and far, of the risks of applying physical principles, analogies, theories and explanations to psychical puzzles. Despite their increasing caution about the psychical applications of physical theories, they never gave up the hope that some theories and ideas in physics might constitute the basis for more satisfactory interpretations of psychical phenomena and this partly depended on new understandings of the physics of the ether, energy and matter as they unfolded in the early

---

[180] Conwy Lloyd Morgan, 'Supernormal Psychology', *Nature*, vol. 35 (1886–7), pp. 290–2, p. 292.

[181] J. B. Stallo, *The Concepts and Theories of Modern Physics* (New York: D. Appleton and Co., 1884), p. 128.

[182] Stallo, *Concepts and Theories*, p. 128.

[183] Ernst Mach, 'The Propensity Toward the Marvellous', *Open Court*, vol. 14 (1900), pp. 539–50, p. 541.

[184] Mach, 'Propensity', p. 543. He undoubtedly meant Crookes, Zöllner and Wallace.

1900s. What proved more useful than such theorising in the long term were the general physical principles and analogies that, well into the twentieth century, peppered physical–psychical scientists' arguments against the impossibility of psychical effects and the futility of psychical investigation. These rhetorical strategies had a limited effect on one significant audience – physical scientists – whose interest in psychical research seems to have declined after the early 1900s. Physical–psychical scientists acknowledged that persuading colleagues of psychical research's strengths required an additional strategy: the transformation of sites of psychical inquiry into spaces that more resembled those of experimental physics – spaces where some psychical effects promised to be more satisfactorily registered, controlled, measured and, in general, evidenced than they had been previously. It is to the origins and development of this strategy that we turn in the following chapter.

4   Psychical Investigation as Experimental Physics

Few of William Crookes's scientific colleagues were more enchanted by his public performances of experimental physics than William Thomson. Few events proved more enchanting to this eminent Scottish physicist than Crookes's presidential address to the Institution of Electrical Engineers in 1891. Peppered with a string of visually stunning experimental demonstrations, Crookes took stock of nearly two decades of study into the mysterious behaviour of electrified molecules in near-vacuum conditions, including their radiation-like properties.[1] Thomson was so impressed that he nominated Crookes for the Copley Medal, the prestigious award conferred by a much older scientific organisation of which Thomson was president: the Royal Society of London. A year after Crookes's address, he put his case to Rayleigh by emphasising that the chemist and physicist had created "really the greatest things so far, in developing the physics of the 19th century".[2] What surprised Thomson was that such intellectual milestones "all arose from his 'investigating' mesmeric attraction!!"[3]

Thomson's surprise is understandable. About a decade earlier, he had spoken publicly of "that wretched superstition of animal magnetism, and table-turning, and spiritualism, and mesmerism, and clairvoyance, and spirit-rapping", and evidently believed that such "superstition" could neither be the subject of scientific investigation nor the inspiration for physical discoveries.[4] Rayleigh's reply to Thomson's proposal is unknown, but even if he did lend his weighty support, Crookes had to wait until 1904 for his Copley Medal. Rayleigh would have been less surprised than Thomson by the apparent connection between physical experiment and psychical investigation. Like Crookes, he had been interested in using the resources of experimental physics to investigate the

---

[1] William Crookes, 'Electricity *in Transitu*: From Plenum to Vacuum', *Journal of the Institution of Electrical Engineers*, vol. 20 (1891), pp. 4–49.
[2] William Thomson to Lord Rayleigh, 13 February 1892, Rayleigh Family Papers, Terling Place, Terling, Essex.
[3] Thomson to Rayleigh, 13 February 1892.
[4] Thomson, 'Six Gateways of Knowledge', p. 258.

physical phenomena of spiritualism and did so in his private residence, where rooms for seances were not far from a physical laboratory.[5]

Rayleigh's interest in spiritualism owed a great deal to the strong "impression" made upon him by an article that Crookes had published in an 1874 issue of a periodical that Crookes edited: the *Quarterly Journal of Science* (henceforth *QJS*).[6] This article was the last in a series describing Crookes's investigations into the phenomena of spiritualism, which included the movement of untouched objects, spirit-rapping and other intelligent sounds, moving lights, phantom human forms and disembodied luminous hands that wrote messages. Rayleigh's positive response illustrates the significant role that experimental enquiry played in shaping attitudes towards psychical investigation per se. Chapter 3 showed that theoretical and conceptual developments in the physical sciences certainly stimulated much private and public debate about the plausibility of psychical effects, but the results of what were variously called 'experimental', 'scientific' and 'accurate' tests of psychical phenomena were more significant determinants of the scientific profile of psychical investigation. Physical–psychical scientists were acutely aware of this, as suggested in Lodge's argument of 1902 that the foundation of a "psychical laboratory" modelled on a physical equivalent was needed if the physical phenomena of psychical research were "ever to become recognised as a branch of orthodox physics".[7]

Rayleigh's impression of Crookes's publications was more widely shared, not least among many of those who later founded the SPR. In July 1874, during his early spiritualistic investigations, Rayleigh's brother-in-law Henry Sidgwick told his wife that Crookes's articles bolstered the "idea of the weight of the evidence in favour of the phenomena".[8] But what made it difficult for Sidgwick to accept that Crookes was either "affirming a tissue of purposeless lies, or

---

[5] Rayleigh, *Life of Rayleigh*, p. 67n. Rayleigh's most intriguing use of physics was a hermetically sealed glass retort within which he left a pencil and paper for 'spirits' to write on. The apparatus, which still exists, has yet to show an inscription. On Rayleigh's domestic laboratories see Donald Opitz, '"Not Merely Wifely Devotion": Collaborating in the Construction of Science at Terling Place', in Annette Lykknes, Donald L. Opitz and Brigitte van Tiggelen (eds.), *For Better or For Worse? Collaborative Couples in the Sciences* (Heidelberg: Birkhäuser, 2012), pp. 33–56; Simon Schaffer, 'Physics Laboratories and the Victorian Country House', in Crosbie Smith and Jon Agar (eds.), *Making Space for Science: Territorial Themes in the Shaping of Knowledge* (Basingstoke: Macmillan, 1998), pp. 149–80.

[6] Rayleigh, 'Presidential Address', p. 278; Crookes, 'Notes of Enquiry'. See also William Crookes to Lord Rayleigh, 27 January 1874, Folder 4, Box #17, Series 4, R-USAF.

[7] Lodge, 'Address by the President', p. 47.

[8] Henry Sidgwick to [Mary] Sidgwick, 11 July 1874, in A[rthur] S[idgwick] and E[leanor] M[ildred] S[idgwick], *Henry Sidgwick: A Memoir* (London: Macmillan and Co., 1906), pp. 290–1, p. 290.

a monomaniac" was that at precisely this time Crookes was displaying considerable experimental acumen in the form of Royal Society exhibitions of his research into the apparent repulsive force of radiation.[9] Sidgwick had fewer problems than did Thomson with the idea of experimental physics being relevant to psychical investigation. Only weeks before his remarks about Crookes, he told Myers that the idea of the famous American medium Katherine Fox-Jencken "rapping away" in Rayleigh's house "using all the resources of his laboratory is too tempting a prospect".[10]

Throughout the last third of the nineteenth century, physical–psychical scientists displayed enormous confidence in the relevance of their experimental resources and skills to the development of psychical inquiry. In the first of his *QJS* papers, Crookes presented a virtual manifesto for the proper scientific study of spiritualism in which both attributes played a prominent role. The kinds of "experimental proof" that "science has a right to demand before admitting a new department of knowledge into her ranks", he insisted, depended critically on the careful use of the delicate instruments used in physical and chemical laboratories, but here deployed in establishing psychical "facts" that could not always be trusted to the "unaided senses" in seance conditions where fraud and mental "excitement" were often present.[11] Spiritualist claims about a mysterious power levitating heavy objects, for example, could only be accepted if the same power supposedly moved a "delicately-poised balance".[12] Together with "a long line of learning" that furnished skills in identifying the "dangers", uncertainties and certainties of inquiries, instruments gave the "scientific man" a "great advantage" over the "ordinary observer" in weighing spiritualists' claims.[13]

Crookes's manifesto made the resources and skills of experimental physical sciences relevant by sidestepping the fact that even the most 'physical' phenomena of spiritualism exhibited intelligence – properties that he (like Maxwell and most other physicists) agreed were beyond the formal boundaries of the physical sciences.[14] This sidestepping was explicit in Lodge's address to physicists at the 1891 meeting of the British Association. He anticipated objections that psychical investigations were psychological

[9] Sidgwick to Sidgwick, 11 July 1874, p. 290.
[10] Henry Sidgwick to Frederic W. H. Myers, 4 June [1874], Add. MS.c.100[262], Henry Sidgwick Papers, Trinity College, Cambridge. Katherine Fox became Katherine Fox-Jencken in 1872 when she married the English barrister Henry D. Jencken.
[11] Crookes, 'Spiritualism Viewed', p. 318.
[12] Crookes, 'Spiritualism Viewed', p. 319. This echoed Faraday, 'On Mental Education', p. 471.
[13] Crookes, 'Spiritualism Viewed', p. 319.
[14] See Crookes's remarks in [Anon.], 'The Psychological Society of Great Britain', *Spiritualist*, vol. 15 (1879), p. 235. See also Barrett, 'Address by the President', p. 332.

rather than physical, but reassured his auditors that since the results of such investigations had a "physical side" then it was a "proper subject for physical enquiry" where physics had to "lead, not to follow".[15] Physical analogies, concepts and theories were obvious ways of illuminating this "side", but they also helped to legitimate ways of achieving the same goal via the hardware of physical investigation, and thereby create a psychical kind of experimental physics.

Arguments for the importance of physical experiment in psychical investigation were often made in the context of attempts to mobilise the resources of the physical laboratory in illuminating the reality, nature and provenance of psychical effects. This chapter analyses four of the most elaborate and revealing of these attempts made by physical–psychical scientists between the 1870s and the 1890s. What chiefly distinguished them from the vast majority of nineteenth-century tests of psychical effects was the presence of scientifically trained participants and the use of a host of devices for studying psycho-physical phenomena with puzzling mechanical, optical, magnetic, electrical, thermal and acoustical properties. As we shall see, however, the resulting transformations of seances into sites of physical enquiry, and sites of physical enquiry into seances, proved less effective than physical–psychical scientists hoped. While some evidence produced in this way persuaded physical–psychical scientists of the reality of specific effects and the need for further inquiries, it proved difficult for them to amass the quantity of evidence required to stimulate greater interest and conviction among professional scientific colleagues. This was partly because the experiments depended too strongly on human (psychic) subjects who were hard to find, whose powers were capricious and who most scientific practitioners associated with fraud. But it was also because leading physical–psychical scientists made the pragmatic decision that their skills and resources in experimental physics could better serve their professional goals by deploying them in purely physical questions.

From Psychic Force to the Radiometer

William Thomson was only partially right in claiming a mesmeric provenance for *all* of Crookes's researches on molecular physics. He probably remembered that in the mid-1870s he had refereed many of the papers on the subject that Crookes had submitted to the Royal Society and that, much to his irritation, one such paper had suggested that the unanticipated movements of a delicate object suspended in near-vacuum conditions

[15] Lodge, 'Address', p. 555.

might be due to an aura around human hands that Reichenbach claimed as a manifestation of od.[16] What had annoyed Thomson about the Reichenbach reference was that it reminded him of the "delusions" into which he believed Crookes had been led, mainly because of seemingly dishonest witnesses to mesmeric and spiritualistic phenomena.[17] But the psychical phenomena that Thomson saw as irrelevant and pernicious to Crookes's physical discoveries had long been seen by Crookes as potential sources of discovery in the physical sciences. This blurring of distinctions between physical and psychical investigation was starkly apparent in, and made possible by, the laboratory that he had installed in his family home near Regent's Park.

When, in July 1870, Crookes published his manifesto for a new experimental approach to spiritualism, he had already found what he called a potential "key to these strange phenomena" in the form of Daniel Dunglas Home, whose reputation for powerful mechanical effects at a distance made him an obvious subject for physical tests.[18] Crookes had been convening seances with Home in his London residence since at least April 1870, when this globe-trotting medium was in the middle of one of his longest stints in Britain and eager to have new, high-profile tests of his abilities.[19] By this time, Crookes had been attending seances with other mediums for about a year and these had persuaded him of the existence of a "power in some way connected with gravitation".[20] His participation in the London Dialectical Society's seances would have encouraged him to regard Home as a potential "key" to spiritualism because opposing David Brewster and others who had accused Home of deception and fraud were Dialectical Society witnesses of such scientific calibre as Varley and Lindsay, who testified to the genuineness of many of Home's powers.[21] With such an impressive history of cooperating with scientific enquirers, Home would have struck Crookes as an

---

[16] William Thomson, 'Report on Mr. Crookes's Paper "On Attraction and Repulsion Resulting from Radiation"', 16 June 1875, RR/7/369, Royal Society Archives; William Crookes, 'On Repulsion Resulting from Radiation. Part II', *Philosophical Transactions of the Royal Society of London*, vol. 165 (1875), pp. 519–47, p. 526.

[17] Thomson, 'Report on Mr. Crookes's Paper'.

[18] Crookes, 'Spiritualism Viewed', p. 320.

[19] Medhurst, *Crookes and the Spirit World*, pp. 149–50. The following discussion uses R. G. Medhurst's edition of Crookes's notes on the Home seances because it includes material that Crookes did not include in his own published notes: William Crookes, 'Notes of Séances with D. D. Home', *PSPR*, vol. 6 (1889–90), pp. 98–127.

[20] William Crookes to John Tyndall, 22 December 1869, in Medhurst, *Crookes and the Spirit World*, p. 233.

[21] The investigations in which Crookes participated are described in *Report on Spiritualism*, pp. 159–60, 213–17.

excellent experimental subject with whom to achieve his goal of conducting tests "more calmly than is possible when in a party of enthusiasts".[22]

The domestic location and the personnel of these early Home seances were deliberately chosen to create the convivial psychological 'atmosphere' widely believed by spiritualists to favour the exhibition of psychic power. Held in the evening, the sittings were staged at Crookes's residence and the homes of prominent London spiritualists, and among those around the main seance tables were Crookes's wife and brother, some of Crookes's scientific colleagues (including Robert Angus Smith and Alfred Russel Wallace), and many individuals who had sat with Home on previous occasions (notably Lord Adare).

Crookes's published and unpublished notes on these events suggest that until spring 1871 little happened that would have surprised those attending performances by Home or any other leading mediums. Whilst fully conscious or semi-entranced, Home appeared to be able to move tables, bells and other objects without touching them; to levitate himself; to produce a cool breeze and faint disembodied hands; and to channel professed spirits of the dead who rapped out encoded messages on furniture and responded to questions given vocally by participants. These effects were capricious: some seances witnessed the appearance of very few manifestations, and the mysterious bodily force by which Home appeared to produce them was, according to Crookes, subject to "unaccountable ebbs and flows".[23] Very few of the physical effects that Crookes witnessed in these early seances suggested to him the need for scientific tests more elaborate than testing the possible thermal and electrical properties of the mysterious breeze with a thermometer and gold leaf electrometer respectively, or exploring the effect on manifestations of putting coloured filters over the gas lamps illuminating the seance room.[24] The outcomes of these particular tests were not recorded.

By spring 1871, however, Crookes's experiences of Home had given him confidence in the fruitfulness of more searching instrumental tests. He was impressed that Home performed most of his seances by gaslight, and that the medium allowed him a good deal of freedom in scrutinising the effects, as well as his body and the seance room.[25] The favourable impression that he had of Home's character was bolstered by his failure to

[22] Crookes to Tyndall, 22 December 1869, in Medhurst, *Crookes and the Spirit World*, p. 233.

[23] William Crookes, 'Experimental Investigation of a New Force', *Quarterly Journal of Science*, vol. 1 (New Series) (1871), pp. 339–49, p. 340.

[24] Medhurst, *Crookes and the Spirit World*, p. 153.

[25] Crookes, 'Experimental Investigation', pp. 340–1. See also Crookes's contributions to [Anon.], 'Discussion of Professor Lodge's Paper', *JSPR*, vol. 6 (1893–4), pp. 336–45.

detect any accomplices, wires or other means of defrauding witnesses in any of the seances he convened or attended.[26] But it was Home's capacity to repeatedly exhibit two specific psychical effects that proved most important to Crookes's experimental ambitions: the medium seemed to be able to cause an accordion to expand, contract and play tunes whilst holding it from the end furthest from the keys, and sometimes without holding it at all; and on many occasions, he could cause the weight of the seance table to increase and decrease on Crookes's command, the weight being measured by Crookes with a small spring balance hooked under the table.

The weight tests were particularly appealing to Crookes. They promised to give him additional reasons for challenging the assumption that the movements of objects were due to ordinary mechanical forces exerted by participants: his close observations of Home and other seance participants suggested that they had rarely made any mechanical contact with seance tables and on those occasions when there was such contact, they could not have caused the significant weight changes merely by pushing or lifting.[27] The weight tests were also attractive to Crookes because, as his "friend" and fellow chemist Alexander Boutlerow had shown in tests of Home in late 1870, the effect was measurable with simple mechanical instruments.[28]

To fulfil his manifesto commitment to subjecting spiritualistic phenomena to accurate measurement and "proper tests", Crookes needed to transform the domestic seance into a space that bore key attributes of physical experiment.[29] This was made easier since he persuaded Home to give seances in his dining room, which was near his private laboratory.[30] Moreover, in the seances that he gave in this room between late May and late June 1871, Home was happy to perform in the presence of mechanical and other kinds of apparatus and of scientific participants, notably William Huggins, Edward Cox and Crookes's laboratory assistant Charles Gimingham. Personally trained by Crookes, Gimingham's roles in the seances merely extended those he fulfilled in his employer's purely physical enquiries – observing, measuring and recording effects and undertaking other investigative tasks that Crookes was unable to execute himself owing to other professional and commercial commitments.[31]

[26] For Crookes's positive view of Home's character see William Crookes to Alexander Boutlerow, 13 April 1871, quoted in Fournier d'Albe, *Life of Sir William Crookes*, pp. 196–8.

[27] For example, Medhurst, *Crookes and the Spirit World*, pp. 163, 170–1 and 178.

[28] Crookes, 'Some Further Experiments', p. 475.

[29] Crookes, 'Spiritualism Viewed', p. 320.

[30] Medhurst, *Crookes and the Spirit World*, p. 189.

[31] On Gimingham see Gay, 'Invisible Resource'.

There were two forms of apparatus in the test seances. The first was a cylindrical wire cage, almost as tall as the gap between the seance table and the floor, connected via long wires to an electric battery in the laboratory. By asking Home to hold the accordion inside the cage, Crookes aimed to study the possible effect of electric currents on the musical instrument's striking movements, although this test proved inconclusive. Furthermore, by asking Home to reproduce the effect with the cage under the table, Crookes sought to severely restrict how far Home could surreptitiously use his feet or hands to achieve the effect. The second apparatus was a balance, which was employed to measure Home's apparent ability to change the weight of objects independently of any known force. This comprised a wooden board 36 inches long, one end of which rested on a fulcrum near the edge of the seance table, the other end being suspended from a self-registering spring balance, itself hung from a tripod.

In the test seance whose results eventually formed the basis of his second *QJS* paper on spiritualism, Crookes emphasised that in the accordion test, Home sat near the table, with Crookes and his wife resting their feet on his as an additional form of control.[32] Home took the accordion from the end furthest from the keys and was asked to hold it with one hand in the cage under the table, his other hand resting on the table. Soon the accordion was heard to play successive notes and a tune, and Gimingham and Huggins reported seeing such startling sights as the instrument expanding and contracting, and playing sounds whilst the hand touching the instrument was "quite still".[33] A particularly striking development was when Home let go of the instrument and it carried on playing, suspended in the air.

More striking still, however, were the results of the balance experiment. With Huggins and Crookes observing the wooden board, Home rested his fingers near the fulcrum. The spring balance was seen to register successive increases and decreases, and at one stage the weight of the board reached six pounds more than its normal weight of three pounds. To counter objections that Home could have caused this startling result by secretly pressing down on the end of the board, Crookes jumped up and down on it and observed that this increased the weight by only two pounds. In addition to this crude but clear calibration exercise, Crookes emphasised that Home was only permitted to position his hands near the fulcrum so that any pressure applied there could not have produced the significant weight increases.

---

[32] Medhurst, *Crookes and the Spirit World*, pp. 172–5.
[33] Crookes, 'Experimental Investigation', p. 343.

The version of these seances that Crookes published in the July 1871 issue of the *QJS* was carefully designed to make a scientific case for the existence of some kind of new force exuded by Home, a force that he provisionally identified with Edward Cox's "psychic force", whose "correlation" with known forces he refused to "hazard the most vague hypothesis" about, owing to its close connection with "rare physiological and psychological conditions".[34] Evidently mindful of Tyndall's warning about mixing matters of the heart and mind in questions of phenomena bordering the supernatural, his paper was also designed to counter suspicions that he was a closet spiritualist whose judgement had been clouded by "[r]omantic and superstitious ideas".[35] This version emphasised his "critical acumen" and the frequency with which he had experimented on the "capricious" force, and included supportive letters from Huggins and Cox, and also omitted potentially embarrassing references to exchanges with disembodied spirits that occurred in these test seances.[36]

Crookes undoubtedly used the same rhetorical strategies in papers on the psychic force that he sent to the Royal Society in June 1871, but neither these nor the *QJS* versions resulted in strongly favourable responses. Reactions to the *QJS* paper in spiritualistic and general intellectual quarters were mixed, but in scientific and medical circles they were generally more negative.[37] While some scientific commentators believed that Crookes had successfully distanced himself from spiritualist theories of 'spirits' and made a case for further investigation, others saw flaws in experimental design (including those allowing vibrations to produce the movements of the board), contradictions and gaps in the observations, and a failure to repeat the tests a satisfactory number of times.[38] All of these raised suspicions that Crookes had deluded himself or been duped by a medium who was not as spotless as Crookes seemed to think.

[34] Crookes, 'Experimental Investigation', p. 347.
[35] Crookes, 'Experimental Investigation', p. 347.
[36] Crookes, 'Experimental Investigation', p. 340. An account of one such seance with 'spirits' is published in Medhurst, *Crookes and the Spirit World*, pp. 172–5.
[37] [James Burns], Editorial Comments, *Medium and Daybreak*, vol. 2 (1871), p. 231; C. W. Pearce, 'Mr. Crookes's Experiments', *Spiritualist*, vol. 1 (1869–71), p. 190; [Anon.], 'Psychic Force', *Saturday Review*, 15 July 1871, p. 83; [Anon.], 'A Scientific Testing of Mr. Home', *Spectator*, vol. 44 (1871), pp. 827–8.
[38] Other critical responses include [Anon.], 'On a New Force, Falsely So Called', *Medical Times and Gazette*, vol. 2 (1871), pp. 99–100; J. P. Earwaker, 'The New Psychic Force', *Nature*, vol. 4 (1871), pp. 278–9; P. H. Van der Weyde, 'On the Psychic Force', *Scientific American*, 23 September 1871, p. 197. More positive were George Fraser, 'The New Psychic Force', *Nature*, vol. 4 (1871), pp. 279–80; 'A Fellow of the Royal Astronomical Society', 'Psychic Force, &c', *English Mechanic and World of Science*, vol. 13 (1871), p. 539; and Cromwell F. Varley to William Crookes, 8 July 1871, published in [Anon.], 'Correspondence Between Mr. Cromwell F. Varley and Mr. William Crookes', *Spiritual Magazine*, vol. 6 (New Series) (1871), pp. 350–3.

Among the scientific criticisms that Crookes took most seriously were those from the physicists George Gabriel Stokes and Balfour Stewart. As one of the Royal Society's secretaries to whom Crookes had sent his psychic force papers, Stokes wielded much power as an "arbiter" of British physical sciences.[39] From previous experience of Stokes's refereeing, Crookes was acutely aware that the physicist could seriously help or hinder his ambition to secure scientific attention for the new research. But having read Crookes's manuscript, Stokes signalled his low opinion of the work by refusing to witness the effects in Home's presence and by suggesting that the medium could have easily caused the puzzling weight increases by careful application of pressure on the board.[40] For Stokes, whose strong Christian faith underpinned a private abhorrence of dubious mediators of potentially unclean spirits, mediums were simply not legitimate features of sound physical enquiry. The disappointment that Crookes felt in response to Stokes's resistance turned to exasperation when he discovered that the physicist's verdict had inspired rumours that his experiments had been officially rejected by the Royal Society as lacking *"scientific precision"*.[41]

Stewart's problem was with Crookes's state of mind rather than experimental design. In a leading article in *Nature*, he echoed many scientific commentators' praise for Crookes's courage in tackling "mysterious" phenomena but regarded the evidence of mesmerism, and in particular the power of an individual to induce subjective impressions in another, to be strong enough to doubt the chemist's conclusions.[42] It was more likely that Crookes had only imagined the manifestations and instrumental measurements of psychic force because of Home's subtle influence over him.

Between June and October 1871, Crookes made three major changes to his weight test apparatus that aimed to eliminate the specific problems identified by these two formidable scientific adversaries. Probably inspired by Robert Hare's apparatus of the 1850s, the first change added, near one end of the board exactly over the fulcrum, a water-filled glass vessel in which a perforated copper bowl was

---

[39] See David B. Wilson, 'Arbiters of Science: George Gabriel Stokes and Joshua King', in Kevin C. Knox and Richard Noakes (eds.), *From Newton to Hawking: A History of Cambridge University's Lucasian Professors of Mathematics* (Cambridge University Press, 2003), pp. 295–342.

[40] See Stokes's remarks in Crookes, 'Some Further Experiments', p. 478.

[41] Crookes, 'Some Further Experiments', p. 481. Crookes's disappointment with Stokes was echoed in other scientific quarters: see, for example, [Anon.], 'Psychic Force', *English Mechanic and the World of Science*, vol. 14 (1871), p. 85.

[42] Stewart, 'Mr. Crookes'. Stewart's concern about Home's mesmeric influence was shared by Barrett in 'On Some Phenomena', p. 87.

suspended.[43] Addressing Stokes's criticism that small impulses applied to the fulcrum could cause large weight changes, this change eliminated direct mechanical contact between the hands and the board. Shaking the heavy stand holding the vessel or dipping hands into the water caused no "appreciable" deflection of the spring balance.[44]

The second change responded to Stewart's criticism insofar as it aimed to eliminate the subjectivity of impressions of mechanical force. This involved mechanising the registration of the movements of the spring balance indicator by allowing the indicator to trace curves on smoked glass plates that moved relative to the indicator via a clockwork mechanism. By taking over the registration process, this mechanism also enabled Crookes and other leading investigators to focus more closely on holding and watching Home. The third major change, implemented after the results of the first two, involved replacing the board arrangement entirely with a still more delicate apparatus that Crookes had mentioned in his 1870 manifesto: a phonautograph. This was a circular parchment membrane whose movements in response to the mechanical effect of the pulsing psychic force were magnified and registered via an indicator moving across a smoked glass plate. The crucial feature of the phonautograph was that it promised to detect pulses of psychic force when the medium was standing away from the apparatus (Figure 4.1).

The seances featuring the new apparatus were organised in much the same way as the previous series, except that Crookes's wife Ellen now played a more significant role as a close observer of Home's body, and an unnamed non-professional female medium replaced Home for some of the phonautograph experiments.[45] The first experiment run during the seances involved Home placing his hands in the water vessel, which eventually resulted in a series of irregular curves on smoked glass plates, one indicating the exertion of a force of about 0.7 pounds. Similar curves arose from subsequent experiments, in which the glass vessel was removed entirely and Home asked simply to make gentle contact with the board, and then to stand three feet away from it. Far subtler traces resulted from the final experiments, in which Home and the unnamed female medium held their hands near the phonautograph. In each case, Crookes satisfied himself that he had good control of his psychical

[43] Crookes accepted that Hare's apparatus was similar to his own, rather than being the inspiration for it: Crookes, 'Some Further Experiments', p. 477. Crookes may have known about Hare's apparatus long before 1871. Hare's work was discussed in Epes Sargent's *Planchette* (1869), a spiritualist book admired by Crookes: see William Crookes to Epes Sargent, 17 April 1874, Mss.Acc.420, Boston Public Library.

[44] Crookes, 'Some Further Experiments', p. 486.

[45] Medhurst, *Crookes and the Spirit World*, pp. 194–203.

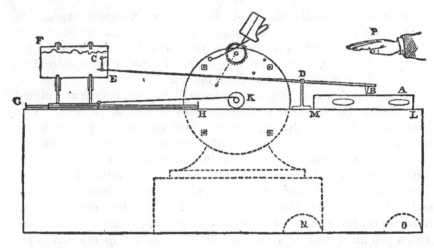

4.1 Crookes's automatic machine for registering psychic force. The force emanating from a medium's hand (P) impacts on a drum (ALM) made from thin parchment. The movement of the parchment causes the lever (CB) to pivot about D, and this causes the pointer at C to inscribe curves on a smoked glass plate (FE), which is pulled to the right by a clockwork mechanism (K). From William Crookes, 'Some Further Experiments on Psychic Force', *Quarterly Journal of Science*, 1 (2nd Series) (1871), pp. 471–93, p. 489. Reproduced by permission of Leeds University Library, Special Collections.

subjects, whether through close observation of their movements or by ensuring that their hands and feet were held by or were in contact with those of another participant. These controls underpinned Crookes's conviction that the curves could not have been produced by any known forces and confirmed "*beyond doubt*" the existence of some kind of force associated with the body that could alter the weight of material bodies.[46]

Once again, Crookes turned these experiments into papers for the Royal Society and the *QJS*, and again, these papers omitted potentially embarrassing references to the spirit-rapping that we know took place.[47] But none of these papers proved significantly more successful than the earlier ones. Some scientific commentators accepted that his *QJS* paper was scientifically more credible than the previous instalment and that it justified the need for further enquiry, but others saw no improvement.[48]

[46] Crookes, 'Some Further Experiments', p. 490.
[47] See, for example, Medhurst, *Crookes and the Spirit World*, pp. 194–203.
[48] Sympathetic responses were [Anon.], 'Further Experiments by Mr. Crookes', *Spiritualist*, vol. 1 (1869–71), p. 177 and [Anon.], 'Psychic Force', *English Mechanic and World of*

In response to the Royal Society submission, Stokes warned Crookes that Home could have caused significant mechanical forces simply by moving his hands whilst they were submerged in the water vessel, a verdict that evidently carried weight with the Society's Committee of Papers, which officially rejected Crookes's psychic force manuscripts in January 1872.[49]

By this time, Crookes's professional reputation had already suffered a huge public battering from scientific critics (and even some close friends and colleagues), who severely questioned his scientific competence per se, not just his abilities to safeguard against well-known sources of deception arising from studies of spiritualism, mesmerism and od.[50] Much of the controversy that followed the second *QJS* paper on psychic force focussed more on these general issues than on the details of his new experiments. What was painfully clear to Crookes from this debate was that no increases in the delicacy of apparatus or controls over mediums removed the fact that the prime experimental subjects remained a liability because of their reputation for deception.[51] Yet even before the controversy over psychic force peaked at the end of 1871, Crookes was exploring a way of exhibiting psychic forces independently of mediums, seances, communications with professed spirits and spiritualism altogether. This would prove to be Crookes's most telling attempt to make experimental physics matter to psychical investigation.

Towards the end of his second *QJS* paper on psychic force, Crookes made the provocative suggestion that psychic force did not require "access to known psychics" and was "probably possessed by all human beings", although this supposition could only be verified with a "more delicate apparatus" that registered fractions of a grain rather than the cruder instruments he had been using.[52] Within months, he had made significant steps towards this non-spiritualist goal when he told Huggins that "I have got an indicator so delicate that it will work without a medium" and showed a "new force, or a new form of a known force".[53] Frustratingly, Crookes's surviving correspondence and laboratory notebooks are silent on the exact

Science, vol. 14 (1871), p. 85. A particularly hostile one was P. H. Van der Weyde, 'On Mr. Crookes's Further Experiments on Psychic Force', *Journal of the Franklin Institute*, vol. 92 (1871), pp. 423–6.

[49] Crookes, 'Some Further Experiments', pp. 478–9; George G. Stokes to William Crookes, 18 January 1872, Letter Book, MS/426, p. 356, Royal Society Archives.

[50] One such close friend and colleague was the chemist John Spiller: John Spiller, 'Mr. Crookes's "Psychic Force"', *Echo*, 6 November 1871, p. 1.

[51] This was an argument in 'Fellow of the Royal Astronomical Society', 'Psychic Force, &c'.

[52] Crookes, 'Some Further Experiments', pp. 491–2. He had made a similar suggestion to Stokes earlier in 1871: Crookes, 'Some Further Experiments', p. 478.

[53] William Crookes to William Huggins, 6 November 1871, in Fournier d'Albe, *Life of William Crookes*, p. 227. Cf. William Crookes, 'Mr. Crookes' Psychic Force', *Echo*, 10 November 1871, p. 2.

nature and provenance of this "indicator", but it is likely that it arose from a significant, albeit short-lived, convergence of one of Crookes's physical researches with spiritualism.[54]

The physical research was a painstaking attempt to establish the atomic weight of thallium.[55] Having secured a Royal Society fellowship (in 1863) and much professional fame from the co-discovery of this chemical element, Crookes was now trying to further raise his reputation as an accurate investigator by focussing on one of its major physical attributes. As several historians have shown, Crookes needed to produce the atomic weight measurement in a good vacuum because minute changes in air pressure compromised the accuracy of this gravimetric work.[56]

In the midst of this work, Crookes had observed an anomalous physical effect: at pressures as low as his mercurial pressure gauge would register and where the effects of convection currents and vapour films were believed to be insignificant, the human body, sunlight and other sources of heat outside the iron vacuum chamber he had built seemed to decrease the measured weight of bodies inside it, while cold sources increased their weight. This seemed anomalous because heat and light were not believed to exert *direct* mechanical action or to interfere with the force of gravity. To achieve more consistent results, Crookes and Gimingham built a series of increasingly sensitive evacuated glass instruments which showed a surprisingly strong repulsive action of warm bodies, and a corresponding attractive action of cold bodies, on delicate material indicators that either pivoted about a central axis or were suspended from torsion fibres. It was the capacity of these instruments to respond to a force somehow associated with heat that made them so promising as detectors of psychic force. They embodied Crookes's declaration of 1870 that the delicate instruments of the physical sciences were the ultimate court of appeal in judging the mechanical effects of spiritualism.

There is no *direct* evidence to suggest that Crookes's prior quest to show psychic force without mediums inspired the particular instrumental strategies that he and Gimingham adopted after the discovery of the anomalous heat action. But it is likely that Crookes's parallel psychical and physical investigations were nurturing each other at this stage – the psychical investigations dramatising the existence of exciting new forces

[54] The earliest surviving notebook from the 1870s is from the period July 1875 to December 1877 but contains no references to od or spiritualism: William Crookes, Laboratory Notebook IV, MS / 0410, William Crookes Papers, Science Museum Group Collection.

[55] William Crookes, 'Researches on the Atomic Weight of Thallium', *Philosophical Transactions of the Royal Society of London*, vol. 163 (1873), pp. 277–330.

[56] Brock, *William Crookes*, chapter 9; Robert K. DeKosky, 'William Crookes and the Quest for the Absolute Vacuum', *Annals of Science*, vol. 40 (1983), pp. 1–18.

to be discovered, and the physical investigations suggesting new possible ways of persuading scientific audiences of their reality. The fertile effect of psychical ideas on Crookes's physical research is apparent from his correspondence with Gimingham and the Irish chemist James Emerson Reynolds. In October 1871, he urged Gimingham to explore the effect on one of his delicate instruments of "hot water[,] spirit lamp[,] magnesium wire[,] your fingers, &c.[,] as well as magnets[,] crystals &c", and about a year later was still "very anxious to ascertain if there are any causes" of the deflection of a kyanite needle suspended in a highly exhausted apparatus "which are *not* due to heat".[57]

Crookes's choice of candidate sources for the effect, and his emphasis on causes besides heat and those associated with the human body, is telling: they suggest a strong debt to Reichenbach's od, a force which the German chemist had, in the late 1860s, linked to the movement of material bodies at a distance, as well as magnetism, heat, light, vitality and crystalline action.[58] He shared his interest in Reichenbach with Reynolds, and this exchange may well have resulted in Reynolds lending Crookes a "little instrument" that showed, more decidedly than anything Crookes had then built, an unexpected repulsive force produced by the hand and other sources of heat.[59] Crookes's and Reynolds's interest in od clearly owed much to the fact that Reichenbach and his greatest English champion (William Gregory) were respected chemists, and their exchanges illustrate the seriousness with which they took Reichenbach's insistence on the inclusion of od within the "domain of physics".[60]

A telling insight into the psychical uses to which Crookes put his instruments is provided by the statistician Francis Galton, who, in spring 1872, attended seances with Home and Katherine Fox at Crookes's house. What intrigued Galton was that at one point during his encounter with Fox, Crookes brought out instruments comprising "needles" suspended in evacuated glass bulbs, and that the needles were both attracted and repelled by the finger, especially if, as Galton told his cousin Charles Darwin, the "operator" was "bright and warm and comfortable after

---

[57] William Crookes to Charles H. Gimingham, 8 October 1871 and 9 November 1872, MS/0409, William Crookes Papers Science Museum Group Collection.

[58] Reichenbach discussed the mechanical action of od on pendulum bobs in *Die odische Lohe und einige Bewegungserscheinungen als neuendeckte Formen des odischen Prinzips in der Natur* (Vienna: Wilhelm Braumüller, 1867), pp. 39–80.

[59] William Crookes, 'On Attraction and Repulsion Resulting from Radiation', *Philosophical Transactions of the Royal Society of London*, vol. 164 (1874), pp. 501–27, p. 505. Reynolds's instrument comprised a glass flask in which a deal slip was suspended from a thin fibre. See also Crookes to Gimingham, 9 November 1872.

[60] Reichenbach, *Researches on Magnetism*, p. 3.

dinner".[61] More intriguingly, Fox reputedly had a greater effect on the needles than either Crookes or Galton. The "grand discovery" that Galton claimed Crookes thought he had made was undoubtedly that there was a new force exuded by the body, but Crookes knew that to persuade the scientific world of this was going to require much more work.[62] Having suffered so much scientific hostility towards his psychic force research, he needed to regain the confidence of senior colleagues in his experimental abilities, a goal towards which his painstaking work on thallium's atomic weight would soon move him. Achieving this goal via a new instrumental display of psychic force was going to be difficult because such an apparatus needed convincingly to show the new force independently of known physical forces and of the bodies and spaces of Crookes's domestic spiritualistic enquiries.[63]

The intense work that Crookes and Gimingham devoted to the anomalous mechanical action of heat from spring 1872 eventually gave them considerable control over the effect, but no convincing evidence that it was anything other than something arising from known invisible and visible forms of radiation (notably, heat and light). By August 1873, when he submitted his first paper on the anomalous heat effect to the Royal Society, Crookes was confident that the quality of the vacuum he and Gimingham had achieved within his delicate indicators justified the argument that convection currents were not plausible causes, but he was also somewhat disappointed to report that he had not "yet been able to get distinct evidence of an independent force (not being of the nature of heat and light)".[64] What finally persuaded him that his delicate instruments were not detecting psychic, od or any other "independent" force was that with delicate instruments (notably a pith index suspended within a glass tube) exhausted by Gimingham's powerful version of the Sprengel vacuum pump, inorganic sources were as effective as living bodies in producing the mechanical effect. Tubes of warm water, for example, produced the same results as fingers.

In a second paper accepted for the Royal Society's *Philosophical Transactions*, Crookes summarised this significant conclusion in the

---

[61] Francis Galton to Charles Darwin, 28 March 1872, cited in Karl Pearson (ed.), *The Life, Letters and Labours of Francis Galton*, 3 vols. (Cambridge University Press, 1914–30), vol. 2, p. 63.

[62] Galton to Darwin, 28 March 1872.

[63] Darwin well understood this objective. He told Galton that the "question" of a new bodily power would be "settled at once" if Crookes could build an apparatus that could be sold by an instrument maker for everyone to try for themselves: Charles Darwin to Francis Galton, 21 April [1872], Frederick Burkhardt et al. (eds.), *The Correspondence of Charles Darwin, vol. 20 1872* (Cambridge University Press, 2013), pp. 168–9, p. 169.

[64] Crookes, 'Attraction and Repulsion', p. 523.

passage that Thomson criticised for illustrating the author's spiritualistic "delusions":

Many persons believe that there is a peculiar emanation or aura proceeding from the human hand and Baron von Reichenbach considered that he had proved this to be the case. Were this true it was not impossible that the emanation would affect the pith index. I have been unable, however, to detect the slightest action exerted by my or any other person's hand which I could not entirely explain by heat.[65]

This was one of the few places where Crookes publicly revealed the entanglement of his psychical and physical researches, but it was more ambiguous than many readers might have assumed. He was indeed announcing that his instruments had failed to detect od or a similar bodily force and that any effect originally ascribed to such a force was due to the anomalous mechanical action of radiation, but he was not ruling out the possibility that od existed and was detectable by other means, a position which Stone, Varley and other physical–psychical scientists endorsed.[66]

The ambiguity of Crookes's conclusion about od is supported by the fact that long after 1875 he defended his original claims about psychic force and believed in the existence of a "residuum of fact" pointing to "other kinds of force than what physicists are accustomed to deal with".[67] He did so despite the fact that he had not been able to achieve with psychic force what he would with the anomalous mechanical action of radiation: to convince senior scientific colleagues of its objective reality and to embody the effect in instruments that were not contingent on particular spaces and persons.

The series of papers on radiation that Crookes submitted to the Royal Society from 1873 onwards impressed referees of the calibre of Thomson and Maxwell, who praised the care with which Crookes had been able to isolate, replicate and control the radiation effect and approved the work for publication in the Society's prestigious *Philosophical Transactions*.[68] Thomson judged that Crookes had made a discovery of "transcendent interest and first rate importance", but Crookes's hopes of achieving something even remotely similar with psychic force had started dimming

---

[65] Crookes, 'Repulsion Resulting from Radiation', p. 526; Thomson, 'Report on Mr. Crookes's Paper'.

[66] Crookes later privately defended Reichenbach: William Crookes to Alfred R. Wallace, 24 May 1877, ff. 139–40, ARW-BL.

[67] Crookes, 'Address by Sir William Crookes', p. 30. Citation from William Crookes to Robert Bulwer-Lytton, 19 November 1891, V16/280, Bulwer-Lytton Papers, Knebworth House Archive, Hertfordshire.

[68] On Maxwell's and Thomson's responses see Brock, *William Crookes*, pp. 162–72; DeKosky, 'William Crookes and the Quest for the Absolute Vacuum'.

in 1873.[69] Although his instrumental approach to the subject had raised the hopes of Barrett, Rayleigh and others for a genuinely scientific approach to spiritualism, he could no longer rely on the most powerful and reliable source of the psychic force: Home.[70] Until mid-1873, the medium had continued giving seances for Crookes and other British-based spiritualist enquirers who were impressed with his usual manifestations of raps and levitating objects as well as rarer effects such as handling hot coals and displaying ghostly human figures.[71] However, after this time, Home was unavailable for seances owing to deteriorating health and an increasing devotion to foreign travel.[72] Most other spiritualist mediums were evidently not attractive alternatives because, unlike Home, they lacked "extraordinary powers" and a willingness to be subjected to close scrutiny, and behaved in ways likely to threaten a newly restored scientific reputation.[73]

For Crookes, the radiation researches were proving so important to his professional objectives and so time-consuming that, as well as burdensome commitments to a host of other scientific, journalistic and commercial enterprises, they took priority over the spiritualistic inquiries that partly inspired them and whose credibility hinged on them. As he explained to one prominent American spiritualist in early 1875: "I *dare* not (for my reputation's sake) omit prosecuting physical research, that is the only thing which will cause my spiritualistic publications to be looked at by men of science".[74] By the end of the year, Crookes had another good reason to think that his strategy was working, since his radiation researches had secured him the Royal Society's prestigious Royal Medal.[75]

The radiation researches, however, were far from being the unproblematic sources of scientific credibility that this honour implied. They prompted heated debate about whether radiation alone or trace gases in 'evacuated' vessels coerced by radiation caused the mechanical effects,

---

[69] William Thomson to George G. Stokes, 18 February 1874, RR/7/294, Royal Society Archives.

[70] See [Barrett], 'Science and Spiritualism'; Rayleigh, 'Presidential Address', p. 278. Crookes's instrumental tests may have inspired a simple mechanical apparatus that Rayleigh built to test the movement of tables: see Eleanor Sidgwick to Gerald Balfour, 7 February [1876], GD433/2/449/2/7–8, Papers of the Balfour Family of Whittingehame, National Records of Scotland.

[71] Medhurst, *Crookes and the Spirit World*, pp. 203–20.

[72] Home, *D. D. Home*, pp. 206–30.

[73] Crookes quoted in [Anon.], 'Discussion of Professor Lodge's Paper', p. 345.

[74] William Crookes to Robert Dale Owen, 2 January 1875, L122, Robert Dale Owen Papers, Indiana State Library. I owe this reference to Bill Brock.

[75] [Anon.], 'Anniversary Meeting', *Proceedings of the Royal Society*, vol. 24 (1874–5), pp. 70–121, pp. 91–3.

and which explanation Crookes advocated.[76] One of the most spectacular instruments that Crookes and Gimingham built for the research was the radiometer. Comprising an evacuated bulb in which black and white vanes spun around a central pivot when exposed to heat or light, the radiometer was built and marketed by instrument-makers for the wealthier and more scientifically minded consumer. It advertised Crookes's ability to bottle and export a puzzling effect well beyond his laboratory, an effect that at least one scientific observer believed was even more wonderful than the physical phenomena of spiritualism.[77] But the radiometer's puzzling behaviour exposed it to enormous scientific scrutiny and would spark further controversies over the precise cause of the effect and over Crookes's continuing and apparently delusive attachment to mysterious forces.

Tying Mediums with Electricity

In his second (1875) *Philosophical Transactions* paper on the mechanical action of radiation, Crookes explained that to eliminate the possibility that the movements of a delicate torsion apparatus were due to static electricity, he had built an apparatus suggested by Cromwell Varley but found no evidence of the supposed connection.[78] This was not the first time that year that Crookes had borrowed the electrician's suggestions for an apparatus that might limit interpretations of mysterious effects. In early March 1875, over a week before submitting the radiation paper to the Royal Society, he had published an account of his adaptation of Varley's 'electrical test' of the spiritualist medium Florence Cook.[79] Like Varley's electrical test, this seemed to yield evidence in favour of the sensational powers of another young female medium – Annie Eva Fay – whose genuineness had been severely questioned inside and outside spiritualist circles. Less ambitious than Crookes's psychic force investigations – they were designed to evidence the absence of fraud rather than the presence of a new force – the electrical tests further highlight the extent to

[76] Brock, *William Crookes*, chapters 9 and 12.

[77] After attending a Royal Society soirée in 1874 featuring Crookes's experiments, Maxwell reported to his friend Tait that the repulsive effects associated with radiation "whip spirits all to pieces". James Clerk Maxwell to Peter G. Tait, late April 1874, in Harman (ed.), *Scientific Letters and Papers of James Clerk Maxwell*, vol. 3, pp. 66–7, p. 66.

[78] Crookes, 'On Repulsion Resulting from Radiation', p. 546.

[79] William Crookes, 'A Scientific Examination of Mrs. Fay's Mediumship', *Spiritualist*, vol. 6 (1875), pp. 126–8. Harrison promoted the results of Varley's test as Cromwell F. Varley, 'Electrical Experiments with Miss Cook When Entranced', *Spiritualist*, vol. 4 (1874), pp. 134–5.

which physical–psychical scientists saw the physical sciences as a resource for tackling the problems of psychical experiment.

Both Crookes and Varley had published the results of their electrical tests in the *Spiritualist*, a leading spiritualist weekly whose editor, William H. Harrison, diverged from many spiritualists in the vigour with which he championed the use of scientific instruments in better understanding mediumship and establishing the laws underlying spirit manifestations.[80] The choice of publishing venue suggests that Crookes and Varley did not think the results of these tests were appropriate for the scientific audiences at whom Crookes had aimed his psychic force researches. I want to suggest two main reasons for this. First, the *Spiritualist* was one of the most vigorous public platforms for accounts of the principle mediums involved, and it welcomed the electrical tests as crucial interventions in its debates on these mediums' abilities. Second, and more significant, the performances of Cook and Fay were highly controversial, and Varley and Crookes were evidently not prepared to risk their professional reputations by engaging scientific audiences with studies on phenomena that even scientists sympathetic to spiritualism found hard to accept.

Born in London in 1856, Florence Cook rose to become one of the most extraordinary materialisation mediums of the late Victorian period.[81] By the early 1870s, several mediums seemed to be able to materialise 'spirit forms' that resembled human faces and hands, and were anything from solid to vaporous to touch. The faces often spoke and the hands sometimes wrote and carried objects. By December 1873, when Crookes had been attending Cook's seances for nearly 20 months, the 17-year-old girl could, when entranced, fully materialise the spirit of a young woman that she claimed controlled her.[82] The spirit called itself 'Katie King', a familiar visitor to late-nineteenth-century seances who claimed to be the daughter of Henry Owen Morgan, the seventeenth-century Welsh-born Lieutenant-Governor of Jamaica. Clothed in white robes and headdress, it breathed, had a pulse, walked, talked, joked and interacted with seance-goers. An important feature of Cook's performances was the use of a darkened cabinet. Often an improvised wooden construction or a room flanking the main seance area, the 'cabinet' was where some mediums sat during seances and was often required by them for channelling the subtle energies required to produce startling

---

[80] For example, William H. Harrison, 'The Work of a Psychological Society', *Spiritualist*, vol. 1 (1869–71), pp. 206–7.

[81] On Cook see R. G. Medhurst and K. M. Goldney, 'William Crookes and the Physical Phenomena of Mediumship', *PSPR*, vol. 54 (1964), pp. 25–157, pp. 48–89; Owen, *Darkened Room*, chapter 3.

[82] Medhurst and Goldney, 'William Crookes', pp. 63–4.

manifestations. To prevent this delicate process being upset, seance-goers were usually prohibited from entering or shining a light into the cabinet.

A measure of exactly how controversial Cook's performances were is suggested by the January 1874 number of the *QJS*, where Crookes summarised his observations of spiritualistic phenomena. There was an account of Home's "phantom" human form but no mention of 'Katie King'.[83] Fully and partially formed materialised spirits seemed too incredible and material even for sympathetic spiritualist enquirers. One attendee at Cook's seances explained that the "difficulty" he had with a materialised spirit form was that it seemed "so thoroughly material and flesh-and-blood like", while Barrett refused to pass judgement on the objective reality and provenance of such "amazing" visitors from the unseen universe.[84] In the case of 'Katie King', the problem of her materiality was compounded by the striking facial resemblance between her and Cook. For many spiritualists, this was not surprising given their belief that spirits borrowed most of their physical and psychological attributes from mediums, but for others the similarity was deeply suspicious and encouraged the method of tethering Cook to her chair or the floor of her cabinet to prevent her from masquerading as her spirit form.[85] This latter method fuelled rather than settled the burning question of Cook's relationship to 'Katie', since many were alarmed to find that after her seances the restraints had been loosened or cut, although this did not resolve the question of whether Cook had *consciously* pretended to be 'Katie'.[86]

Other participants in Cook's seances resorted to more direct methods of exposing what they believed to be fraudulence. During one widely reported seance in early December 1873, a spiritualist called William Volckman grasped 'Katie' and declared it to be the medium. A struggle ensued between Volckman and other participants which led to 'Katie' fleeing to the cabinet and the seance being prematurely terminated. Volckman's actions were not entirely self-motivated. He was an intimate friend of the medium Agnes Guppy, who, like Cook and many other women, saw spiritualist performance as a way of making a career that would provide her with an income, a conspicuous public identity, and powers over male and female visitors to her domestic sphere otherwise denied to her. A medium wielding materialisation powers herself, Guppy

---

[83] William Crookes, 'Notes of an Enquiry into Phenomena Called Spiritual', *Quarterly Journal of Science*, vol. 3 (New Series), pp. 77–97, p. 90. Crookes clearly intended his instrumental evidence for psychic force to stand independently of these notes.

[84] Charles Maurice Davies, *Mystic London; Or, Phases of Occult Life in the Metropolis* (London: Tinsley Brothers, 1875), p. 346; Barrett, 'Phenomena of Spiritualism', p. 1019.

[85] [William H. Harrison], 'Spirit Forms', *Spiritualist*, vol. 3 (1872–3), pp. 451–4.

[86] Medhurst and Goldney, 'William Crookes', pp. 51–2.

seems to have been so jealous of the performances of Cook and other rivals that she sought to ruin their careers.[87] The Volckman 'exposure' divided spiritualists. Many agreed that Volckman's actions were justified in the cause of truth, while Cook's closest supporters (including relatives and *Spiritualist* contributors) were outraged that Volckman had violated his voluntary agreement to comply with the rules of the seance and widely accepted codes of conduct of gentlemen towards women.[88]

Severely shaken by the incident, Cook later recollected that she turned to Crookes, one of her most distinguished sitters, and requested that he subject her to observations and experiments that would determine whether or not she was an "impostor".[89] For her manager, the wealthy spiritualist patron Charles Blackburn, Cook's reputation had been seriously damaged and an intervention by spiritualistic investigators of Crookes's and Varley's stature was desirable. Irrespective of the controversial nature of materialisation, Crookes and Varley clearly found Blackburn's invitation morally very appealing. Impressed by what he had already seen in Cook's seances, Crookes considered lending the "weight of his testimony" to the act of "removing an unjust suspicion" on a young woman who was willing to submit to careful tests and who, by virtue of her being "young, sensitive, and innocent", was unlikely to deceive careful observers.[90] Moreover, he and Varley clearly wanted to demonstrate that a scientific approach to this question, *pace* Volckman and his allies, did not have to violate the specific conditions under which 'Katie' appeared and those of civil society. Indeed, it was Crookes's apparently gentlemanly behaviour, his acceptance of 'Katie's' requests, and refusal to play tricks of a kind made notorious by Tyndall that eventually persuaded the medium and spirit to allow him to make more conclusive scientific tests, which involved relaxing the rules about entering the 'cabinet' and a greater degree of physical intimacy between the scientist and his subjects.[91]

The main purpose of Varley's electrical test was to establish the bodily relationship between Cook and 'Katie', a question that Crookes had already started to address with an observation of 'Katie' at the precise

[87] Medhurst and Goldney, 'William Crookes', pp. 57–61; Owen, *Darkened Room*, pp. 66–7.

[88] J. C. Luxmoore, 'The Outrage at a Spirit Circle', *Spiritualist*, vol. 3 (1872–3), p. 491; Henry Edward Thompson, 'Grasping a Spirit', *Medium and Daybreak*, vol. 4 (1873), pp. 598–9.

[89] Cook quoted in [Anon.], 'Miss Florrie Cook', *Two Worlds*, vol. 10 (1897), pp. 173–4, p. 185.

[90] William Crookes, 'The Outrage at a Spirit Circle', *Spiritualist*, vol. 4 (1874), p. 71.

[91] [Anon.], 'Miss Florrie Cook', p. 185; William Crookes, 'Spirit Forms', *Spiritualist*, vol. 4 (1874), pp. 158–9, p. 159. Crookes revealed his obedience to 'Katie's' "orders" in William Crookes to Mr. Noyes, 10 March 1874, Rayleigh Family Papers, Terling Place, Terling, Essex.

moment that he had heard what he believed to be Cook's sobbing sounds from within the cabinet.[92] The test was effectively a remote sensing device that permitted investigators to gauge what Cook was doing without violating the crucial condition of not entering or illuminating the cabinet. Staged in the London home of one of Cook's closest supporters, John Chave Luxmoore, the equipment comprised a small battery of electrical cells, wires and a "regular cable testing apparatus".[93] The latter comprised standardised electrical resistance coils and a 'reflecting' galvanometer that measured tiny currents in terms of the position of a spot of light on a graduated scale, and was the very equipment on which Varley had built his scientific reputation.[94] Indeed, in declaring that "Miss Cook took the place of a telegraph cable under electrical test", Varley wanted to extend his hard-won authority as an investigator of hidden electrical circuits from oceanic depths to mediums' cabinets.[95] He was also allying himself with an increasing number of medical practitioners, physiologists and electricians who believed that the human body was a mass of electrically resisting material whose properties could be measured with electrical instruments.[96]

Affixed to Cook's wrists were electrical contacts made from sovereigns and blotting paper moistened with an alkaline solution. Wires led from the contacts, under thick curtains dividing a room functioning as her darkened cabinet, to the gas-lit room where the cable-testing equipment and participants were situated. To render the movements of the galvanometer spot more meaningful, Varley calibrated it before the seance commenced. He noted that once connected, Cook's bodily resistance produced a deflection of 220 divisions; that the drying of the blotting paper produced a steady decline in this reading; and that shorting the circuit produced a decided deflection of 300 divisions. More significant, when Cook moved her hands the galvanometer spot moved between 15 and around 30 divisions, but when the circuit was broken the spot moved over 200 divisions. Varley now had the measurements required to establish whether Cook escaped from his electrical tether.

During the 45-minute test, Varley occupied himself with the galvanometer readings, Crookes and Luxmoore observed 'Katie', and Harrison recorded vocalised observations. Varley's suspicions were aroused at around 36 minutes when the movement of 'Katie's' arms corresponded

---

[92] Crookes, 'Outrage at a Spirit Circle'.   [93] Varley, 'Electrical Experiments', p. 134.

[94] Varley's reputation is evident from numerous electrical manuals of the period: see, for example, Latimer Clark and Robert Sabine, *Electrical Tables and Formulae* (London: E. & F. N. Spon, 1871), pp. 43–4.

[95] Varley, 'Electrical Experiments', p. 134.

[96] Morus, 'Physics and Medicine', pp. 688–9.

with a spot movement of 17 divisions, but were removed by what he considered a few "excellent" tests several minutes later: the spot did not move more than one division when 'Katie' touched Crookes's head, wrote on and threw some paper, and repeatedly opened and closed her fingers.[97] He was evidently equally impressed by the fact that the galvanometer spot had never shown any of the large movements associated with circuit breakage. Varley left it to *Spiritualist* readers to draw their own conclusions from his report, which offered proof that Cook and 'Katie' were bodily distinct and implied that Cook was a genuine medium. For some spiritualists, the test was a welcome vindication of a cherished medium, but there remained many others who criticised it for being beyond the expertise and pocket of most spiritualists, who had long held that simple tests using sight and touch were more meaningful than those with "scientific appliances".[98]

Harrison and Crookes were, predictably, among the most enthusiastic champions of Varley's test. A former telegraph operator himself, Harrison sought to reassure *Spiritualist* readers with his own supportive testimony: he saw no tell-tale wires trailing behind 'Katie' or any deflection of the galvanometer spot when she dipped her hands into a bowl of alkaline solution (which would have shorted the circuit).[99]

For Crookes, the test confirmed the conclusions he was drawing from a long series of test seances that Cook had been giving in his house since December 1873.[100] As in the Home seances, the domestic setting gave Crookes greater control over his experimental subject and her performance space. He could more effectively safeguard his investigations against potential trickery, which included searching and locking rooms, and delegating to his wife Ellen some of the responsibility for observing and holding the medium, and the obviously delicate duty of checking Cook's body for hidden props.[101] With his locked library functioning as the darkened cabinet and his adjacent laboratory as the location for seance participants, Crookes's tests also aimed to establish the bodily relationship of Cook and 'Katie' by more direct means. Owing to the trust that they had in him, Cook and 'Katie' allowed Crookes the rare privilege of entering the cabinet during the seance, to see and touch the figure in the gloom that was purportedly Cook, and to illuminate her and 'Katie' with the light of a phosphor lamp. By seeing

---

[97] Varley, 'Electrical Experiments', p. 135.

[98] Benjamin Coleman, 'Spirit Forms', *Spiritualist*, vol. 4 (1874), p. 177; citation from [James Burns], 'Electrical Tests with Miss Cook When Entranced', *Spiritual Magazine*, vol. 9 (New Series) (1874), pp. 161–8, p. 168.

[99] [William H. Harrison], 'Miss Cook's Mediumship', *Spiritualist*, vol. 4 (1874), pp. 133–4.

[100] Crookes told the American spiritualist Epes Sargent that he organised over forty such seances: see note 43.

[101] Ellen Crookes's role is documented in Medhurst and Goldney, 'William Crookes', pp. 63–70.

'Katie' only seconds after touching and seeing what he believed to be Cook, Crookes was even more confident in the medium's honesty. 'Katie' also allowed him to take a long series of photographs in the seance, which Crookes used to argue for the "absolute certainty" of the medium and the spirit having separate bodies (Figure 4.2).[102] Like Varley's, Crookes's conclusion conspicuously avoided the question of 'Katie's' alleged spiritual provenance. Many spiritualists took the verdict as further evidence for 'Katie's' reality *qua* spirit of the dead, but it did not give Crookes the proof of the survival of deceased personalities that he had been seeking for several years.[103]

The context within which Crookes publicly praised Varley's test was his adaptation of it for testing Annie Eva Fay. Five years Cook's senior, Ohio-born Fay came to London in 1874 and gave public performances of her sensational abilities to move untouched objects while tied to a chair.[104] Fay was hardly unique in her apparent telekinetic powers, but the regularity and power with which she displayed them drew the attention of Crookes, Myers, Henry Sidgwick and other scientific enquirers into spiritualism. What made her so controversial, however, was that she performed for a fee (and such mediums were generally suspected of being more tempted to defraud their audience) and behaved in other ways that put her closer to the world of popular stage magic than the domestic seance. Few would have been surprised that the leading 'anti-spiritualist' magicians of the day, John Nevil Maskelyne and George Cooke, claimed to have made an "exact re-production of all Miss Fay's tricks".[105] In the midst of private seances with her, Sidgwick told Myers that he thought the decision to advertise her as an "entertainer" rather than a "medium" was "degrading and humiliating", and he found her father–manager to have "Tradesmanlike Dignity".[106] Sidgwick's distaste ultimately prompted him to focus his spiritualistic enquires elsewhere, but Myers, Crookes, Rayleigh and others had more confidence in her character and persuaded her to give them a series of private seances.

---

[102] William Crookes, 'The Last of Katie King. The Photographing of Katie King by the Aid of the Electric Light', *Spiritualist*, vol. 4 (1874), pp. 270–1, p. 271. Crookes described rather than reproduced these photographs because Cook and 'Katie' only allowed them to be taken if they were circulated privately: Medhurst and Goldney, 'William Crookes', p. 149.

[103] William Crookes to Madame B[oydanof], 1 August 1874, in [Anon.], 'Sir William Crookes on "Invisible Intelligent Beings"', *Light*, vol. 20 (1900), p. 223. Boydanof's identity is suggested by Fournier d'Albe, *Life of Sir William Crookes*, p. 180.

[104] For Fay see Barry H. Wiley, *The Indescribable Phenomenon: The Life and Mysteries of Annie Eva Fay* (Seattle, WA: Hermetic Press, 2005).

[105] John Nevil Maskelyne, *Modern Spiritualism: A Short Account of Its Rise and Progress, with Some Exposures of So-Called Spirit Media* (London: Frederick Warne and Co., 1875), p. 121.

[106] Henry Sidgwick to Frederic W. H. Myers, n.d., Add. MS.c.100[136], Henry Sidgwick Papers, Trinity College Library, Cambridge.

4.2 The only known 'double' photograph that Crookes made to compare
the heights of Florence Cook and 'Katie King'. The figures, including
Crookes arm-in-arm with his experimental subjects, were photographed
at the curtain dividing Crookes's laboratory (in the foreground) from his
library (which functioned as Cook's darkened room). Reproduced by
permission of the Estate of Emil Prinz zu Sayn-Wittgenstein-
Berleburg, Private Collection, Gerd H. Hövelmann Marburg.

The most elaborate of these seances involved the electrical test and
these took place in Crookes's house on several evenings in February 1875
(Figure 4.3). Similar to the Cook seances, Fay sat on a chair in a locked
and darkened library whilst seance participants sat in his gas-lit laboratory,

THE MEDIUM AND DAYBREAK.

A WEEKLY JOURNAL DEVOTED TO THE HISTORY, PHENOMENA, PHILOSOPHY, AND TEACHINGS OF

SPIRITUALISM.

[REGISTERED AS A NEWSPAPER FOR TRANSMISSION IN THE UNITED KINGDOM AND ABROAD.]

No. 258.—VOL. VI.]   LONDON, MARCH 12, 1875.   [DOUBLE SHEET—PRICE 1½d.

A SCIENTIFIC SEANCE.—THE ELECTRICAL TEST FOR MEDIUMSHIP.

By what means is the investigator to determine that the phenomena which he observes are indeed spiritual; that is, produced by a power other than that furnished by the volitions of someone present? In the ordinary affairs of life, actions can be generally traced to the actors, but with spiritual manifestations of some kinds the case is very different. Some of the most important of these, as indicating a source of action independent of mediums and sitters, usually occur in darkness, when it is impossible to control the conduct of everyone present. True, hands may be held all round, or wrists may be tied together, but there are so many ways of escaping from bondage, and so many tricks indulged in by the practitioners of manual dexterity, that though the sitters may be morally

the question. The man of science is called in, and he demonstrates—by a process of investigation—that muscular force or the action of the sitters could in no wise cause the movement; and so it is rendered certain that they are due to some other agency. This important demonstration is beautifully afforded in Mr. Crookes's published "Researches," part I, which contains sixteen diagrams of the apparatus and methods used by him in his numerous experiments with Mr. Home. A more difficult task than that of proving that an object being moved in a dark room, the act is due to a so-called spiritual agency. This was the work which Mr. Crookes undertook on Thursday evening week. We were invited to witness the experiments for the benefit of the readers of this journal, and the following is the account of what took place :—

The medium selected was Mrs. Fay, and the result will prove a source of satisfaction to many who have witnessed her public seances. The genuineness of Mrs. Fay's mediumship has been

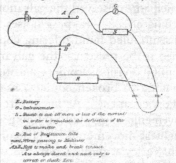

DIAGRAM OF MR. CROOKES'S LIBRARY.

DIAGRAM AND EXPLANATION OF THE ELECTRICAL APPARATUS FOR TESTING MEDIUMSHIP.

certain that all is genuine, yet the stranger who hears the story may ask—How am I to know that someone did not loose hands or play some trick which his fellows could not detect? Though these objections do not in the least invalidate the genuineness of the physical manifestations, yet they are an obstacle to their being received by all as an experimental demonstration.

Natural phenomena of many kinds are familiar to ordinary observers, which are so little understood that few can give a satisfactory definition of them. To individualise knowledge, so to speak, is the work of science. A table moves when several hands are placed lightly thereon. This movement may be due to some force other than muscular pressure, but the sitters may be divided on

widely questioned—as, indeed, has been the probity of every other medium—more particularly because she permitted herself to be advertised and exhibited in showman fashion. The phenomena occur at her seances with such pre-arranged regularity, that many cannot escape the suspicion that the experiments are a series of tricks, inscrutable to the public, but capable of imitation by experts.* Others again boast that they can permit themselves to be tied and then perform "all her tricks." At the present moment the showman who worked her seances at Hanover Square is now imitating

* Those accustomed to investigate with well-developed mediums, are favoured with an almost equal certainty and regularity of the phenomena. The objections raised against mediums are often unnecessary, and sometimes malicious.

divided from the library by a curtain. The only significant change that Crookes made to Varley's apparatus was that the medium was asked to hold onto two brass handles covered in cloths soaked in salt water and fixed to a table near the laboratory entrance. Any hand movements greater than an inch were immediately registered as suspicious movements of the galvanometer spot. Crookes also went to much more effort than did Varley to raise the scientific credibility of his experimental spaces. In addition to staging the tests in the laboratory now renowned for many scientific discoveries, he secured the assistance of several unnamed scientific participants (later identified as Galton, Huggins and Rayleigh).[107] Two such participants were also used to calibrate the apparatus: they attempted to fool the galvanometer by connecting the brass handles with a damp handkerchief, but neither could do so without causing tell-tale movements of the galvanometer spot. For Harrison, this powerfully captured Varley's and Crookes's implicit argument that the authority of the electrical test could be transferred from telegraphy to spiritualism: even if the medium had been an "accomplished electrician", they could not have fooled an instrument capable of detecting the tiny currents emerging from 2000-odd miles of Atlantic cable.[108]

During the test seances, Crookes stood near the curtain whilst other participants observed the galvanometer. Despite the occurrence of some startling effects – rapping noises, a hand-bell ringing in the library and a disembodied hand offering copies of books to the authors who happened to be present – the galvanometer spot never moved more than about 60 divisions from its starting point and neither sank to zero (corresponding to broken contact) nor flew off the scale (corresponding to a short circuit).[109] As far as Crookes was concerned, none of the spot movements suggested anything suspicious about Fay or, indeed, her "slightest movement".[110] His conclusion was not much less ambiguous than Varley's about Cook: he was more explicit in denying that the medium could have produced the effects fraudulently but was as reticent as the telegraph engineer as to the provenance of the medium's abilities.

---

[107] Galton's identity is revealed in Edward Cox, *The Mechanism of Man: An Answer to the Question of What Am I?*, 2 vols. (London: Longmans and Co., 1876–9), vol. 2, p. 447. Huggins's identity is implied by Crookes in his 'Scientific Examination', p. 127. Rayleigh's identity is revealed in Rayleigh, *Life of Lord Rayleigh*, p. 68.

[108] William H. Harrison, 'Electrical Tests Popularly Explained', *Spiritualist*, vol. 6 (1875), pp. 135–6, p. 136.

[109] The books were Edward Cox's *What am I? A Popular Introduction to Psychology and Philosophy*, Francis Galton's *The Art of Travel, or, Shifts in Contrivances Available in Wild Countries*, and William Huggins's edition of Heinrich Schellen's *Spectrum Analysis in Its Application to Terrestrial Substances and the Physical Constitution of Heavenly Bodies*.

[110] Crookes, 'Scientific Examination', p. 128.

Crookes's *Spiritualist* account of the Fay tests would be his last substantial publication on spiritualism for over a decade. He certainly did not abandon the subject, as testified by the private seances he continued to attend with Cook and other mediums and the ongoing inclusion of spiritualistic and more broadly occult subjects in his *QJS*.[111] He and Varley had persuaded some spiritualists that the authority of cable-testing equipment had been transferred to the seance and yielded powerful evidence for Cook's and Fay's genuineness, but most spiritualists maintained that simpler tests of mediumship carried at least as much authority.[112]

Much more troublesome, however, were critics who questioned the integrity of the test and suspected that mediums could still have evaded it. In 1881 the American spiritualist F. F. Cook told an audience of leading British spiritualists that the electrical tests had "no scientific value" in the light of the "seriously compromised" reputations of Cook and Fay, even though this did not undermine the truths of spiritualism.[113] Some of F. F. Cook's auditors would have been painfully aware of allegations about these mediums. In 1875, Florence Cook's honesty was called into question when Mary Rosina Showers, a medium with whom she often gave joint seances and whose honesty Crookes had already questioned, told Fay that she had used fraudulent means to exhibit a materialised spirit, while in 1880 Cook herself was found impersonating a spirit form, although many spiritualists defended her by denying that it was conscious fraud.[114] Doubts about Fay circulated from the mid-1870s onwards and caused Crookes more difficulties: Myers thought she was an "undoubted cheat" despite accepting the robustness of the electrical test, and in 1877 Crookes was forced to defend her against allegations that she was an impostor who had evaded the test.[115]

[111] For Crookes's later seances see Medhurst and Goldney, 'William Crookes', pp. 66–74. Examples of this material in the *QJS* (which changed its name to *Journal of Science* in 1879) are: [Anon.], 'Human Levitation', *Quarterly Journal of Science*, vol. 5 (New Series) (1875), pp. 31–61; [Anon.], 'Occultism Reconsidered', *Journal of Science*, vol. 4 (1882), pp. 404–9, 441–6; 'R. M. N.', 'Psychography', *Journal of Science*, vol. 7 (1885), pp. 143–9. His next major publication was Crookes, 'Notes of Séances'. However, this was a transcription of selected notes he took during Home seances in the early 1870s.

[112] Examples of positive responses are: [James Burns], 'A Scientific Séance – The Electrical Test of Mediumship', *Medium and Daybreak*, vol. 6 (1875), pp. 161–3; Epes Sargent, *Proof Palpable of Immortality* (Boston: Colby and Rich, 1876), p. 100; Alfred Russel Wallace, *Miracles and Modern Spiritualism* (London: James Burns, 1875), pp. 181–4.

[113] F. F. Cook, 'The Relations of Spiritualism to Science', *Light*, vol.1 (1881), pp. 130–1, 138–9, p. 131. See also G. H. Reddalls, 'Spirit Apparatus', *English Mechanic and World of Science*, vol. 21 (1875), p. 283 and 'Sigma' [John T. Sprague], 'Modern Spiritualism', *English Mechanic and World of Science*, vol. 19 (1874), p. 122.

[114] Brock, *William Crookes*, chapter 11; Medhurst and Goldney, 'William Crookes', pp. 80–3, 105–23.

[115] Medhurst and Goldney, 'William Crookes', pp. 93–4, 103–5.

For Crookes, the cases against Cook, Fay and Showers and associated criticisms of the electrical test were outweighed by serious concerns about his professional and moral integrity. The physical intimacy that he had with young female mediums was accepted by many spiritualists, his relatives (including his wife) and others as a necessary part of creating the harmonious psychological conditions within which these instruments of spiritualism could produce powerful manifestations. It was certainly tolerated by Eleanor Sidgwick, Rayleigh, Myers and others, who, inspired by Crookes's tests of Cook, collaborated on their own tests of materialisation mediums.[116] But for many spiritualist believers, critics and uncommitted enquirers, Crookes's behaviour and writings indicated a degree of attraction to the beauty and charm of female instruments of research that was improper for a married man and fatal for a scientific experimenter. They inspired rumours and more concrete allegations that he had been duped or, worse, deliberately concealed mediumistic trickery in exchange for sexual favours.[117] It was hardly surprising that in late 1875, when these stories posed considerable threats to his moral and scientific standing, he admitted to Home that he was sick of spiritualists' "calumny, slander, backbiting and abuse" and threatened to "cut the whole Spiritual connection".[118]

Far from cutting his connections with spiritualism, Crookes maintained his belief in Cook's and Fay's honesty despite claims to the contrary. But the threat posed by these mediums' reputations to his professional scientific ambitions prompted a new strategy for engaging with psychical topics. From the late 1870s onwards, his public statements on spiritualism, including those defending his earliest investigations, rarely mentioned the Cook and Fay tests.[119] As he told Lodge in 1909, when the younger physicist found himself accused of being driven by "immoral motives" in studying Eusapia Palladino, "hints and rumours" were the "tax" to be paid for studying female mediums and this was why he appealed to his evidence for Home rather than Cook when defending his position.[120] Ultimately, Crookes, like Lodge and other professionally ambitious physical–psychical scientists, was only prepared to pay so much

---

[116] Sidgwick, 'Results of a Personal Investigation', p. 48. For discussion of these investigations see Gauld, *Founders of Psychical Research*, pp. 107–14.

[117] Medhurst and Goldney, 'William Crookes'; Owen, *Darkened Room*, pp. 228–32.

[118] William Crookes to D. D. Home, 3 November 1875, SPR.MS 28/122, Daniel Dunglas Home Papers, Society for Psychical Research Archive, Cambridge University Library; William Crookes to D. D. Home, 24 November 1875, quoted in Home, *D. D. Home*, p. 218.

[119] See, for example, Crookes, 'Address by Sir William Crookes', p. 30; [Anon.], 'Sir William Crookes on Psychical Phenomena', p. 397.

[120] William Crookes to Oliver Lodge, 5 July 1909, SPR.MS 35/357, OJL-SPR.

of this "tax", even when their personal experiences of particular mediums had been largely positive and there remained some potential for further scientific discovery. When, as was the case with Crookes from the mid-1870s, professional and intellectual rewards could be gained from studying purely physical (but nonetheless troublesome) instruments that did not incur this tax, then it was time to switch resources away from psychical investigation.

Magnetic Sense or Nonsense?

If Oliver Lodge ever read Crookes's *Philosophical Transactions* papers on the mechanical action of radiation, he would have agreed that Reichenbach's od had been absent from these experiments but not killed off. In 1889, during his presidential address to the fledging Liverpool Physical Society, he saw it as one of the "likely-looking avenues" awaiting exploration by the emerging "workers in physics", whose field of labour the Liverpool organisation was helping to define.[121] Alongside the photography of ultra-red radiation, phosphorescence and the newly discovered Hertzian waves, the effect of magnetism on living organisms, which included Reichenbach's claims regarding physiological sensitivity to magnetism, had the potential to lead to anticipated and unanticipated discoveries. While admitting that Reichenbach's claims had not been confirmed by "subsequent observation", Lodge still regarded it as one of the more "physical" aspects of psychical investigation that were accordingly fit subjects for physical enquiry.[122]

Reichenbach's claims had been the subject of many medical and scientific investigations since the late 1840s, but this work had not been conclusive and underpinned much scientific scepticism towards his claims.[123] By the early 1880s, a key effect associated with Reichenbach's work – the sensitivity of the human body to the magnetic field – was still deemed plausible by such scientific savants as William Thomson and George F. FitzGerald. The source of the "subsequent observation" to which Lodge referred, however, was undoubtedly the SPR's 'Reichenbach Committee'. Similar to other SPR committees, this was launched in

---

[121] Oliver Lodge, 'Presidential Address to the Liverpool Physical Society', *Proceedings of the Liverpool Physical Society*, vol. 1 (1889–92), pp. 1–8, pp. 3 and 5. On the Liverpool Physical Society see Rowlands, *Oliver Lodge*.

[122] Lodge, 'Presidential Address to the Liverpool Physical Society', p. 6. Lodge, 'Address', p. 555.

[123] See discussion of Stone in Chapter 3. In 1879, the British amateur astronomer John Rand Capron announced that his visual tests of magnetic luminosity had failed: J. Rand Capron, *Aurorae: Their Characters and Spectra* (London: E. and F. N. Spon, 1879), pp. 165–6.

January 1882, but unlike its siblings, it was dominated by physical–psychical scientists – Barrett, Stone and Walter Coffin – rather than Myers, Gurney and other members of the Sidgwick group. Published in the SPR's *Proceedings*, the Reichenbach Committee's work was one of the most conspicuous *collaborative* experimental initiatives pursued by physical–psychical scientists, and its trajectory and legacy is the focus of this section.

The Reichenbach Committee's main goal was to make conclusive tests of Reichenbach's claim that some "sensitive" persons experienced visual, thermal and other sensory responses to an unknown "external cause" present in magnets, crystals, the human body and other sources.[124] Although it deemed od to be no more acceptable than mesmeric fluids and other physical hypotheses from which the SPR tended to distance itself, and acknowledged objections that Reichenbach's results were probably due to fraud, the imagination and "hysterical illusion", the Committee maintained that there was sufficient evidence for some of the effects described by the German chemist to justify a "strenuous and exhaustive attempt at their reproduction".[125]

Another source of justification was the potential of Reichenbach's work to connect psychical research to physics. One aspect of his work – the luminosity of the magnetic field – seemed to represent an "actual physical phenomenon" that could be captured independently of potentially unreliable humans whose optical sensitivity enabled them to see the light, and this had a "high scientific interest" beyond the SPR.[126] The latter scientific audiences certainly included physicists. This is suggested by the fact that the Committee's reports included FitzGerald's tentative Maxwellian explanation of magnetic luminosity and William Huggins's suggestions regarding more successful ways of photographing the effect, but also from Barrett's distillation of the Committee's work for the leading British physics journal, the *Philosophical Magazine*, and his attempt to connect it to Rayleigh's recent study of the limits of vision in poor light.[127] Despite its hopes of showing that magnetic luminosity was physically objective, if invisible to most people, the Committee maintained an interest in broader and related issues of the magnetic sensitivity of the human

---

[124] [Anon.], 'Preliminary Report of the "Reichenbach" Committee', *PSPR*, vol. 1 (1882–3), pp. 99–100, p. 99.
[125] [Anon.], 'Preliminary Report', p. 100.   [126] Barrett et al., 'First Report', p. 230.
[127] William F. Barrett, 'Note on the Alleged Luminosity of the Magnetic Field', *Philosophical Magazine*, vol. 15 (5th Series) (1883), pp. 270–5; William F. Barrett to Lord Rayleigh, 22 December 1883, Folder 2, Box #17, Series 4, MS 63, R-USAF. See also Lord Rayleigh, 'On the Invisibility of Small Objects in Bad Light', *Proceedings of the Cambridge Philosophical Society*, vol. 4 (1880–3), p. 324.

body, which inevitably raised questions about the direct effect of magnet-ism on human physiology rather than of light.

By leading the Committee, Barrett, Coffin and Stone were able to bring together a range of skills associated with the scientific study of magnetism, electricity and the human body. By the late 1870s, Barrett was convinced, from his own experiences of mesmerism and spiritualism, from Crookes's evidence of psychic force, and from other sources that living things exuded an obscure form of radiant energy, although it is not clear whether he (like Crookes) ever associated this with od.[128] Like many nineteenth-century scientists, Barrett was impressed by Reichenbach's 'sensitives', whose high social status and apparent soundness of mind justified the need for further scientific enquiries, which Barrett's friend Alfred Russel Wallace had encouraged him to initiate in 1876.[129] Barrett's interest in Reichenbach certainly dovetailed with his purely physical researches. In the 1860s, he had written much on invisible vibrations and the capacity of some individuals to perceive sounds and lights to which others were insensible, and, the following decade, studied the apparent non-effect of magnetism on the brain and popularised the researches of Faraday and others on the connections between magnetism and light.[130]

The origins of Walter Coffin's interest in Reichenbach are harder to trace, but the Committee's decision to mobilise physical instruments and measurements in the study of magnetic sensitivity was entirely consistent with the strategies he advocated for studying the human body and its relationship with known and unknown forces. Only a few months after the Committee was founded, he supported Stone's and Kilner's call for the use of physical instruments of measurement in clearing up mysteries sur-rounding electrical and magnetic therapies.[131] In 1874, he had cham-pioned Varley's electrical test of mediumship and two years later joined Varley, Harrison and Desmond Fitzgerald on the fledgling Experimental Research Committee of the leading spiritualist organisation of the day, the British National Association of Spiritualists, which staged its own (incon-clusive) instrumental tests of the anticipated weight loss of mediums during

---

[128] Barrett, 'Phenomena of Spiritualism', p. 1019.
[129] Alfred R. Wallace to William F. Barrett, 18 December 1876, in James Marchant (ed.), *Alfred Russel Wallace: Letters and Reminiscences* (London: Cassell and Co., 1916), p. 197.
[130] Barrett, 'Light and Sound'; William F. Barrett, 'A Fragment of Faraday's Electrical Discoveries', in *Science Lectures for the People. Science Lectures Delivered in Manchester. Third and Fourth Series* (Manchester: John Heywood, 1873), pp. 286–303; William F. Barrett, 'On the Points of Contact between Magnetism and Light', *Telegraphic Journal*, vol. 4 (1876), pp. 301–2, 319–20.
[131] See Coffin's remarks in [Anon.], 'Discussion on Dr. Stone's Paper', *Journal of the Society of Telegraph Engineers*, vol. 11 (1882), pp. 118–28, pp. 123–4.

materialisation seances.[132] Coffin's connections with Varley and Harrison probably acquainted him with their unsuccessful attempt of 1875 to photograph "odic flames" near magnets.[133]

Barrett may have been the Reichenbach Committee's chairman and chief reporter but Stone was clearly its main inspiration. Its experiments were effectively continuations of those that Stone had conducted over the previous eight years. In the 1870s, he tried unsuccessfully to detect a physiological response to magnetism by placing his head between the poles of a powerful electromagnet situated in a darkened room.[134] He also reported no evidence of luminosity associated with magnetism, or of a magnetic effect on fluorescent materials and highly sensitive photographic plates.[135] In the early 1880s, Stone followed up these tests in St Thomas's Hospital, whose electrical department he headed. There, his investigations of the therapeutic effect of a powerful electromagnet on a neuralgia sufferer yielded no conclusive evidence for magnetic sensitivity.[136] Stone's interest in magnetic sensitivity survived these further disappointments. He was one of many physicists who considered magnetic sensitivity eminently plausible in light of the new relationships being elucidated between the human body and electricity and, like Barrett, did not dismiss seemingly "trivial" evidence for magnetism's genuine influence on the body.[137] Further investigations of magnetic sensitivity had the potential to bring such "trivial" facts into the realms of scientific knowledge and to curb what Stone and Barrett agreed was the circulation of erroneous ideas about the physiological effect of magnetism by professional and quack medical practitioners.[138] The moral objective of the Committee's leaders was entirely consistent with the SPR's general goal of policing knowledge of obscure phenomena.

Perhaps the most telling indication of Stone's confidence in the future of Reichenbach research was that he allowed the Committee to borrow

[132] Minutes for 9 May 1876, British National Association of Spiritualists, Minute Book No. 1, College of Psychic Studies Archives; William H. Harrison, 'Weighing a Medium During the Production of Spiritual Manifestations', *Spiritualist*, vol. 12 (1878), pp. 210–16. For discussion see Richard Noakes, '"Instruments to Lay Hold of Spirits": Technologising the Bodies of Victorian Spiritualism', in Iwan Rhys Morus (ed.), *Bodies/Machines* (Oxford: Berg, 2002), pp. 125–63, esp. pp. 151–4.

[133] William H. Harrison, 'New Experiments on Odic Flames from Magnets', *Spiritualist*, vol. 7 (1875), pp. 97–8.

[134] Stone, 'On Hysteria and Hystero-Epilepsy', p. 100.

[135] Stone, 'Reichenbach's Experiments'.

[136] Stone, 'On Hysteria and Hystero-Epilepsy'.

[137] Stone, 'On Hysteria and Hystero-Epilepsy', p. 102.

[138] William F. Barret, 'Is There a "Magnetic Sense"?', *Dublin University Review*, vol. 1 (1886), pp. 23–34, p. 34; William H. Stone, 'The Physiological Bearing of Electricity on Health', *Journal of the Society of Telegraph Engineers*, vol. 13 (1884), pp. 415–32, esp. pp. 418–19.

one of his large electromagnets and to stage its experiments in his home in Dean's Yard, Westminster – the same place where, from 1882 to 1887, the SPR rented rooms for its administrative headquarters.[139] For all his enthusiasm, however, Stone was conspicuously absent from the Committee's work undoubtedly because, following a severe stroke in October 1882, he was mentally and physically incapacitated.[140] Stone's illness left him permanently weakened and this dealt the Committee a blow from which it never fully recovered.

Held in early January 1883, the Committee's experiments were similar to Stone's of 1874 in two important respects: first, they tested whether individuals could see luminosity near an electromagnet in a "perfectly darkened" room, the electromagnet being excited from a battery and commutator in an adjacent room whose entrance was covered with "darkening screens"; second, the individuals tested included trained scientific investigators, thus responding to criticisms that witnesses to Reichenbach's phenomena tended to be persons of unsound mental constitution.[141] Over 45 people were tested, including Committee and other SPR members, and each person sat in the darkened room for 2½ to 3 hours (partly to allow visual accommodation to the darkness). Of these, only three reported seeing "luminous appearances", but their performances in subsequent tests seemed equally impressive.[142] In these tests, Coffin and two other Committee members randomly and silently switched the electrical circuit and recorded anything said by the test subjects, whilst Barrett and three others sat in the darkened room. All three subjects claimed that from the upward-pointing poles of the electromagnet streamed an unsteady luminosity roughly in the shape of an inverted cone, and which could be deflected by the breath. One subject, George Albert Smith, whose mesmeric and thought-reading powers were being investigated by the SPR, particularly impressed the Committee because on 14 consecutive occasions his reports of luminosity were simultaneous with the electromagnet being excited.

The Committee was satisfied that this first round of experiments constituted a "*primâ facie* case" for magnetic luminosity "visible only to certain individuals".[143] Similar to the SPR's arguments for telepathy, its case for luminosity rested on a probabilistic argument against the effect being due to chance coincidence and on a general satisfaction with the

---

[139] [Anon.], 'Meeting of Council', *JSPR*, vol. 1 (1884), p. 33. The SPR relocated to rooms about one kilometre away in Buckingham Street in 1887: [Anon.], 'Meetings of Council', *JSPR*, vol. 3 (1887–8), pp. 149–50.

[140] William H. Stone, 'Some Effects of Brain Disturbance on the Handwriting', *Saint Thomas's Hospital Reports*, vol. 12 (1882), pp. 67–75.

[141] Barrett et al., 'First Report', pp. 230 and 232.

[142] Barrett et al., 'First Report', p. 231.   [143] Barrett, 'First Report', p. 236.

experimental conditions. The Committee emphasised that the probability against Smith rightly guessing the state of the magnet on successive occasions was several million to one. It was also satisfied that it had eliminated sources of potential trickery and experimental error. Smith could not have determined the state of the magnet from magnetic materials secreted about his person because a search of his clothes revealed no such materials, which, even had they been found, would not have greatly assisted Smith at his distance from the electromagnet. It was also concluded that Smith was unlikely to have surreptitiously used a compass because he had entered the darkened room ignorant of the nature of the experiment and had persuaded the Committee of his "good faith".[144] Barrett also insisted that Smith could not have used the faint ticking sounds accompanying magnetisation because Smith was unlikely to have had knowledge of this effect and, even if he had, he would have been seen placing his ear against the magnet to make fraudulent use of it.[145] In this sense, knowledge of magnetism was thought to be no more useful in achieving deception in this trial than electrical knowledge in the Varley–Crookes tests.

The Committee was disappointed to report that they had not been able to repeat these performances under what they believed to be "identical circumstances", and suggested that this might be due to "physiological conditions" exceeding their knowledge and control.[146] Neither had Huggins succeeded in helping them to produce photographic evidence of luminosity.[147] These results would have severely weakened the Committee's hopes of capturing a purely physical effect, but they did not stop Barrett from investigating the less purely 'physical' and more physiological question of Smith's sensitivity to magnetism. In these studies, Smith claimed that the magnetism from an electromagnet affected his eyes and temples, and used this effect to correctly identify, on 21 successive occasions, when a current to an electromagnet was made or broken. Barrett invited Smith to help him repeat this in his physics laboratory in Dublin. There, Smith and "several others" failed to experience anything due to magnetism, but on 10 out of 12 occasions Smith correctly determined when the magnetism was excited, even though Barrett was convinced he could not have known this from visual or auditory clues.[148]

[144] Barrett, 'First Report', p. 233.
[145] Barrett, 'First Report', p. 233. Barrett had researched this effect in the 1870s: William F. Barrett, 'On the Molecular Changes That Accompany the Magnetisation of Iron, Nickel and Cobalt', *Philosophical Magazine*, vol. 47 (4th Series) (1874), pp. 51–6.
[146] Barrett, 'First Report', p. 235.   [147] Barrett, 'First Report', p. 237.
[148] William F. Barrett, 'Note on the Existence of a "Magnetic Sense"', *PSPR*, vol. 2 (1883–4), pp. 56–60, p. 58.

Barrett pushed hard to gain scientific interest in the Committee's work. Apart from reports in the SPR's *Proceedings* and the *Philosophical Magazine*, he presented its findings in *Nature* and at the Royal Dublin Society.[149] He had limited success in persuading scientific audiences of the robustness of the Committee's methods and of the need for further enquiry. Some agreed that Reichenbach's claims now rested on a securer basis and warmly welcomed further results, but the editors of the *Philosophical Magazine* (who included William Thomson) seem to have expressed their doubts by strategically inserting, immediately before his paper, research by the Swiss biologist Emile Yung on the capacity of even healthy experimental subjects to experience visual, auditory and other forms of hallucination in response to subtle suggestions from experimenters.[150] *Philosophical Magazine* readers were thus being invited to question whether Barrett had eliminated the possibility that he, like Reichenbach before him, had unwittingly provided clues regarding the nature of the experiment which made his subjects only imagine what they claimed they could see.

A far weightier response came several years later in the form of new research by two American scientists – the psychologist Joseph Jastrow and the zoologist George Nuttall – working from the psycho-physical laboratory at Johns Hopkins University. In a paper eventually published by the fledgling American branch of the SPR, they expressed more widely shared misgivings about the quality of Reichenbach's original work, but also questioned whether the Reichenbach Committee's methods had really prevented experimental subjects knowing the condition of the magnet "by other means".[151] Once again, their experiments tested the physiological response of a subject near an electromagnet, with the source of electrical power (in this case, a hand-cranked Gramme dynamo) being in another room. A key difference was that the subject was asked to place their head between the electromagnet's poles and to record what they believed the magnetic state to be, whilst the distant circuit operator switched (or did not switch) the current according to a random sequence unknown to the subject. To eliminate any chance of tell-tale sounds from the dynamo being communicated along the connecting wires, mercury-filled cups

---

[149] William F. Barrett, 'On a "Magnetic Sense"', *Nature*, vol. 29 (1884), pp. 476–7; [Anon.], 'Societies and Academies', *Nature*, vol. 30 (1884), pp. 161–4, p. 163.

[150] Positive responses include [Anon], 'Reichenbach and the Psychical Research Society', *Journal of Science*, vol. 5 (1883), pp. 313–19; John T. Sprague, 'Luminosity of Magnets', *English Mechanic and World of Science*, vol. 37 (1883), p. 233. Emile Yung, 'On the Errors of Our Sensations: A Contribution to the Study of Illusion and Hallucination', *Philosophical Magazine*, vol. 15 (5th Series) (1883), pp. 259–70.

[151] Joseph Jastrow and George Nuttall, 'On the Existence of a Magnetic Sense', *Proceedings of the American Society for Psychical Research*, vol. 1 (1885–9), pp. 116–26, p. 118.

were interposed along the wires. To deal with the problem of coincidence, the experimenters devised a method in which pure guesswork would result in exactly half of the judgements being correct; significantly more than this would suggest a genuine magnetic sensitivity.

After putting themselves and eight healthy students through hundreds of tests, Jastrow and Nuttall concluded that, since the total number of incorrect judgements was close to what would have been achieved by guesswork, there was no satisfactory evidence for magnetic sensitivity. In terms of scale and robustness Jastrow and Nuttall's research was evidently intended as a significant challenge to the Committee's modest study of Smith's magnetic sensitivity. Countering the Committee's specific evidence for magnetic luminosity, Jastrow and Nuttall called on the services of one distinguished participant in their experiments: the American astronomer William H. Pickering. Having stared at the electromagnet in a dark room, Pickering failed to see any luminosity, and emphasised that as someone whose eyesight could distinguish faint stars he considered himself a "good subject".[152]

Pickering's conclusion highlighted one of Barrett's main problems: people of normal physical and mental constitution were relatively unproblematic as experimental subjects, but they failed to see magnetic luminosity or experience any physiological effect of magnetism; those with abnormal physical and mental constitutions were more problematic as experimental subjects, but they seemed to confirm Reichenbach's findings. As in the case of Crookes and psychic force, Barrett needed to demonstrate the capacity to perceive magnetic luminosity in a far wider range of people, rather than just those deemed medically abnormal. Smith certainly fell into this latter category because, as Barrett well knew, in the early 1880s he had impressed the SPR with his thought-reading abilities. But it was this very capacity that undermined his reliability as a subject for the Committee. Barrett may have denied that Smith could respond to his "unexpressed will", but there were many who surmised that his responses to magnetism might be the result of his genuine capacity to read the minds of Committee members.[153]

One person who may have shared this view was Myers, who, in his posthumous work on psychical research, upheld the SPR's evidence of Smith's telepathic powers, but drew the ambiguous conclusion that the

[152] William H. Pickering, 'A Research on the Reality of Reichenbach's Flames', *Proceedings of the American Society for Psychical Research*, vol. 1 (1885–9), p. 127.

[153] Barrett, 'Note on the Existence', p. 56; J. Brown, 'The Psychical Society's Experiments on Reichenbach's Phenomenon', *English Mechanic and World of Science*, vol. 37 (1883), p. 246; William H. Harrison, 'Recent Psychical Researches', *Medium and Daybreak*, vol. 14 (1883), pp. 310–11.

value of the Committee's experiments depended on "whether the subjects had any means, direct or indirect, of knowing when the current was made or broken".[154] Eleanor Sidgwick agreed because, in 1890, she privately doubted Barrett's reassurance that Smith could not have heard – consciously or unconsciously – the magnetic tick.[155] The enormous respect that Myers's and Sidgwick's opinion commanded in the SPR makes it easier to understand why the Committee was disbanded in 1884, long before other SPR committees were wound up.[156] For Myers, Gurney and the Sidgwicks, the other committees were simply proving more effective in delivering the kind of evidence for psychical effects that they believed the SPR needed to present in public. But the brevity of the Reichenbach Committee's life derived from the SPR's leadership ambivalence *and* from the changing circumstances of its main drivers: Stone was incapacitated; Coffin seems to have decided to devote himself more to dentistry than to psychical research; and, despite his private conviction of the existence of magnetic luminosity, Barrett's heavy professional commitments made it very difficult for him to pursue any SPR investigations, let alone produce fresh experimental evidence for the Reichenbach phenomena.[157]

Despite its limited short-term impact, the Reichenbach Committee seems to have had greater significance in the longer term. It inspired investigations made by French and Dutch psychical researchers between the 1880s and early 1900s yielding new evidence for sensitivity to magnetic luminosity.[158] A more telling impact, however, occurred in 1910 when Silvanus Thompson published a paper in the *Philosophical Transactions* presenting new evidence of the physiological effect of an alternating magnetic field.[159] The article partially achieved what one

[154] Myers, *Human Personality*, vol. 1, p. 483. In 1890, Myers had also derided the "dubious" belief of French occultists in the magnetic sense: Myers, 'Subliminal Consciousness', p. 340.

[155] Eleanor Sidgwick to William Bateson, 14 March 1890, MS Add. 8634/C.32, William Bateson Papers, Cambridge University Library.

[156] The Committee published only one full report and disappeared from the SPR's publications after 1884.

[157] In 1910, Barrett revealed that he had made "scores of additional experiments with much more elaborate apparatus & precautions" in a purpose-built dark room, but these seem to have been inconclusive: William F. Barrett to Silvanus P. Thompson, 13 April [1910], No. 24 in SPT-IC. By the early 1890s, Barrett was finding it so difficult to find time for original psychical investigations that Myers had to beg him to "start some *experimental* work!": Frederic W. H. Myers to William F. Barrett, 8 November 1891, SPR.MS 3/A4/77, WFB-SPR. Coffin's final contribution to the SPR was the donation of an electromagnet, although this does not appear to have been used: [Anon.], 'Meetings of Council', *JSPR*, vol. 3 (1887–8), pp. 65–8, p. 66.

[158] Nahm, 'Sorcerer of Coblenzl', pp. 398–401.

[159] Silvanus P. Thompson, 'A Physiological Effect of an Alternating Magnetic Field', *Philosophical Transactions of the Royal Society of London*, vol. 82B (1909–10), pp. 396–8.

early scientific commentator on the Committee's work had desired: to shift the question of magnetic sense from psychical to "physical" research.[160] Given Thompson's close friendship with Barrett and his armchair interest in the SPR's work, it was arguably more indebted to the Committee than he revealed.[161] The only prior investigation of the effect that Thompson cited was that of Varley and Lindsay, to which he referred in a biography of William Thomson that he was then finishing.[162]

For Thompson, it was the development of powerful new generators of magnetic fields that promised to yield stronger effects than those found by the Reichenbach Committee. Having observed a "faint visual effect" when experimenting with an alternating electromagnet in the mid-1890s, Thompson was evidently prompted by the Thomson biography to return to the question more systematically (Figure 4.4).[163] His experiment differed from the SPR's in several crucial respects. He employed a vastly more powerful magnetic field, he relied on his own judgement rather than those of potentially dubious subjects, and he examined the effect in the dark and daylight.[164] He concluded from a series of trials that on placing his head between the electromagnet's pole, a "faint flickering illumination, colourless, or of a bluish tint" could be seen in the dark and with the eyes closed, but that a definite flicker remained in daylight or with the eyes open.[165]

Just as Crookes shifted the discovery of the mechanical effect of radiation away from the od context that partly inspired it, so Thompson distanced his evidence of a magnetic sense from its psychical context, and this proved effective in attracting favourable attention from scientific and medical practitioners, many of whom shared with him other cases of the physiological effect of magnetism.[166] It certainly impressed fellow

---

[160] [Anon.], 'Reichenbach and the Psychical Research Society', p. 314.
[161] Barrett had used an alternating magnetic field in his Dublin experiments on Reichenbach: Barrett to Thompson, 13 April [1910].
[162] Thompson, *Life of William Thomson*, vol. 2, p. 1104. Thompson was also aware of the enormous electromagnet that Lord Lindsay had built for this purpose: Andrew Jamieson to Silvanus P. Thompson, 12 April 1902, SPT/P/I/112/28, Silvanus Philips Thompson Papers, Institution of Engineering and Technology Archives.
[163] Thompson, 'Physiological Effect', p. 397. Thompson was studying the effect at least as early as 1906 when he tried unsuccessfully to persuade Lodge to follow suit: Silvanus P. Thompson to Oliver Lodge, 27 February 1906, MS Add. 89/104, OJL-UCL. Lodge declined on the grounds that the head might not have evolved a "protective sense of pain": Oliver Lodge to Silvanus P. Thompson, 3 March 1906, MS Add. 89/104, OJL-UCL.
[164] The quoted field strength at the centre of the coil was 1000 CGS units.
[165] Thompson, 'Physiological Effect', p. 397.
[166] He received letters on the subject from the Danish physicist Kristian Birkeland, the English psychiatrist James Crichton-Browne and the Scottish cardiologist James Mackenzie Davidson: letters no. 36, 63a and 94, SPT-IC.

4.4 Silvanus Thompson's photograph of himself inside the powerful apparatus that he built to study the effect of an alternating magnetic field on vision. From James Crichton-Browne to Silvanus P. Thompson, 14 April 1910, 63a, Silvanus Thompson Papers, Imperial College London. Reproduced by permission of Archives of Imperial College London.

physicist and Thomson hagiographer Joseph Larmor, who, from a Maxwellian analysis of the optical impact of alternating magnetic fields on the electrically conducting material in the eyes, concluded that there was "nothing very surprising" about the effect.[167] Yet, as the Dutch and French investigations into magnetic sensitivity suggest, Thompson's publication did not persuade everybody that the question of the magnetic sensitivity with less powerful magnetic fields and with individuals possessing abnormal sensory powers – on which Reichenbach and the SPR had focussed – had been settled. Indeed, the Dutch investigations bolstered the final appeal that Barrett made for further Reichenbach studies the year before he died.[168]

Physical as Psychical Laboratories

The Dean's Yard rooms where the SPR's Reichenbach Committee staged its investigations were the closest the SPR got to having specific sites for the experimental investigation of the more physical aspects of psychical phenomena. The Society's sites of investigation tended to be domestic spaces bereft of the kind of physical apparatus that we have explored in this chapter, partly because its leading figures had concentrated mainly on telepathy and other more purely psychological aspects of psychical phenomena that did not require particularly elaborate apparatus. It was also because the SPR's leadership was, in the wake of many inconclusive test seances and evidence of mediumistic trickery, losing faith in the physical phenomena of spiritualism as a fruitful line of enquiry. In the absence of mediums as powerful or apparently reliable as Home, there seemed little need to build on the example of Crookes and others and establish a "psychical laboratory".[169] Yet that is exactly what Lodge called for in a paper privately circulated to SPR members in 1894.

The director of an academic teaching and research laboratory since 1881 and, more recently, the supporter of the idea of a national laboratory for pursuing the routine quantitative work of physics, Lodge's choice of topic would not have surprised some readers. More than most SPR members, Lodge understood the enormous significance that laboratories now commanded in the sciences, whether as spaces for researching

---

[167] Joseph Larmor to Silvanus P. Thompson, 5 April 1910, No. 189 in SPT-IC.

[168] William F. Barrett, 'Sur la luminosité du champ magnétique et de certaines personnes qui, d'après le baron Reichenbach, serait perçue par les sensitifs', in *L'état actuel des recherches psychique d'après le travaux du IIme Congrès International tenu à Varsovie en 1923* (Paris: Les Presses Universitaires de France, 1924), 169–73. Barrett was referring partly to the work of Floris Jansen: see Nahm, 'Sorcerer of Coblenzl', pp. 399–400.

[169] Lodge, 'Experience of Unusual Physical Phenomena', p. 357.

natural phenomena, for developing standards of measurement and testing instruments, or for inculcating the skills of experimental investigation. The critically important role that laboratories had in the production of reliable scientific knowledge was the reason why, in 1902, he judged them to be essential to making the "physical aspect" of psychical research part of "orthodox physics".[170]

Lodge's proposals for a psychical laboratory recognised the fact that the prime "instrument" of psychical research was a human being and so required a "humane and cautious treatment" of a "kind quite different" from that commonly adopted with physical and chemical apparatus.[171] Yet the non-living instruments did not have to differ radically from those in scientific laboratories, and especially in the relatively new laboratories of human physiology and experimental psychology. For example, one major requirement was a large self-registering balance which, by placing the medium on a chair at one end, could illuminate the puzzle of the mechanical reaction experienced by the medium during telekinetic phenomena. Measurements of such bodily properties as weight, temperature and pulse could be automatically recorded on electrical apparatus outside the laboratory. To bring the laboratory closer to the kind of domestic environment within which most mediums preferred to perform, most apparatus could be carefully "concealed under a superficial aspect of comfort and ordinary homeliness".[172] Cameras hidden in walls could continuously take images while the laboratory was lit by ultra-violet light – which was less harmful to spirit manifestations than visible light. And the seance table could be inconspicuously adapted to probe the question of mechanical reaction by resting it on an array of wooden blocks, the forces on which could be measured via changes in the electrical resistance of carbon paper placed beneath them.

There were two significant contexts for Lodge's paper on a psychical laboratory. The first was that it was an appendix to his report on investigations undertaken in July and August of 1894 into the most powerful physical medium since Home: Eusapia Palladino.[173] Born in Bari in 1854, Palladino had come to the attention of European scientists in the late 1880s. In test seances staged with her in 1892–4, individuals of such intellectual weight as the aristocratic physiologist Charles Richet, the

---

[170] Lodge, 'Address by the President', p. 47.
[171] Lodge, 'Experience of Unusual Physical Phenomena', p. 357.
[172] Lodge, 'Experience of Unusual Physical Phenomena', p. 359.
[173] On Palladino and the SPR see Gauld, *Founders of Psychical Research*, pp. 221–45. For other investigations into her mediumship see Christine Blondel, 'Eusapia Palladino: la méthode experimentale et la "diva des savants"', in Bensaude-Vincent and Blondel, *Savants face à l'occulte*, pp. 143–71; and Sommer, 'Psychical Research'.

psychologist Cesare Lombroso and the astronomer Giovanni Schiaparelli were strongly impressed by her abilities, when entranced, to alter her own weight, move untouched objects and create a host of other acoustical, mechanical and optical effects.

Richet's conviction that these phenomena could not be put down to trickery prompted him to invite Myers, Lodge and his wife, the Sidgwicks and others to test seances with Palladino at his Mediterranean retreat on Île Roubaud. Participants in these seances were struck by the fact that Palladino managed to produce these effects under conditions that seemed to preclude fraud: the seance room was thoroughly searched, Lodge's wife carefully examined the medium's clothes and, crucially, Palladino's hands and feet were firmly held during the sittings.

Lodge was especially impressed by Palladino's ability to wind up, play and swing a musical box suspended from the ceiling beyond her reach, and by what appeared to be a temporary vital "prolongation" from her body (a manifestation of what Richet christened 'ectoplasm') that

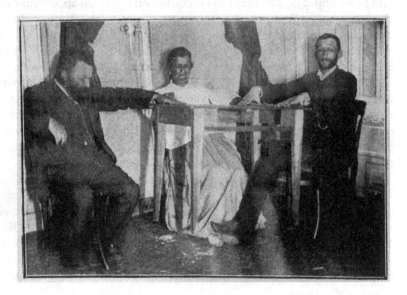

4.5 Two of Lodge's fellow psychical investigators, the Polish psychologist Julien Ochorowicz and the French medical doctor Charles Ségard, experience Eusapia Palladino's power of levitating tables at the Carqueiranne seances in 1894. Detail from Albert De Rochas, *L'extériorisation de la motricé* (Paris: Bibliothèque Chacornac, 1906), plate V, p. 191. Reproduced by permission of Universal History Archive/ Universal Images Group via Getty Images.

touched him and other participants.[174] These effects underpinned his measured verdict that, having treated her as somebody who was "liable to deceive" both consciously and unconsciously, he had repeatedly seen, heard and touched objects moving by means other than "normal" (which included the fraudulent).[175] Lodge's conclusion about the genuineness of the telekinetic effects was only shared by Myers and Richet (and not the Sidgwicks), and it was partly for this reason that he confined his report to the SPR's *Journal* rather than its public-facing *Proceedings*. Nevertheless, his claim to have made a "*primâ facie* case for investigation" was evidently shared by all investigators, since they joined up for a further series of seances with Palladino in Richet's chateau in· the Provençal town of Carquieranne and in Myers's house in Cambridge (Figure 4.5).[176]

It was in the Carquieranne seances that Lodge began exploring the effectiveness of instruments in understanding Palladino's powers, and particularly the extent to which she might be the source of the energy and seat of mechanical reaction of the telekinetic effects. This was hardly innovative, because in 1892 Richet and other scientists had convened test seances in Milan where they used large balances to measure Palladino's capacity to change her own weight and that of other bodies, although the results were inconclusive.[177] Lodge was predictably interested in Richet's use of a hand dynamometer in the Carquieranne seances, which showed that on one occasion when Palladino was under the influence of her spirit control (the ubiquitous 'John King'), she managed to exert nearly four times her "normal" strength of about fifty pounds, which was also far more than the maximum of the strongest participant (Lodge).[178] Lodge accepted that this was not significant given Richet's experiences of the abnormal strength shown by hysterics, but he certainly did not abandon the idea of instrumental ways of solving the critical question of where the force and energy of telekinetic effects originated.

The Palladino seances dramatised many of the critical investigative problems faced by students of the physical phenomena of spiritualism, and these constituted the second significant context for Lodge's paper on the psychical laboratory. By the 1890s, leading SPR figures had been

---

[174] Lodge, 'Experience of Unusual Physical Phenomena', p. 334. For Richet and ectoplasm see Robert M. Brain, 'Materialising the Medium: Ectoplasm and the Quest for Supra-Normal Biology in *Fin-de-Siècle* Science and Art', in Anthony Enns and Shelley Trower (eds.), *Vibratory Modernism* (Basingstoke: Palgrave Macmillan, 2013), pp. 115–44.

[175] Lodge, 'Experience of Unusual Physical Phenomena', pp. 325 and 334.

[176] Lodge, 'Experience of Unusual Physical Phenomena', p. 336.

[177] Hereward Carrington, *Eusapia Palladino and Her Phenomena* (London: T. Werner Laurie, 1910), pp. 29–33. The scientists also included the physicist Giovanni Ermacora and the astronomer Giovanni Schiaparelli.

[178] Lodge, 'Experience of Unusual Phenomena', p. 327.

arguing for some time that the evidence for these kinds of phenomena was severely weakened by the strong possibilities of observational error, memory lapses and mediumistic trickery. The problems had been powerfully articulated in a paper by Eleanor Sidgwick in the May 1886 number of the SPR's *Proceedings*.[179] Based partly on critical reflections on her own extensive but largely disappointing experiences of seances, the paper doubted that satisfactory safeguards against fraud had been taken in the "vast majority of recorded cases" of such phenomena, which included the means to compensate for the difficulty of achieving the "power of continuous observation".[180] Powerful empirical support for Sidgwick's warnings came later in 1886, when SPR members Richard Hodgson and S. J. Davey showed how easy it was to persuade people of the genuineness of 'slate-writing' using sleight of hand.[181] With self-registering instruments, Lodge sought to tackle the lapses in observational continuity that such trickery exploited.

Lodge's confidence in an instrumental solution owed a great deal to the widespread belief that mechanical devices were not susceptible to the subjectivities and other defects of human judgement.[182] He may have known about earlier attempts, notably by Crookes and the British National Association of Spiritualists, to create automatic registers of psychical phenomena. But the main source of inspiration was undoubtedly one that Lodge never mentioned in public but which spectacularly illustrates the convergence that many physical–psychical scientists sought between psychical research and experimental physics. This was a series of experimental seances held in April 1894 in University College Liverpool's laboratory for teaching advanced undergraduates in physics. Designed to determine whether mediums could levitate or move a table without touching it, these seances were directed by Lodge but principally the work of his private laboratory assistant Benjamin Davies.[183]

Bereft of a formal academic education, Davies received scientific training from Lodge and thereby rose from the lowly position of an assistant in the Liverpool teaching laboratory to someone to whom Lodge could easily delegate complex research and pedagogical tasks while he was away from the laboratory.[184] Davies's experimental skill proved especially

---

[179] Sidgwick, 'Results of a Personal Investigation'.
[180] Sidgwick, 'Results of a Personal Investigation', pp. 64 and 70.
[181] Richard Hodgson and S. J. Davey, 'The Possibilities of Mal-Observation and Lapse of Memory from a Practical Point of View', *PSPR*, vol. 4 (1886–7), pp. 381–495.
[182] On this belief see Lorraine Daston and Peter Galison, *Objectivity* (New York: Zone Books, 2007), chapter 3.
[183] Benjamin Davies, 'Experiments on Levitation', *Light*, vol. 36 (1916), pp. 186–7, 194–5 and 202–3, p. 186.
[184] Clow, 'Laboratory of Victorian Culture', chapter 4; Roberts, 'Training of an Industrial Physicist', chapter 2.

invaluable in the early–mid-1890s when he contributed significantly to Lodge's painstaking attempts to determine whether the ether of space was dragged by matter moving rapidly past it.[185] Staged in a room not far from where Davies would hold the experimental seances, the ether-drag experiments bordered the levitation trials in more than an architectural sense. As we saw in Chapter 3, by the early 1890s Lodge was entertaining the possibility that the apparent power of mediums to move untouched objects might involve some kind of energy transfer through the ether but that this might require the ether to have properties very different from those of gross matter. This latter possibility was strongly suggested by the ether-drag experiments, which had failed to find any viscosity between ether and matter. The results of levitation experiments would not, therefore, have been seen as irrelevant to those on the ether, even if they did not provide clues to the ether's possible constitution.

The levitation experiments were partly inspired by Davies's personal interest in psychical research which, undoubtedly stimulated by Lodge, led to private seances at the home of a Liverpool medium called Mr Duke.[186] Davies was impressed by Duke's apparent ability to repeatedly tilt a heavy table without appearing to exert any effort. Duke achieved the result when sitting at the end furthest from the tilting end and sometimes with his mediumistic daughter sitting on the tilting end. Keen to help Davies investigate the effect (and in particular to study the source of the mechanical reaction), Duke and his daughter agreed to give further seances in the Liverpool laboratory surrounded by instruments devised by Davies to continuously measure and register the forces on a table when one end tilted (Figure 4.6).

It is not clear how much Lodge was involved in the design of these experiments, but it had two features that would resurface in his ideas about a psychical laboratory. The first was an electrical "pressure apparatus" attached to one end of the seance table, and on which the mediums were asked to place their hands.[187] The apparatus comprised a wooden board resting on four springs within a trough, and between the springs and the bottom of the trough was a piece of carbonised cloth through which an electric current was passed. Pressure on the board caused changes in the electrical resistance of the cloth and, consequently,

[185] Oliver Lodge, 'Aberration Problems: A Discussion Concerning the Motion of Ether Near the Earth, and Concerning the Connection Between Ether and Gross Matter; with Some New Experiments', *Philosophical Transactions of the Royal Society of London*, vol. 184 (1893), pp. 727–804. For discussion see Hunt, 'Experimenting on the Ether'.
[186] Davies's interest by this time is evident in Benjamin Davies, 'Unusual Physical Phenomena', *Liverpool Daily Post*, 10 January 1895, p. 3.
[187] Davies, 'Experiments', p. 186.

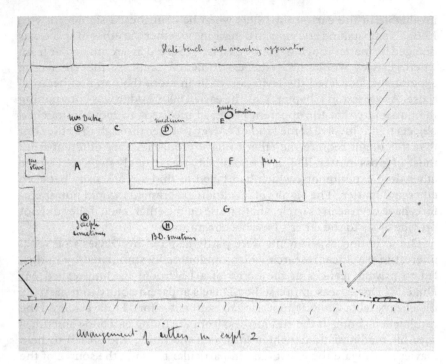

4.6 Benjamin Davies's drawing of the practical physics laboratory in University College Liverpool where he staged his experimental seances. It shows the positions of Davies ('B. D.'), the medium Mr Duke, and Duke's wife and son Joseph around the seance table; Davies's pressure table; and the slate bench where the recording apparatus was placed. From Benjamin Davies, 'SPR' notebook, item 8, Box 17, BD-NLW. Reproduced by permission of the National Library of Wales.

changes in current strength. The second feature was "in common use in all laboratories", and comprised an electromechanical device that automatically recorded the current changes as continuous curves on a sheet of moving smoked paper fed from a rotating cylinder.[188] Similar to Crookes's approach to one of his psychic force devices Davies calibrated and tested his apparatus by determining the effect on the instrument of a known mechanical force: in this case, he produced a "standard curve" representing his deliberate tilting of the table via the pressure board and to which he compared subsequent curves.[189]

[188] Davies, 'Experiments', p. 187.   [189] Davies, 'Experiments', p. 194.

Several experimental seances were held with Duke and his daughter, all of which witnessed the mediums tilting the table under the gaze of Davies, his wife and Lodge's private secretary Alfred Briscoe, who closely watched for any suspicious bodily movements. What particularly impressed Davies was that the tilts occurred during moments when, according to the curves, "no appreciable force" had been exerted on the pressure apparatus, or at least only fractions of those signified by the standard curve.[190] In the absence of convincing evidence for a direct force on the pressure apparatus, Davies concluded that the forces producing the tilts, whether they were among the known mechanical or other physical forces or some unknown force, had to act directly on the table and independently of the mediums' bodies. It was not impossible for the mediums to have acted directly on the table: just as two magnets could repel each other through non-magnetised material, so some kind of force exuded by the mediums' hands could have bypassed the pressure apparatus and pushed the table. However, Davies judged that this was "extremely unlikely" given the difficulty of achieving it without some tell-tale sign being recorded on the pressure apparatus.[191]

Like the ether-drag experiments that Lodge and Davies were pursuing nearby, the table-tilting experiments were difficult to interpret in terms of what Lodge called "simple mechanics".[192] But by mid-1894, exploring the exciting question of a more complex mechanics via psychical research seemed far less promising than via the ether. Davies's plans to repeat the table-tilting experiments and remove unspecified "doubts" that he had about them were thwarted because Duke's departure from Britain and subsequent premature death deprived him (as Home's death had deprived Crookes) of a powerful and cooperative experimental subject.[193]

An obvious question is why Lodge and Davies did not try to repeat the experiments with another medium, notably Palladino, whose powers had so impressed Lodge in the months after the Duke trials. One reason was clearly lack of time. From the spring of 1894, they directed most of their experimental efforts to the question of electromagnetic wave propagation and detection, work that resulted in the invention of a new form of detector (the 'coherer') and, eventually, new clues about the complex mechanics of the ether.[194]

---

[190] Davies, 'Experiments', p. 195.   [191] Davies, 'Experiments', p. 202.
[192] Lodge, 'Interstellar Ether', p. 862.
[193] Davies, 'Experiments', p. 202. He and Lodge believed these doubts would be partly addressed by having a medium sit in a swing suspended from the ceiling, thus reducing disturbances from vibrations.
[194] This is apparent from Lodge's laboratory notebooks from the mid-to late 1890s: Oliver Lodge, Research Notebooks, nos. 8–9, MS.3.17–18, Oliver Lodge Papers, University of Liverpool Library.

Even without these demands on their time, Palladino's behaviour would have made her a difficult choice of experimental subject. Given her stringent demands about performing in homely environments, she is unlikely to have agreed to hold seances in a physical laboratory.[195] Even had she agreed to this, by late 1895 she would have caused Lodge much professional embarrassment. The credibility of his evidence for her genuineness was virtually destroyed by the sensational discovery, in the Cambridge seances of July and August 1895, that she had consciously resorted to trickery. Privately, Lodge maintained that fraudulence could not satisfactorily explain what he and others had seen in Île Roubaud and that it was not wise to assume that she had always duped her sitters.[196] Two other physicists who participated in the Cambridge seances – Rayleigh and J. J. Thomson – privately agreed that the case against her was not decisive, but they, as well as Lodge, accepted the SPR's decision to distance itself from Palladino.[197] This was a decision reflecting the SPR's policy of dissociating itself from mediums proven to be deceptive, and whose trickery "tainted" all the evidence for their genuineness.[198] As a physicist whose ongoing goals included building his professional reputation and the scientific profile of psychical research, Lodge could not afford to present his evidence for Palladino in public, however promising she had been as a way of extending experimental physics into the seance.

Wanting Opportunities?

By the late 1890s, physical–psychical scientists had achieved only a limited success in mobilising the resources of experimental physics in studying the more physical aspects of psychical phenomena. They had persuaded themselves and some others (including some fellow scientists) that their approaches had yielded positive evidence for the genuineness of certain effects and demonstrated the need for further enquiries. The longer-term impacts of their experimental investigations were figural and literal. The most widely debated of these investigations – Crookes's

---

[195] This is discussed in Hamilton, *Immortal Longings*, pp. 213–21.

[196] See Lodge's contribution to [Anon.], 'General Meeting', *JSPR*, vol. 7 (1895–6), pp. 131–8, pp. 134–5.

[197] Rayleigh, 'Presidential Address', p. 282; Thomson, *Recollections and Reflections*, pp. 151–2. However, later investigations of Palladino in Genoa, Turin and Paris yielded more positive results and informed the SPR's decision in 1908 to stage new tests of her.

[198] Oliver Lodge, 'Address by Professor Lodge', *Light*, vol. 17 (1897), pp. 162–8, p. 166. On this policy see Henry Sidgwick, 'Eusapia Palladino', *JSPR*, vol. 7 (1895–7), pp. 230–1. This policy forced Barrett to suspend publication of his first book on spiritualism until the publication of the stronger evidence of the later Palladino tests: Barrett, *Threshold of a New World of Thought*, pp. v–ix.

experiments on psychic force – was praised as much for symbolising a definite shift towards a thoroughly scientific phase in spiritualism as directly inspiring new instrumental approaches to psychical phenomena.[199] The latter included attempts by several French psychologists in the 1890s and early 1900s to build apparatus purporting to register the mechanical effect of "magnetic" fluids exuded by the human body.[200] Lodge's partially realised ideas about a psychical laboratory also symbolised psychical researchers' ambitions for scientific credibility and may have inspired American physicist Robert W. Wood's idea of taking X-ray photographs of Palladino's ectoplasmic extrusions in dark seances of 1910.[201] Countering these champions of early attempts to merge experimental physics and psychical investigation were scientists such as Stokes and William Thomson, whose grave misgivings about 'borderland' phenomena were evidently unaffected by what Crookes, Varley and Barrett had done because in the 1890s they were still ascribing most psychical effects to mediumistic deception and self-delusion.

Many physical–psychical scientists accepted much criticism of their experimental strategies and in response tried to improve their methods and instruments. This criticism did not weaken their belief in the importance of studying the more physical kinds of psychical phenomena and of the potential of new tools in the physical sciences to succeed where older instruments had failed or proved inconclusive. Not surprisingly, many of their reflections on the prospects of further investigations focussed on the difficulty of finding powerful and reliable 'physical' mediums, of whom Home was exemplary. This is what Crookes meant when, in 1884, he told an ailing Home that "[m]y belief is the same as ever, but opportunities are wanting".[202] He was being somewhat disingenuous. He was indeed steadfast in his belief about psychic force, but was unable to reinforce his position more because of brightening "opportunities" outside spiritualism than dwindling ones inside. By the 1880s, he had decided that these outside opportunities – which included researches on radiant matter,

---

[199] Myers, *Human Personality*, vol. 1, p. 6; Charles Richet, *Traité de métapsychique* (Paris: Librairie Félix Alcan, 1922), pp. 34–42.

[200] For example, Hipployte Baraduc, *La force vitale: Notre corps vitale fluidique. Sa formule biométrique* (Paris: Georges Carré, 1893), pp. 65–73; [Ernest] Bonnaymé, *La force psychique. L'agent magnétique* (Paris: Librairie du Magnétisme, 1908). Crookes's approach was being emulated well into the 1920s: see Fritz Grunewald, *Physicalisch-mediumistische Untersuchungen* (Pfullingen: Johannes Baum, 1920).

[201] Hereward Carrington, *Personal Experiences in Spiritualism* (London: T. Werner Laurie, 1913), pp. 257–69; Hereward Carrington, *The American Seances with Eusapia Palladino* (New York: Garrett Publications, 1954), pp. 40–2, 191–5. Carrington suggested in the latter work that Wood was unable to fully implement his ideas, partly because Palladino disliked the scientific apparatus.

[202] Crookes quoted in Home, *D. D. Home*, p. 218.

commercial ventures in electric lighting and government-sponsored studies of London's water supply – had to be prioritised for meeting intellectual, professional, financial and moral objectives.[203]

Crookes was certainly not alone in thinking that exploiting opportunities for making new discoveries in physics was a more effective way of building a scientific reputation and therefore raising the confidence of others in statements on psychical phenomena.[204] To better understand why the experimental projects discussed in this chapter had such a limited success, we cannot treat them in isolation from the often competing and reinforcing commitments that physical–psychical scientists were juggling in their lives and careers.

***

This chapter has effectively extended the argument of the last one: psychical experimentation as well as psychical theorising preoccupied late-nineteenth-century physical scientists, and inspired creative uses of physical science, to a greater extent than we have assumed. These efforts were primarily directed towards producing evidence in favour of psychical phenomena, but not exclusively. To raise the confidence of scientific and other critics that they had the skills and resources to draw these positive conclusions, physical–psychical scientists often engaged in experimental work to debunk claims made about psychical phenomena. In the 1870s, Harrison deployed his considerable photographic skill in the exposure of fake spirit photographs, while Varley exploited the electrical resources of his private telegraphic consultancy to rubbish spiritualists' claims that the body exuded electricity and magnetism, and that these forces explained seance manifestations.[205]

More spectacularly, in 1888, Crookes seized on an opportunity to reassure scientific audiences that, irrespective of their thoughts on psychic force, he still knew how to distinguish genuine from bogus forces. The opportunity was a paper by the French chemist Jules Thore, who claimed evidence of a new force associated with the human body and which caused the rotation of "delicately suspended" ivory cylinders shielded from convection currents.[206] In a paper for the Royal Society's *Philosophical Transactions*, Crookes explained that with a "more accurate"

[203] Brock, *William Crookes*, chapters 14–15 and 17.
[204] See Oliver Lodge to Frederic W. H. Myers, 21 October 1890, SPR.MS 35/1309, OJL-SPR.
[205] [William H. Harrison], 'Real and Sham Spirit Photographs', *Spiritualist*, vol. 2 (1872), pp. 75–6; Cromwell F. Varley, 'Electricity, Magnetism and the Human Body', *Spiritualist*, vol. 1 (1869–71), pp. 137–8.
[206] William Crookes, 'On the Supposed "New Force" of M. J. Thore', *Philosophical Transactions of the Royal Society of London*, vol. 178 (1888), pp. 451–69, p. 451.

version of Thore's apparatus he had reproduced all the effects with inorganic sources of heat.[207] Crookes's conclusion, and the example he set of deploying his experimental skills in distinguishing fact and fancy, certainly impressed two SPR scientists assessing one of a plethora of instruments purporting to register a new bodily force.[208]

The need for physical–psychical scientists to be seen distinguishing fact from fancy became increasingly pressing in the 1890s, when the discoveries of X-rays and radium emanations seemed to lend plausibility to old and new claims about the existence of rays, fluids and forces associated with the mind and body. Disputes about the reality and nature of these latter entities focussed heavily on the extent to which their discoverers – including some physical scientists – had only imagined what they claimed was real. For physical–psychical scientists, these disputes raised painfully familiar questions about whether physical scientists were fit to study physical phenomena entangled with life and mind. Crookes was not alone in believing that they were fit. The apparatus that he had built to replicate Thore's results sometimes failed to show the subtle movements he claimed for it and reminded him that the difficulty of demonstrating obscure effects at will plagued experimental physics as well as psychical investigation.[209] As we shall see in Chapter 5, he joined other physical–psychical scientists in arguing that experience coping with troublesome physical instruments furnished an understanding of scientific approaches to the more difficult human instruments of psychical research.

---

[207] Crookes, 'Supposed "New Force"', p. 453. Cf. Crookes's verdict on od in 1875.

[208] Frederick J. M. Stratton and P. Phillips, 'Some Experiments with the Sthenometer', *JSPR*, vol. 12 (1905–6), pp. 335–9.

[209] In 1888, he grumbled to the Assistant Secretary of the Royal Society that he had a "most signal failure" showing the movements at a crowded Royal Society soirée because of the heat emitted by attendees: William Crookes to [Herbert] Rix, 10 April 1888, WMS Autograph Letters Series, Box 11, Wellcome Collection.

5   Expertise in Physics and Psychics

In late 1884, Edmund Gurney temporarily stepped back from his laborious SPR investigations into hypnotism, mesmerism and apparitions to write an incisive analysis of the problem of expertise in psychical research. Published in that leading forum of Victorian intellectual debate, the *Fortnightly Review*, Gurney's article opened by suggesting that the fledgling field of study seemed to be unique in one "unfortunate" way: its "surprising facts" had neither interested nor constituted much of a "direct opportunity" to scientists.[1] The first people to be convinced by the "facts" had not been scientists, but those lacking "intellectual superiority" and "some special aptitude for observation or power of reasoning".[2]

The situation hardly surprised Gurney. Since so many psychical phenomena were difficult to see and control, the "method of direct experimental treatment" was hardly ever appropriate, so practitioners of the "physical sciences" were deprived of the "customary method of vindicating their authority".[3] Yet apparitions, haunted houses and other psychical phenomena had interested and been investigated by vast numbers of scientifically untrained individuals for millennia. In a situation where the "ordinary rules of experimental procedure" were often useless, psychical research had arrived at truths by

something which is both more and less than laboratory and hospital experiences. The method is wider but less precise, more various but less technical; and the application of it demands disengagedness and common-sense rather than any special aptitude; where phenomena cannot be commanded at will (as is the case in some of the more striking departments of our research), the work of investigating them must consist, not in their origination, but in the collecting, sifting and bringing into due light and order, of experiments which Nature has from time to time given ready-made. And the due estimation of these depends, in the

[1] Edmund Gurney, 'The Nature of Evidence in Matters Extraordinary', *Fortnightly Review*, vol. 22 (1884), pp. 472–91, p. 472.
[2] Gurney, 'Nature of Evidence', p. 472.   [3] Gurney, 'Nature of Evidence', p. 477.

broadest sense, on the due estimation of testimony; on what may be called historical, as opposed to experimental, methods of enquiry.[4]

Gurney did not deny that "laboratory" and "hospital" experiences had value in a "definite corner" of psychical research.[5] In a significant concession to the well-known arguments of Carpenter and others that mesmerism and spiritualism were areas where trained scientific expertise carried the most weight, he explained that the physician could rightfully claim authority on the manifestations of hysteria and that somebody of Faraday's technical skill could devise experiments revealing something hidden to the untrained eye: unconscious muscular action causing table-turning. Yet it was not surprising that Gurney should privilege the "common sense" and "historical" over the "experimental" approaches. His ongoing work of collecting, sifting and interpreting thousands of cases of visual and auditory hallucinations was proving more fruitful in evidencing what the SPR called "spontaneous" telepathy than elaborate tests of thought-reading in showing "experimental" telepathy.[6]

The kind of expertise that Gurney and fellow SPR members judged appropriate in psychical research was problematic because of its perceived lack of scientific credibility. Gurney wanted to "protest in the strongest manner against the idea that knowledge, because it is not experimental, is essentially unscientific", and cited the scientific "spirit" shown by "recent anthropology and history" as a useful challenge to those who identified science with a "command of technical appliances" or theoretical and conceptual disputation.[7] Psychical research depended much more on "general sagacity" and "educated common-sense" than these attributes of trained scientific expertise, and on this basis had the potential to develop an independent scientific mode of arriving at truth.[8]

Many leading SPR members agreed with Gurney that expertise in psychical research could not depend solely on that arising from scientific and medical training. In 1889, for example, Henry Sidgwick remarked that the improbability of the kinds of marvel coming under the SPR's remit needed to be weighed in the "rough scales of common sense" rather than those of "exact science", and assessing the value of the testimony for such marvels required abilities in textual criticism and conjuring.[9] Twenty-five years later, the businessman and long-serving SPR secretary J. G. Piddington had similar reasons for declaring that "educated

---

[4] Gurney, 'Nature of Evidence', pp. 481–2.   [5] Gurney, 'Nature of Evidence', p. 478.
[6] The distinction between these forms of telepathy was discussed in Gurney, Myers and Podmore, *Phantasms of the Living*.
[7] Gurney, 'Nature of Evidence', p. 483.   [8] Gurney, 'Nature of Evidence', p. 483.
[9] Henry Sidgwick, 'The Canons of Evidence in Psychical Research', *PSPR*, vol. 6 (1889–90), pp. 1–6, p. 4.

people", rather than more specifically trained scientists, were the "proper court of appeal" in psychical matters.[10] As we saw in Chapter 4, in the early 1870s Sidgwick had been genuinely excited by the prospects of experimental physicists investigating the physical phenomena of spiritualism, but by the late 1880s, the prospects of this area of psychical research seemed bleak owing to a string of disappointing and apparently fraudulent performances by mediums whose forte was the production of physical effects. For the SPR's leadership, this would have given weight to Gurney's argument that some types of psychical phenomena were not fit subjects for the "direct" experimental methods of the physical sciences because they were so difficult to control. Accordingly, the value of physical scientists in the SPR stemmed more from their general intellectual lustre than the particular expertise they could bring to sites of psychical enquiry.

However, the experimental investigations analysed in Chapter 4 suggest that many physical–psychical scientists did not think that there was such a gulf between specialist scientific and psychical forms of expertise. Long after the SPR was founded, they were still defending some "definite corner" of psychical research where they believed phenomena were easier to see and control and where trained scientific expertise had proved, and would continue to prove, valuable.[11] Most critical of Gurney's position was Lodge, who, in his presidential address to the SPR in 1903, took the controversial step of insisting that the organisation could not always rely on the spontaneous kind of psychical occurrences that lent themselves to Gurney's more passive, observational, taxonomic and literary kinds of psychical expertise. As someone who had spent several decades teaching and making a professional reputation from experiment, Lodge was bound to insist that "we shall not make progress in understanding the laws of the phenomena and in disentangling their deeper meaning if we confine ourselves to observation alone. We must experiment, we must endeavour to produce and examine phenomena as it were in a laboratory and submit them to minute investigation".[12]

The difficulties that Lodge and others had experienced in subjecting the often capricious and evanescent physical phenomena of spiritualism and od to the regimes of a laboratory were clearly not good enough reasons to neglect such phenomena, which, for all their problems, had more potential to appeal to those scientific audiences who upheld experiment as an epistemically privileged form of natural enquiry. Neither had these difficulties shifted Lodge significantly from his earlier position that

[10] J. G. Piddington, 'Presidential Address', *PSPR*, vol. 34 (1924), pp. 131–52, p. 139.
[11] Gurney, 'Nature of Evidence', p. 478.   [12] Lodge, 'Presidential Address', p. 9.

while psychical research was relevant to many different scientific and medical fields of enquiry, it involved types of phenomena whose elucidation needed to be led by physicists. In a 1906 issue of the same journal that published Gurney's 1884 article, he suggested that a physicist could take a commanding role in solving problems of certain "rather elusive phenomena" by combining the functions of a lawyer, medical doctor, psychologist and biologist, whether by learning their skills or cooperating with them.[13]

Lodge's defence of the place of the physicist in psychical research was only the latest in a series of interventions by physical–psychical scientists in the long and often fierce debates about what kinds of expertise were relevant to the satisfactory unravelling of psychical puzzles. Historical and sociological studies of scientific controversy have shown that disputes about the reality of phenomena are frequently disputes about the expertise of those declaring evidence for such phenomena.[14] This entanglement was spectacularly exhibited in the controversies sparked by the experimental investigations examined in Chapter 4. The burden of this chapter is to better understand the reasons why these episodes generated such mixed responses from key audiences including spiritualists, psychologists, psychical researchers, conjurors and physicists. We shall see that different kinds of critics converged more than they might have imagined on the key issues: for often radically different reasons, they agreed that it was physical scientists' forms of expertise and use of instruments that ultimately threatened their claims to authority in psychical investigation.

In many ways, we should not be surprised by the identity of the physical–psychical scientists who were most outspoken in their defences of the legitimacy of physical expertise in psychical matters. These were the same individuals who, in the very period when they launched such defences, were especially active in helping to define and raise the profile of the physical sciences through institutional, pedagogical and literary enterprises. Barrett, Crookes, Stone and Varley were founder members (in 1874) of the Physical Society of London, whose official objective was to "promote the advancement and diffusion of the knowledge of physics" among academic physicists, electricians, school teachers and others and

---

[13] Lodge, 'Scientific Attitude to Marvels', p. 460.
[14] The classic studies include H. M. Collins, *Changing Order: Replication and Induction in Scientific Practice* (London: Sage, 1985); Cooter, *Cultural Meaning*; Martin J. S. Rudwick, *The Great Devonian Controversy: The Shaping of Scientific Knowledge Among Gentlemanly Specialists* (Chicago University Press, 1985); James A. Secord, *Controversy in Victorian Geology: The Cambrian-Silurian Dispute* (Princeton University Press, 1986); Steven Shapin and Simon Schaffer, *Leviathan and the Air Pump: Hobbes, Boyle and the Experimental Life* (Princeton University Press, 1985).

to provide a platform for research and pedagogical topics that fell outside the remit of existing scientific societies.[15]

Only three years before he took part in the foundation of the Physical Society, Varley had helped establish the Society of Telegraph Engineers (STE). The STE was founded to promote the advance of telegraphy and electrical science in general, enterprises which the founders believed were of immense industrial, economic, political and cultural significance and yet were not properly represented in existing learned societies.[16] Like other founder members, Varley rejected perceptions that the new society might be too specialist to ensure its longevity.[17] By encouraging papers from "all those sciences where electricity plays an important part", he insisted, the STE had a supreme opportunity to take electrical science from the narrow telegraphic "groove into which it seemed to be drifting" into one truly symbolising its presence in "every operation in nature".[18] By 1889, when the STE rebranded itself the Institution of Electrical Engineers, it had no doubt about its image as an organisation that had escaped from its narrow telegraphic origins. Among those electrical sciences with which the STE eventually engaged were medicine and physiology, and it is not surprising to find that contributors to debates on these subjects included such physical–psychical scientists as Coffin, Desmond Fitzgerald, Kilner and Stone, who were also interested in the question of whether electricity might operate at the subtler, psychological level of nature.[19]

For some physical–psychical scientists, the tasks of the Physical Society and STE to define the subject matter of physics and electrical science also drove what they did as teachers, popular lecturers, textbook writers and contributors to non-specialist periodicals. The definitions that they gave often depended on the specific role they were discharging. In their university lectures and textbooks, for example, Barrett, Lodge and Stewart helped promulgate the idea that the study of physics was the study of matter, motion and energy and accepted that the fledgling scientific

---

[15] Objectives quoted in Lewis, *Promoting Physics*, p. 14. See also Gooday, 'Periodical Physics'.

[16] Appleyard, *History of the Institution of Electrical Engineers*; W. J. Reader (with Rachel Lawrence, Sheila Nemet and Geoffrey Tweedale), *The Institution of Electrical Engineers, 1871–1971* (London: Institution of Electrical Engineers, 1987).

[17] See views of C. W. Siemens and W. H. Preece in [Anon], 'The Society of Telegraph Engineers', *Journal of the Society of Telegraph Engineers*, vol. 1 (1872–3), pp. 19–39.

[18] Cromwell F. Varley quoted in [Anon], 'Society of Telegraph Engineers', p. 34.

[19] See remarks by Coffin, Fitzgerald, Kilner and Stone in [Anon.], 'Discussion on Dr. Stone's Paper', *Journal of the Society of Telegraph Engineers*, vol. 11 (1882), pp. 118–28.

discipline formally sidestepped questions of life and mind.[20] But in research papers, addresses to learned societies and semi-popular writings they explicitly challenged, and encouraged the interrogation of, the disciplinary boundaries reinforced in these pedagogical contexts. Those of Lodge's undergraduate students who attended his opening address to the Liverpool Physical Society in 1889 would have been intrigued to find that in this forum, he represented physics as an enterprise that was much more open to physiology and questions of vitality than was apparent from the lectures and textbooks for his physics courses.[21]

Scourging Spiritualists and Scientists

The conceptual, theoretical and experimental interventions in psychical debates that we explored in Chapters 3 and 4 carried with them implicit arguments that expertise in physical sciences was both relevant to and authoritative in the solution of psychical puzzles. Crookes's spiritualism 'manifesto' of 1870, however, was one of several interventions in which physical–psychical scientists made these arguments *explicit*, and these sometimes involved stark comparisons with 'unscientific' spiritualists. As we shall see in this section, whether or not these arguments were implicit or explicit, they were not persuasive and prompted much critical assessment, by spiritualists and others, of the value of physical expertise in psychical matters per se.

Crookes was certainly not alone among physical–psychical scientists in expressing serious doubts about the scientific credibility of spiritualists' methods and claims. In 1895, Heaviside revealed to Lodge that he had only met two spiritualists in his life and "both talked a lot of bastard science".[22] Barrett was one of the few scientists prepared to direct this kind of concern directly to spiritualists, albeit more diplomatically. When, in 1877, he turned to the spiritualist press to help him amass genuine cases of thought-reading, he took the opportunity to ask the editors to "scourge" those readers who "habitually abuse scientific phraseology".[23]

---

[20] William F. Barrett and W. Brown, *Practical Physics: An Introductory Handbook for the Laboratory* (London: Percival and Co., 1892); Oliver Lodge, *Elementary Mechanics: Including Hydrostatics and Pneumatics* (London: W. & R. Chambers, 1879); Balfour Stewart, *Lessons in Elementary Physics* (London: Macmillan and Co., 1872).

[21] Lodge, 'Presidential Address to the Liverpool Physical Society'; *University College, Liverpool. Calendar for the Session 1886–1887* (Liverpool: Adam Holden, 1886), pp. 73–82.

[22] Oliver Heaviside to Oliver Lodge, 11 January 1895, MS Add. 89/50(ii), No. 91, OJL-UCL.

[23] William F. Barrett, 'A Letter from Professor Barrett', *Medium and Daybreak*, vol. 8 (1877), p. 209.

Insisting that their "ridiculous" appropriation of the terms electricity and magnetism fostered the "derision of scientific men for subjects that deserve patient investigation", the academic professor gladly accepted an invitation to be "didactic" before a new audience and explain the etymology and proper scientific use of the terms.[24]

For some spiritualists, however, these physics lessons were neither convincing nor welcome. For the veteran mesmerist and spiritualist Henry Atkinson, Barrett's abrasive tone was at least as objectionable as his attempt to police the use of words whose meaning was more flexible than the professor was prepared to allow.[25] Barrett's approach would have done little to persuade Atkinson that he differed significantly from older physicists in their unsympathetic attitudes towards spiritualism. Indeed, when, in the 1870s and 1880s, Barrett, Crookes and other physical–psychical scientists first engaged with spiritualists, they faced audiences for whom Faraday, Brewster and, above all, Tyndall captured many of the worst qualities exhibited by scientific investigators of spiritualism. Faraday still embodied the arrogance of the physicist who embarked upon investigations with firm ideas regarding the naturally possible and impossible.[26] He and Brewster represented the blinkered vision of a physicist who refused to accept testimony (including their own) of phenomena that challenged such ideas of the possible.[27]

Tyndall embodied all these vices and a good deal more. Spiritualists were generally more vexed by his conduct in seances and contemptuous attitude towards spiritualism than the perceived materialism of his addresses and essays, which some spiritualists believed revealed a humble acceptance of the difficulty of reducing mind to matter and force.[28] Few spiritualists were more frustrated by Tyndall's behaviour than William H. Harrison, who, as a contributor to scientific and

---

[24] Barrett, 'Letter from Professor Barrett'; William F. Barrett, 'The Words "Magnetism" and "Electricity" – Their Use and Abuse', *Human Nature*, vol. 9 (1877), pp. 430–1, p. 431. Desmond Fitzgerald wrote a similarly 'didactic' article about the conservation of energy: see Desmond G. Fitzgerald, 'The "Conservation of Energy" in Relation to Certain Views of the Theosophists', *Spiritualist*, vol. 12 (1878), pp. 249–51.

[25] Henry G. Atkinson, 'Animal Magnetism', *Human Nature*, vol. 9 (1877), p. 384.

[26] [Anon.], 'Spiritualism Viewed by the Light of Modern Science, by W. Crookes, F.R.S', *Spiritual Magazine*, vol. 5 (New Series) (1870), pp. 375–81; S[ophia] D[e] M[organ], 'Scientists and Spiritualism', *Light*, vol. 7 (1887), pp. 117–18.

[27] [Anon.], 'The Sense of Identity – Materialistic Explanations of Spiritualism', *Spiritual Magazine*, vol. 5 (New Series) (1870), pp. 429–32; [Anon.], 'Spiritualism in Accord with True Science', *Light*, vol. 5 (1885), p. 464; Robert Chambers, *Testimony: its Posture in the Scientific World* (London: William and Robert Chambers, 1859).

[28] [Anon.], 'The Psychological Society on the Fundamental Nature of Matter', *Spiritualist*, vol. 7 (1875), p. 301; J. Page Hopps, 'Professor Tyndall's Excursions into Spiritualism', *Light*, vol. 14 (1894), pp. 67–9; Epes Sargent, *The Scientific Basis of Spiritualism* (Boston: Colby and Rich, 1881), pp. 68–9.

technical periodicals, had long admired the physicist's performances as a lecturer. Tyndall's notorious essay of 1864, 'Science and the Spirits', gave Harrison plenty of reasons to doubt the capacity of *any* physicist to be a serious investigator of or authority on spiritualism. Tyndall's implicit hostility to spiritualist beliefs and breach of the critical seance condition to remain a passive observer evidently prompted him to declare in 1875 that "[s]piritualists who 'understand conditions', will of course learn more about the nature of manifestations than can possibly be done by disbelieving physicists who break conditions, and scientific discovery will be far more rapid in the hands of Spiritualistic investigators, than in the hands of outsiders".[29] And it was Tyndall's evident ignorance of a subject on which he claimed authority that inspired Harrison to complain that "physicists assume that they know all about the manifestations and have them at their command, whereas the intelligence producing them does not intend to be at their beck and call".[30]

Many spiritualists joined Atkinson in questioning whether younger physical–psychical scientists were much better. In their early engagements with psychical topics, Barrett, Crookes and Lodge occasionally seemed to resemble Faraday, Brewster and Tyndall in arrogantly claiming authority on a subject about which they seemed to possess little knowledge and in not deferring to the conclusions of more experienced investigators of the past and present.[31] Eleven years after his tussle with Atkinson, Barrett found himself attacked by leading spiritualists for embodying what they perceived to be the SPR's overly critical, elitist and ultimately ill-judged approaches to mediumistic phenomena. Desmond Fitzgerald, one of a handful of physical–psychical scientists who explicitly defined himself as a spiritualist, warned Barrett in 1886 that the "principles and methods of the exact physical sciences" that he, Myers and others tried to apply to spiritualistic phenomena might have the "ear of a large section of the educated and thinking public" but were unlikely to persuade older spiritualists who boasted "incomparably greater" experience and who therefore constituted the SPR's "teachers".[32]

---

[29] Harrison quoted in [Anon.], 'The National Association of Spiritualists', *Spiritualist*, vol. 6 (1875), pp. 122–6, p. 125.

[30] Harrison quoted in [Anon.], 'Transactions of the National Association of Spiritualists', *Spiritualist*, vol. 8 (1876), pp. 174–7, p. 175. See also [William H. Harrison], 'Professor Tyndall at a Spirit Circle', *Spiritualist*, vol. 1 (1869–71), pp. 156–7.

[31] [Anon.], 'Professor Barrett on "Thought Reading"', *Medium and Daybreak*, vol. 17 (1886), p. 157; [Anon.], 'Spiritual Science. Spiritualism and the British Association', *Medium and Daybreak*, vol. 22 (1891), pp. 547–9; [James Burns], 'Spiritualism and Science', *Medium and Daybreak*, vol. 1 (1870), p. 108.

[32] Desmond G. Fitzgerald, 'Spiritualism and the Society for Psychical Research', *Light*, vol. 6 (1886), pp. 62–3, p. 63. Varley had also declared himself a spiritualist in Cromwell F. Varley to Alfred R. Wallace, 28 January 1869, ff. 47–50, Add. 46439, ARW-BL.

Crookes's spiritualist manifesto of 1870 prompted a form of spiritualist critique that was never levelled at the older physicists. His privileging of mechanical over other forms of test seemed hopelessly inadequate to address spiritualism's psychological and spiritual questions. "Could all the paraphernalia of Mr. Crookes's workshop reveal to him the presence of a spirit?", rhetorically asked a contributor to the *Medium and Daybreak*.[33] The answer was ultimately negative because the "chemist and electrician may be of great service in investigating the nature of the means used and the material phenomena developed by spirit power, but they can never ascend to the cause, which is far above their sphere of action".[34] The answer also reflected the fact that for many spiritualists, the 'higher' psychological and spiritual aspects of spiritualism were ultimately more important than the 'lower' physical and 'phenomenal' aspects, useful as the latter were in attracting newcomers and converting sceptics.[35] Those whose minds could only apprehend the phenomenal aspects or who relied too heavily on material apparatus were of limited use in the science of spiritualism.[36]

The true spiritualistic scientist was one whose chief instrument was a mind attuned to the delicate psychological conditions of seances and which could apprehend spiritualism's higher aspects. The situation was succinctly expressed in the *Medium and Daybreak*, which, as a leading forum of plebeian spiritualism, voiced the more extreme misgivings of spiritualists towards the approaches of elite scientists, doctors and intellectuals towards their subject.[37] In its critique of Crookes's manifesto, it reassured readers that the "rudiments of a 'science of Spiritualism' is dawning, but it is not on the 'scientific' horizon: it is amongst those who, from aptitude or inclination, give their attention to the subject, acquainting themselves with its facts and their modes of working, and who, to a great extent, have been able to determine laws and conditions for the regulation of the phenomena".[38] Spiritualism was far from being the only form of occultism where these aptitudes and inclinations were contrasted with instrumental ways of knowing. In 1910, William

---

[33] [Anon.], 'About Scientific Spiritualism', *Medium and Daybreak*, vol. 1 (1870), pp. 201–2, p. 201. A similar problem was identified in a leading spiritualist newspaper over forty years later: [Anon.], 'Science and Spirit', *Light*, vol. 33 (1913), p. 246.

[34] [Anon.], 'About Scientific Spiritualism', p. 201.

[35] Remarks of N. Kilburn in [Anon.], 'National Jubilee Conference of Progressive Spiritualists', *Medium and Daybreak*, vol. 3 (1872), pp. 341–4, p. 341 and D. D. Home, *Lights and Shadows of Modern Spiritualism* (London: Virtue and Co., 1877).

[36] Hardinge, 'Scientific Investigation of Spiritualism'; William H. Harrison, 'The Scientific Research Committee of the National Association of Spiritualists', *Spiritualist*, vol. 9 (1876), pp. 193–4.

[37] On the *Medium and Daybreak* see Barrow, *Independent Spirits*, chapter 5.

[38] [Anon.], 'About Scientific Spiritualism', p. 201.

Kingsland spoke for most modern Theosophists when he explained that the methods of "Occult Science" were "diametrically opposed to those of Modern Science, for it does not rely upon mechanical contrivances, but works by the development within the individual himself of higher powers and faculties, by an expansion".[39]

Leading physical–psychical scientists were able, however, to commend themselves to spiritualists by what they claimed from, and how they pursued, their psychical investigations. In general, they were praised rather than criticised for the obvious reason that their investigations seemed to support a range of spiritualist teachings, but also because they seemed to have humbly accepted that the methods of the physical sciences were not always appropriate in spiritualistic enquiries. Spiritualists would certainly have welcomed Barrett's concession of 1886 that the "very nature" of psychical phenomena might "prevent our ever obtaining the kind of evidence that physical science demands".[40] Many accepted that, for all the limitations of their training and laboratory apparatus, some of the younger generation of physical–psychical scientists behaved more appropriately in seances than had Tyndall. Even James Burns, hardly the greatest friend of scientists in seances, was moved to comment in 1875 that in his electrical tests of Annie Eva Fay, Crookes was so "considerate and gentle to everyone" that it was no longer clear that all scientists were the "pronounced enemy of spiritualistic experiments, a terror to mediums, and a source of annoyance to the experienced Spiritualist".[41]

The problem for Crookes and other scientists was that "considerate and gentle" conduct could also be seen by spiritualistic supporters and critics alike as a serious threat to scientific objectivity. The sight of Crookes walking arm-in-arm with 'Katie King' prompted one spiritualist enquirer, Charles Maurice Davies, to conclude in 1875 that the "effusive Professor" had "'gone in' for the Double" of Florence Cook with a "prejudice scarcely becoming an F.R.S."[42] The stage magician John Nevil Maskelyne helped himself to this very quote in his own swipe at Crookes's use of poetry to describe 'Katie's' beauty and his clasping of the materialised spirit in his arms, both of which demonstrated a scientist "much too far gone for 'investigation'" of any medium.[43]

Given the extent to which spiritualists privileged 'higher' psychological and spiritual ways of knowing, it is hardly surprising that followers who

---

[39] Kingsland, Physics of the Secret Doctrine, p. vi.

[40] Barrett, 'On Some Physical Phenomena', p. 39.

[41] [Burns], 'Scientific Séance', p. 163. See also James Burns, 'Professor Tyndall and the Spiritualists', Human Nature, vol. 2 (1868), pp. 454–6.

[42] Davies, Mystic London, p. 319.   [43] Maskelyne, Modern Spiritualism, p. 145.

recognised the value of instrumental tests in understanding seance manifestations were in a minority and tended to be those 'scientific' spiritualists such as Coffin, Harrison, Fitzgerald and Varley who believed that there remained physical and physiological puzzles associated with mediumship that could profitably be studied with the skills and instruments of the established sciences. Many were actively involved in some of the experimental investigations discussed in Chapter 4 and in 1876 helped found the Experimental Research Committee of the British National Association of Spiritualists (BNAS). Based primarily in rooms at the BNAS's Bloomsbury headquarters, the Committee occasionally used instruments to study phenomena accompanying the mediumistic trance.[44] The Committee certainly represented one vision of how to "push on Spiritualism as a science", as its chief publicist put it in 1876, but this was a science that was only slightly more dependent on the established sciences than other visions of scientific spiritualism.[45] Ultimately, this attempt to embrace instrumental ways of knowing was too dependent on the support of a small number of spiritualists to be sustainable in the long run. Serious disputes within the BNAS over finances, management and publicity led in 1881 to the resignation, from the BNAS, of Harrison and other members of the Experimental Research Committee, which seems to have dissolved shortly afterwards.[46]

For members of the BNAS's Experimental Research Committee who joined the SPR in the early 1880s, the new organisation may well have looked like a better venue for the kind of scientific spiritualism that they wanted to develop because it attracted vastly more professional scientists than did the BNAS.[47] As Fitzgerald's remarks illustrate, their problem was that they found the SPR's attitude towards spiritualists, and especially spiritualist practices, overly hostile. Barrett had stronger links with spiritualists than did most late-nineteenth- and early-twentieth-century scientists and publicly welcomed spiritualists as "fellow-workers in the great laboratory of nature", but even he believed they had much to learn about the "methods of research" that "*science* demands".[48] But Barrett and other physical–psychical scientists were sufficiently humble to acknowledge that they needed to learn which of these methods best suited psychical research.

---

[44] Harrison, 'Weighing a Medium'. The Committee included Coffin, Fitzgerald, Harrison, Varley and the physician George Carter Blake.

[45] Harrison, 'Scientific Research Committee', p. 193.

[46] On the BNAS's internal conflict see Geoffrey K. Nelson, *Spiritualism and Society* (London: Routledge, 1968), pp. 107–10; Oppenheim, *Other World*, pp. 54–6. For the abandonment of the Committee's experiments see [William H. Harrison], 'Experimental Research in Spiritualism', *Spiritualist*, vol. 19 (1881), p. 162.

[47] These were Fitzgerald, Coffin and the barrister Charles Carleton Massey.

[48] Barrett, 'Society for Psychical Research', p. 52.

It was in disputes with adversaries intellectually far weightier than any encountered among spiritualists that they rose to defend how much these methods owed to the physical sciences.

Although spiritualists generally privileged the expertise of experienced enquirers over trained scientists in the apprehension of spiritual truths, some recognised the great value that professional scientific methods and knowledge could have in raising spiritualism's public profile as a credible scientific enterprise. The evidence of Crookes, Wallace, Zöllner and others for psychic force, materialised spirits, slate-writing and survival may not have revealed anything that spiritualists did not know already, but it was seen to have great value in attracting audiences far beyond the circles of believers and undecided enquirers. Many spiritualists maintained that they could progress their "science" without the help of men of science, but some agreed with Charles Carleton Massey, who, in 1876, suggested to *Spiritualist* readers that the task of disseminating spiritualism was better in the hands of scientists than newspaper editors and what the *Spiritualist* called the "highly cultured section of society": "[w]e want their names", Massey urged, "and we want their brains".[49]

Late-nineteenth- and early-twentieth-century spiritualist, Theosophical and occult texts amply testify to the extent to which the names and brains of scientific men were exploited for psychical purposes. The authors of these texts had no problems deferring to the expertise of physicists when their 'physical' researches revealed phenomena that seemed to confirm spiritualistic, Theosophical and occult teachings.[50] Few scientific announcements prompted more excitement in this literature than the discovery of X-rays. As a form of invisible radiation that could penetrate solid bodies and be detected photographically, X-rays were often debated as the physical basis of clairvoyance, telepathy, spirit photographs and the mesmeric influence.[51] The discovery of radium emanations and the invention of electric and etherial theories of matter proved equally welcome to spiritualists, Theosophists and occultists in the long term because they challenged

---

[49] Charles Carleton Massey, 'Spiritualism and Men of Science', *Spiritualist*, vol. 9 (1876), pp. 21–2, p. 22; [Anon.], 'Spiritual Phenomena and Men of Science', *Spiritualist*, vol. 9 (1876), p. 1. Massey's view was shared by the American spiritualist Epes Sargent: Sargent, *Scientific Basis of Spiritualism*, *passim*.

[50] For example, [Anon.], 'The Unseen Universe. Mr Crookes FRS, on Materialism', *Light*, vol. 7 (1887), pp. 146–7; Annie Besant and Charles Leadbeater, *Occult Chemistry: Clairvoyant Observations on the Chemical Elements*, ed. by A. P. Sinnett (London: Theosophical Publishing House, 1919).

[51] [Anon.], 'The New Light', *Light*, vol. 16 (1896), pp. 102–3; [Anon.], 'Psychic Photography', *Borderland*, vol. 3 (1896), pp. 313–21; J. W. Sharpe, 'Photographing the Unseen', *Light*, vol. 21 (1901), pp. 429–30.

long-held ideas about matter that were the foundations of a materialistic world view to which occultisms had long been opposed. Thus, in 1905, the Theosophist Fio Hara delighted in the way that these developments had forced "Modern Science" to abandon the idea of the "existence of real material particles" and to accept the existence of the less material "real Ultimates of Force and Matter" that had "ever been part of the teaching of the students in the Secret Schools".[52] Similarly, for the spiritualist weekly *Light*, J. J. Thomson's 1909 remarks on the electrical theory of matter, which attributed inertial mass to electrical charge, and electrical charge to forces in an immaterial ether, showed science "pursuing a path which runs parallel with that of the Spiritualist who postulates that all force is in its ultimate nature spiritual, whether it resides latent in the ether or is manifested to the senses as matter".[53]

Tricky Instruments of Psychics

In a vitriolic pamphlet of 1872, Edward Cox, the British lawyer, psychological writer and spiritualist enquirer, prophesied that "[t]ouched by Science", the "speculations of Spiritualism" will "vanish, and the facts that lie at the bottom of it will become a solid and invaluable addition to our knowledge of the physiology and psychology of Man".[54] Of all these potential "facts", few excited Cox more than his theory of psychic force, for which Crookes's experiments with Home seemed to furnish weighty evidence. By the mid-1870s, Cox believed that this theory of a power flowing from the human body explained how objects in spiritualist seances moved without material contact and how the soul or 'psyche' interacted with the material body. The accumulation of evidence for psychic force, via observations and measurements undertaken in a "variety of conditions", was one of many ways in which Cox believed his "branch of the science of Psychology can be advanced".[55]

As Graham Richards has shown, Cox's vision for psychic force, which would soon inspire him to found the short-lived Psychological Society of Great Britain, provides an illuminating insight into the extent to which late-nineteenth-century individuals fought over the definition of the subject matter for the fledgling science of psychology.[56] Cox's psychological

---

[52] Fio Hara, 'The Advance of Science Towards Occult Teachings', *Theosophical Review*, vol. 37 (1905–6), pp. 548–54, p. 551.

[53] [Anon.], 'Wider Outlooks of Science', *Light*, vol. 29 (1909), p. 427.

[54] Edward W. Cox, *Spiritualism Answered by Science: With the Proofs of a Psychic Force* (London: Longman and Co., 2nd ed., 1872), p. 47.

[55] Cox, *Spiritualism*, p. 67.

[56] Richards, 'Edward Cox'. My discussion is indebted to Richards's analysis.

enterprise is a neglected aspect of the wider transitions in psychology during the last third of the nineteenth century from an empirical to an experimental activity, the more durable manifestations of which included physiological psychology and psycho-physics. In Cox's case, psychology was a science that needed to focus on phenomena such as psychic force, which, seemingly no less tangible than electricity and magnetism and therefore no less susceptible to observation and measurement, would enable psychology to ape the experimental protocols of established sciences.

Cox also defined his science of psychology in opposition to competing psychologies of the mid-nineteenth century. He opposed spiritualists' approach to psychology because it attributed to disembodied spirits phenomena that he believed could be more credibly ascribed to immaterial powers of the living, notably what he christened 'psychic force'. However, his psychology overlapped strongly with the spiritualist variety insofar as it was preoccupied with the phenomena of mediumship and was defined in opposition to an older 'psychology' represented by theological dogma about the soul and to the newer physiological form of psychology that threatened to reduce mind to the molecular constituents of the brain. As something acting well beyond the body, psychic force well served Cox's push for an anti-materialist but non-spiritualist psychology.

Cox's pamphlet of 1872 was principally a defence of the evidence of psychic force against one of the fiercest attacks that he, Crookes, Varley and other scientific investigators of spiritualism would ever suffer. Published in the October 1871 number of a venerable forum of intellectual debate, the *Quarterly Review*, this anonymous diatribe was soon revealed to be the work of William Benjamin Carpenter, the major British architect of the physiological approach to psychology that Cox closely associated with materialism.[57] Cox was furious that Carpenter sought to displace the psychic force theory with a related one about unconscious psychological powers that he judged to be no more scientifically credible. This was Carpenter's theory of 'unconscious cerebration' or 'ideo-motor' action, which he had been using to explain mesmerism, Reichenbach's od, table-turning, spirit-rapping and related phenomena since the 1850s. Individuals who appeared to display an ability to read the thoughts of others, to commune with spirits of the dead, to perceive magnetic luminosity, and to channel forces that turned tables and rapped messages were likely to have been victims of their unconscious mental mechanisms. These mechanisms brought into an individual's

[57] [William B. Carpenter], 'Spiritualism and its Recent Converts', *Quarterly Review*, vol. 131 (1871), pp. 301–53; Cox, *Spiritualism*, p. 77.

consciousness knowledge that they convinced themselves they had forgotten or not known (the basis upon which arguments for thought-reading, od sensitivity and spirit communion were made), compelled them to unconsciously exert mechanical forces ascribed to other agents, and to generally weaken their judgement.

The solution to these now "Epidemic Delusions" lay in educating the mind and in deferring to the authority of those who best understood mental mechanisms.[58] In representing spiritualism as an epidemic originating in well-known mental and bodily processes, Carpenter spoke for many medical practitioners, physiologists and psychologists trying to raise the social, moral and cultural value of their forms of expertise.[59]

What especially exasperated Cox was that he was one of those whose authority Carpenter tried to demolish, a strategy prompting Cox's own attempt to undermine Carpenter's arguments, based as they seemed to be on misrepresentations and slander. For Carpenter, there were two main reasons why Crookes, Varley and other recent investigators of spiritualism lacked scientific authority on the subject. First, they had shown no deference to "scientific men" such as himself and other physiological psychologists whose claims to authority rested on much greater experience in "inquiries of the like kind".[60] Second and more significantly, they were, like so many spiritualistic enquirers without a formal scientific education, the victims of an undisciplined mental apparatus. The scientific training that Crookes, Huggins, Lindsay and Varley boasted had only given them highly specialist and merely "technical" forms of scientific expertise that were useless in enquiries demanding a far wider range of scientific abilities. They exemplified the warning he had directed at the general public and to those who shaped its judgement that "a man may have acquired a high reputation as an investigator in one department of science, and yet be utterly untrustworthy in regard to another".[61] The perils of disregarding this warning were especially acute when a scientific practitioner renowned for his abilities in physical investigation turned to an inquiry that was "psychical rather than physical, and involves a knowledge of the modes in which the Mind of the observer is liable to be misled either by his own proclivities or by the arts of an intentional deceiver".[62] Had Crookes and others deferred more to the likes of Carpenter, they would have been better prepared for self-deception and Home's trickery.

[58] Carpenter, 'Spiritualism', p. 351.   [59] Shortt, 'Physicians and Psychics'.
[60] Carpenter, 'Spiritualism', p. 328.   [61] Carpenter, 'Spiritualism', p. 340.
[62] Carpenter, 'Spiritualism', p. 340.

The limited scientific education of physical scientists was woefully demonstrated in their careless approaches to the human and mechanical instruments at the heart of their spiritualistic investigations. The astronomer Lord Lindsay's evidence of Home's ability to perceive magnetic luminosity suggested someone who placed the "same reliance on Mr. Home's statements that he would in the indications of a well-constructed thermometer" instead of testing the value of Home's utterances "as the maker of a thermometer does the correctness of its graduation".[63] Crookes had seriously blundered in his experiments on psychic force because he had calibrated neither his human nor mechanical instruments, seemingly because his judgement had been impaired by an "*avowed foregone conclusion*" that certain spiritualistic phenomena were genuine: he had signally failed to measure the mechanical force that Home could exert on the wooden board before psychic force welled up in the medium, and how much the effects of psychic force could be produced by the non-psychic means of rhythmical vibration.[64]

This was neither the first nor the last time that Carpenter inveighed against physical scientists' forays into his territory. In articles, public lectures, debates at scientific societies and in private correspondence throughout the 1870s he attacked them for failing to heed the warnings of physiologists, psychologists, medical practitioners and others whose particular experiences and training and their general scientific culture earned them, as he warned Barrett privately in 1876, "the rights to be considered as experts" in spiritualism and mesmerism.[65] Although he denied ruling out the possibility of thought-reading, he was incensed that in Barrett's controversial paper at the 1876 meeting of the British Association, the physicist had effectively privileged his own positive trials of the alleged mental faculty over a far larger body of inconclusive or negative evidence by people who better understood the hidden powers of the mind and body. Furthermore, Barrett did not appear to have prevented his young female subjects from exploiting subtle yet handy "unconscious revelations" that he may have conveyed in tone, gesture or facial expression, and seems to have joined the ignominious company of Reichenbach, Gregory, Crookes and other physicists who put "as much faith in tricky girls or women, as they do in thermometers or electroscopes".[66] Carpenter's gendered contrast between unreliable

---

[63] Carpenter, 'Spiritualism', pp. 335–6.   [64] Carpenter, 'Spiritualism', p. 343.

[65] William B. Carpenter to William F. Barrett, 2 November 1876, SPR.MS 3/A4/12, WFB-SPR.

[66] William B. Carpenter quoted in [Anon.], 'The British Association at Glasgow', *Spiritualism*, vol. 9 (1876), pp. 88–94, p. 90; William B. Carpenter, 'Spiritualism', *Spectator*, vol. 49 (1876), pp. 1281–2, p. 1282.

humans and reliable scientific instruments reflected nineteenth-century medical and scientific understandings of sexual difference. By representing women's bodies and minds as less stable than those of men, the male-dominated medical and scientific professions sought to give natural explanations of and justifications for the inequalities of gender.[67]

Carpenter's preoccupation with physicists' handling of tricky humans and physical instruments took an intriguing turn in the form of a provocative intervention in the controversy over Crookes's radiometer.[68] By early 1876, and partly in response to criticism and encouragement from fellow physicists, Crookes and Gimingham had constructed a series of new instruments exploring the possibility that the residual gas in the radiometer bulb played a much more significant role in moving the vanes than Crookes had previously assumed. Their apparatus showed the presence of a 'molecular wind' accompanying the spinning vanes even at pressures so low that the residual gas resisted a powerful electrical spark and so low that, contrary to Crookes's earlier conclusion that the radiation force intensified with rarefaction, the force actually weakened. This supported Crookes's revised conclusion that the radiometer motion was due to a thermometric action between the vanes and bulb mediated by the residual gas.[69] But many with the means to make or purchase a radiometer were not convinced by Crookes's theory. Fuelling debates over the cause of the rotation were the bewildering number of factors affecting the instrument's behaviour, including the size of the bulb, the type of radiation employed and the shape of the vanes.[70]

For Carpenter, the prominence of the radiometer in scientific discourses was a perfect opportunity to renew his attack on Crookes, who, having given his scientific blessing to Annie Eva Fay and other mediums for whose trickery Carpenter claimed there was strong evidence,

---

[67] On the gender ideologies in Victorian medical and scientific studies of sexuality see Cynthia Russett, *Sexual Science: The Victorian Construction of Womanhood* (Cambridge, MA: Harvard University Press, 1989), esp. chapter 4; Ornella Moscucci, *The Science of Woman: Gynaecology and Gender in England, 1800–1929* (Cambridge University Press, 1990); Mary Poovey, *Uneven Developments: The Ideological Work of Gender in Mid-Victorian England* (Chicago University Press, 1988); Elaine Showalter, *The Female Malady: Women, Madness and English Culture, 1830–1980* (London: Virago, 1985), chapter 5.

[68] Brock, *William Crookes*, chapters 9 and 12; Robert K. DeKosky, 'William Crookes and the Fourth State of Matter', *Isis*, vol. 67 (1976), pp. 36–60.

[69] William Crookes, 'On Repulsion Resulting from Radiation. Influence of the Residual Gas', *Proceedings of the Royal Society of London*, vol. 25 (1876–7), pp. 136–40.

[70] These debates took place in leading scientific periodicals of the period during 1875–7. Typical interventions in these debates are 'Treadle', 'Radiometers', *English Mechanic and World of Science*, vol. 22 (1876), p. 558; T. N. Hutchinson, 'Radiometers and Radiometers', *Nature*, vol. 13 (1875–6), p. 324; George Johnstone Stoney, 'On Crookes's Radiometer', *Philosophical Magazine*, vol. 1 (5th Series) (1876), pp. 177–81.

threatened to fuel the spiritualistic epidemic.[71] Carpenter's strategy, published in another high-profile journal of intellectual debate, the *Nineteenth Century*, was to contrast the trajectories of Crookes's radiometer and spiritualistic investigations. Both began with the experimenter committed to mistaken ideas about new forces, namely the direct mechanical action of light and psychic force. But the experimental programmes then took very different courses. In the radiometer work, Crookes had "evinced the spirit of the true philosopher" by thoroughly testing and then abandoning his original interpretation and building an apparatus establishing the effect of residual gases.[72] This effect not only harmonised with mechanical laws held by all physicists but could be verified by "every one who could construct the apparatus".[73] In his psychic force work, however, Crookes had hardly evinced the same spirit because his strong commitment to spiritualistic ideas led him to rest the case for psychic force on dubious testimony and poorly designed, unrepeatable experiments. The true lessons of the radiometer concerned Crookes's psychology as much as the peculiar nature of radiation transfer in rarefied gases, because it indirectly showed how even an expert mind could have unscientific and scientific sides, and the corresponding need for "training and disciplining of the *whole* mind" and of "cultivating scientific habits of thought" in "*every* subject".[74]

The physiological approach to psychology that Carpenter had been spearheading, as well as its associated theories of automatic mental activity, achieved limited success in Britain, not least because experimental physiology was simply not as well established there as it was in other countries and for this reason it was more difficult to extend its methods to psychology.[75] Physiology enjoyed far greater institutional strength in Germany, and it was here that physiological psychology flourished in the hands of Wilhelm Wundt and others.[76] Wundt shared many of Carpenter's positions on

---

[71] Carpenter was delighted to use Home's *Lights and Shadows of Modern Spiritualism*, which included damning evidence against fellow mediums: William B. Carpenter, 'Psychological Curiosities of Spiritualism', *Frasers's Magazine*, vol. 16 (1877), pp. 541–64. Home's book caused Crookes embarrassment because it publicised the damning evidence against Mary Rosina Showers, the medium whom he had tested in 1874 and who had privately confessed her trickery in 1875: see Medhurst and Goldney, 'William Crookes', pp. 121–2.

[72] William B. Carpenter, 'The Radiometer and Its Lessons', *Nineteenth Century*, vol. 1 (1877), pp. 242–56, p. 254.

[73] Carpenter, 'Radiometer and Its Lessons', p. 254.

[74] Carpenter, 'Radiometer and Its Lessons', p. 256.

[75] Danziger, 'Mid-Century British Psycho-Physiology'.

[76] On Wundt see W. G. Bringmann and R. D. Tweney (eds.), *Wundt Studies* (Toronto: C. J. Hogrefe, 1980); Kurt Danziger, *Constructing the Subject: Historical Origins of Psychological Research* (Cambridge University Press, 1990), chapters 2–3.

spiritualism and psychology.[77] He lambasted spiritualism as a form of superstition and upheld the dignity of the conscious rational will over other mental processes. More tellingly, in his notorious critique of the spiritualistic investigations of the Leipzig astrophysicist Johann K. F. Zöllner, he questioned the ability of somebody with authority in "some particular science" to "transfer this quality at his pleasure to other provinces", especially when these latter provinces related to phenomena threatening scientifically established laws.[78] In 1877, Zöllner had convened seances in Leipzig with the American medium 'Dr' Henry Slade, who had recently fled from Britain following a failed prosecution, led by the British anatomist E. Ray Lankester, for fraudulently producing 'spirit' writing within enclosed slates. Zöllner was vastly more impressed with Slade's performances than the medium's British assailants. For him, Slade's abilities to produce slate-writing, tie knots in continuous loops of fabric and interlink two continuous wooden rings could not be put down to self-delusion or trickery: they suggested the medium's genuine ability to access spirits in a fourth dimension from where these seemingly impossible physical feats could be achieved.[79]

Zöllner's investigations prompted a mixed reaction, the most hostile being from academic psychologists and physiologists. Having attended a single seance with Slade, Wundt strongly suspected that Zöllner had been the victim of "jugglery" and that the large amount of control wielded by the medium over seance conditions demonstrated that it was he who had "made the experiments" on his hapless scientific sitter, rather than vice versa.[80] Echoing Carpenter's attack on scientists' failures to calibrate their mediumistic apparatus, Wundt compared the scientist in the Slade seances to an astronomer who "can not freely manage his senses and his instruments".[81]

A key difference between Wundt and Carpenter was that the German psychologist was much more willing to defer to the authority of conjurors in seances.[82] This was precisely the position of many of the leading American architects of the academic discipline of psychology, some of

[77] Sommer, 'Crossing the Boundaries', chapter 4; Wolffram, Stepchildren of Science, pp. 37–43.

[78] Wilhelm Wundt, 'Spiritualism as a Scientific Question', Popular Science Monthly, vol. 15 (1879), pp. 577–93, p. 580.

[79] Zöllner, Transcendental Physics. For analysis see Sawicki, Leben mit dem Toten, pp. 300–10; Sommer, 'Crossing the Boundaries', pp. 214–28; Staubermann, 'Tying the Knot'.

[80] Wundt, 'Spiritualism', pp. 584 and 586.   [81] Wundt, 'Spiritualism', p. 583.

[82] While he relished the fact that some conjurors could replicate many feats performed by mediums, Carpenter maintained that medical practitioners and psychologists were the supreme scientific experts on psychical matters: Carpenter, 'Psychological Curiosities', p. 553.

whom had studied under Wundt. From the 1880s onwards, G. Stanley Hall, Hugo Münsterberg, Joseph Jastrow and others launched fierce public attacks on mesmerism, spiritualism and psychical research as part of a struggle to claim sovereignty over the science of psychology.[83] One of their major anxieties was the widespread belief that psychical research was a scientifically credible form of psychology, a belief reinforced by the fact that it was closely associated with such individuals as Crookes, Lodge, Richet and, crucially, the Harvard psychologist and philosopher William James. The psychical interests of James, another leading architect of the academic discipline of psychology, exemplified the problematic status of psychical research in the invention of this field per se.[84] On the one hand, academic psychology needed to distinguish itself from what it perceived to be psychical research's serious methodological flaws; on the other hand, it could not simply sidestep the questions pursued by psychical research because these were the ones that fascinated the publics before whom academic psychology needed to demonstrate its authority.

American academic psychologists' complex relationship with psychical research was reflected in their uncertainty about what defined expertise in the subject. As Deborah Coon has argued, they "wanted the authority to dictate" who could study psychical phenomena, even if they did not think all psychologists were qualified to do so.[85] This well characterises Münsterberg, who, in 1899, sought to police theorising on telepathy, apparitions and other psychical phenomena but revealed that he had not attended telepathy trials or seances because his scientific training had "spoiled" him for a task that was better suited to a prestidigitator or a detective.[86] Despite his experiences in psychological research, he still felt closer to the likes of Zöllner, Richet and Crookes, whose laboratory routines had given them a "continuous training of an instinctive confidence in the honesty of their co-operators" which was fatal when applied to human subjects.[87] Those scientists who relied heavily on "physical instruments" were among the least reliable because, as he warned in a Carpenteresque way in 1910, this "shifted the attention away from the woman and her inexhaustible supply of tricks".[88]

---

[83] Coon, 'Testing the Limits'.

[84] On James and psychical research see Sommer, 'Psychical Research'; Taylor, *William James*, chapter 4.

[85] Coon, 'Testing the Limits', p. 148.

[86] Hugo Münsterberg, 'Psychology and Mysticism', *Atlantic Monthly*, vol. 83 (1899), pp. 67–85, p. 78.

[87] Münsterberg, 'Psychology and Mysticism', p. 78.

[88] Hugo Münsterberg, *American Problems: From the Point of View of a Psychologist* (New York: Moffat, Yard and Company, 1910), p. 141.

Physical–psychical scientists seem to have paid less attention to such general scientific critiques of their claimed authority in spiritualism than to those, like Carpenter's, where doubts about their authority were linked to charges against specific claims. One such latter-day critic whom they did take seriously was the British physiologist Ivor Tuckett. Between 1911 and 1913, he published blistering attacks on Barrett, Lodge and other leading SPR researchers for making woefully inadequate claims about telepathy, survival and the physical phenomena of spiritualism. His close analysis of the SPR's reports on Leonora Piper, published from the late 1880s onwards, persuaded him that the powers of telepathy and communion with the dead that many SPR members claimed for her were probably due to a combination of chance, and of skill in using information accessible by normal means. Lodge's report on her mediumship was particularly weak because it clearly did not allow for the possibility that Piper's allegedly "supernormal" knowledge of his deceased relatives was acquired by fishing for information, exploiting clues in his utterances and other surreptitious uses of normal powers.[89]

Equally troublesome for Lodge, Barrett and others was Tuckett's emphasis on the trickery to which they knew some key psychic subjects had at some stage resorted, including Palladino and the Creery sisters, whose extraordinary performances in thought-reading had galvanised Barrett's final push for a psychical research organisation in 1881.[90] These revelations gave weight to Tuckett's argument that Lodge, Richet and other investigators of Palladino had shown a "chivalrous faith" in her genuineness and accepted her conditions "in the most docile manner", thereby allowing themselves to become victims of self-deception and fraud, an allegation that had already been the subject of popular speculation and derision (Figure 5.1).[91] Unsurprisingly, the Creery case gave Tuckett another reason to echo Carpenter's serious misgivings about physicists' experimental protocols. The physiologist had no doubt that had they been "trained in experimental psychology" physicists would never have allowed a strong "*will to believe*" in the reality of psychical effects to turn them into the instruments and victims of a mediumistic experiment.[92]

---

[89] Ivor Tuckett, *The Evidence for the Supernatural: A Critical Study made with 'Uncommon Sense'* (London: Kegan Paul, Trench, Trübner & Co., 1911), pp. 335. 'Supernormal' was Myers's term for human faculties or phenomena that transcended ordinary experience but could not be assumed to be 'supernatural': Myers, Human Personality, vol. 1, p. xxii.

[90] Tuckett, 'Psychical Researchers'. Barrett never doubted the robustness of his evidence for the Creery sisters. Oppenheim's argument (in Oppenheim, *Other World*, p. 360) that in this case Barrett had lost his "critical stance of the neutral scientist" overlooks the fact that Barrett's belief related to his *early* tests, not the later ones where he accepted that cheating had occurred.

[91] Tuckett, 'Psychical Researchers', p. 186.   [92] Tuckett, 'Psychical Researchers', p. 204.

5.1 In November 1909, shortly after Eusapia Palladino arrived in New York for tests with American psychical researchers, the *New York American* poked fun at Lodge's earlier investigations of her. Hoodwinked by 'Madame Fakerino', the 'Royal Professor of Occult Science' is reprimanded by his royal patron and ridiculed in the press. From F[rederick] Opper, 'The Kink of Denmark', *New York American*, 26 November 1909. Reproduced by permission of Ohio State University.

Tricky Instruments of Physics

In his response to Tuckett, Lodge resorted to a strategy used by many in controversies over spiritualism and psychical research who found their authority challenged: question the adversary's authority. Just as he hoped Tuckett would, after becoming "better informed", join the "small group of painstaking and critical explorers in this obscure region", so in 1871 Crookes doubted whether Carpenter's knowledge of this region was as exhaustive as he claimed, not least that relating to the movement of bodies without contact, where his cherished theory of unconscious cerebration seemed to break down.[93] Barrett and Stewart expressed similar intellectual and moral concerns about Carpenter's claims to authority. Carpenter was clearly one of the targets when, in 1877, Barrett warned of the "flimsy explanations, varnished with half-truths, that pass muster at the hands of those psychologists who arrogate to themselves the sole right of instructing the public on this subject" and when, in framing his own positive evidence for thought-reading five years later, Stewart criticised "certain physiologists" for following the poor methodological principle of distrusting evidence that might force an amendment to or extension of an existing scientific generalisation or law.[94]

In these kinds of responses to criticism, physical–psychical scientists claimed to be experts in psychical matters solely on the basis of greater experience of investigating psychical phenomena. Crucially, they believed this experience gave them skills in safeguarding themselves against the trickery of human subjects – skills that a background in the physical sciences alone did not furnish. They would certainly have agreed with the German–English philosopher and SPR member Ferdinand C. S. Schiller, who, responding to Münsterberg's attack on the "competence" of scientists in psychical research, pointed out that while such individuals were not "born experts in psychical research" they became so by a "pretty severe training, in the course of which they may often fall into error" but which eventually gave them an "instinctive insight into the possibilities of fraud".[95] As we saw earlier, one obvious aspect of this expertise that

---

[93] Oliver Lodge, '"Uncommon Sense" as a Substitute for Investigation', *Bedrock*, vol. 1 (1912–13), pp. 333–50, p. 335; Crookes, *Psychic Force*, p. 6. Lodge's warning echoed what he said during the controversy over another 'obscure region' – cathode rays – years earlier. Defending the particle theory of the rays over the rival electromagnetic wave theory, he argued that the "theoretical views" of an experimenter who "lives among phenomena is to be listened to" even if they disagree with our "well-based prepossessions": Oliver Lodge, 'On the Rays of Lenard and Röntgen', *Electrician*, vol. 36 (1896), pp. 438–40, p. 439.

[94] Barrett, 'Demons of Derrygonelly', p. 700n; Balfour Stewart, 'Note on Thought-Reading', *PSPR*, vol. 1 (1882–3), pp. 35–42, p. 36.

[95] Ferdinand C. S. Schiller, 'Psychology and Psychical Research', *PSPR*, vol. 14 (1898–9), pp. 348–65, p. 359.

physical scientists did not possess by reason of their training related to the psychological conditions of experiment. Barrett was acutely aware of this problem. In his presidential address to the SPR in 1904, he maintained that these conditions, and the difficulty of achieving them, led to the notorious replication problems in psychical experiment that so antagonised sceptics. What these sceptics seemed to forget was the "profound difference" between physical and psychical experiment, which was that success in psychical experiment depended on conditions, many of which were unknown, and some of which related to the mental state of participants.[96] However, Barrett had earlier insisted that even the mental "sympathy" that a psychical researcher needed to show towards their experimental subjects could be justified in terms of a comparable approach in the physical sciences: it was "compatible with a calm judgement and clear and accurate observation" that he and so many other academic physicists promoted in their teaching.[97]

The parallel that Barrett drew between expertise in psychical and physical experimentation was weak compared to those that he and other physical–psychical scientists drew on other occasions. These stronger parallels were underpinned by a genuine belief that experimental physics offered salutary lessons for psychical investigation, and engendered a degree of tolerance towards tricky human subjects whose patient study promised to be scientifically fruitful. A key rhetorical strategy that many physical–psychical scientists adopted was to challenge the sharp distinction between physical and human instruments made by many psychological critics. What the latter deemed to be the vices of only having experience using physical instruments were turned into virtues. The physical–psychical scientists who did this were painfully aware that in physical research instruments often behaved unpredictably, effects were difficult to see or control, and progress was often hampered by complex environmental conditions. These made it more difficult to achieve the widely shared expectation that sources of error had been identified, measured, eliminated or managed.[98]

A subtle instance of tricky instruments of physics being given psychical significance occurred in 1877 when Crookes was responding to Carpenter's invidious comparisons between his radiometer and spiritualistic researches. Unsurprisingly, the chemist and physicist denied the physiologist authority on the physical problem of the radiometer and rejected Carpenter's charge that his mind had been scientific in one

---

[96] Barrett, 'Address by the President', p. 331.
[97] Barrett, 'On Some Physical Phenomena', p. 41.
[98] On this expectation see Gooday, 'Instrumentation and Interpretation', esp. 413–20.

case but not the other. Crookes clearly believed that the experimental virtues that led to success with the radiometer had been present in his spiritualistic researches. A key lesson that he drew from the radiometer controversy was the importance of following "residual", unanticipated or slightly anomalous effects, even when such effects were hard to reproduce or challenged what Carpenter upheld as the "'common sense of educated mankind'".[99] It was precisely the "man of disciplined mind and of finished manipulative skills" rather than the "untrained physicist or chemist" who was best able to cope with these kinds of experimental and theoretical difficulty, and could proceed to make important discoveries.[100]

Although this lesson seemed only to be relevant to residual phenomena in the physical sciences, Crookes had long believed that it had a far wider significance. Those with technical and specialist scientific skills could also be trusted to tease out 'residual' truths about spiritualism and, engaging with the same debates on scientific and technical education that had long pre-occupied Carpenter, they were precisely the individuals that had helped give Britain its industrial strength.[101] The psychical lessons of the radiometer were certainly not lost on Lodge and Stone, who agreed that it showed the virtues of persevering with recalcitrant and potentially deceptive effects that were contemptuously rejected by representatives of common sense.[102]

A more explicit attempt to make a virtue of experiences with physical instruments appeared in Barrett's address to the London Spiritualist Alliance in 1894. Engaging with spiritualists' anxieties about the kinds of aggressive scientific investigator that James Burns had expressed nearly 20 years earlier, he tried to reassure his audience that physicists under-stood better than most kinds of scientific practitioner how their very presence in sites of enquiry could adversely affect experimental outcomes and why, *pace* Münsterberg, intrusive methods of investigation were counter-productive. "In every physical process we have to guard against disturbing causes", he explained, but if, for example,

Professor S. P. Langley, of Washington, in the delicate experiments he is now conducting – exploring the ultra red radiation of the sun – had allowed the thermal radiation of himself or his assistants to fall on his sensitive thermoscope, his results would have been confused and unintelligible. We know that similar confused results are obtained in psychical research, especially by those who fancy the sole function of a scientific investigator is to play the part of an amateur detective; and accordingly what they detect is merely their own incompetency to deal with

---

[99] Crookes, 'Another Lesson', pp. 886–7.   [100] Crookes, 'Another Lesson', p. 886.
[101] Crookes, *Psychic Force*, p. 13.
[102] Oliver Lodge, 'Correspondence', *JSPR*, vol. 6 (1893–4), pp. 270–2; William H. Stone, 'The Radiometer', *Popular Science Review*, vol. 17 (1878), pp. 164–81, esp. pp. 180–1.

problems the very elements of which they do not understand and seem incapable of learning.[103]

For some of Barrett's auditors, his argument would have carried much authority given his well-known lectures on the extraordinary sensitivity of flames, Thomas Alva Edison's heat detector and on other instruments that picked up obscure vibrations emitted by the body.[104]

Lodge agreed with Barrett that a more sympathetic treatment of mediums found some justification in the way physicists coped with instruments. Only a month after Barrett's address, he presented his formal report on the Palladino investigations to the SPR, part of which dealt with the question of her potential fraudulence. He judged it "scientific" to accept the possibility that she was capable of conscious and unconscious trickery: even though he was confident that participants held her hands and feet firmly throughout the seances, he acknowledged that some "occurrences" (especially those very near the medium) might seem like the result of her surreptitiously freeing a hand or foot.[105] But he also deprecated the "hasty and ill-founded accusation" of deliberate imposture because it often reflected a poor understanding of Palladino's performance, including the fact that when entranced she was not conscious of her actions.[106] Such interpretations were perilous in psychical *and* physical research. Doubtless drawing on material he used in lectures to his students at Liverpool, he encouraged fellow psychical researchers to treat Palladino

not as a scientific person engaged in a demonstration, but as a delicate piece of apparatus wherewith they are making an investigation. She is an instrument whose ways and idiosyncrasies must be learnt, and to a certain extent humoured, just as one studies and humours the ways of some much less delicate piece of apparatus turned out by a skilled instrument-maker. A bad joint in a galvanometer circuit may cause irregular and capricious and deceptive effects, yet no one would accuse the instrument of cheating. So also with Eusapia: it is obviously right to study the phenomena she exhibits in their entirety, so far as can be done with such a complicated mechanism, but charges of fraud should not be lightly and irresponsibly made.[107]

[103] Barrett, 'Science and Spiritualism', p. 585.

[104] William F. Barrett, 'Sensitive Flames as Illustrative of Sympathetic Vibration', in *Science Lectures at South Kensington*, 2 vols. (London: Macmillan and Co., 1879), vol. 2, pp. 183–200; William F. Barrett, 'Mr. Edison's Inventions', *Electrician*, vol. 2 (1878–9), pp. 76–7.

[105] Lodge, 'Experience of Unusual Physical Phenomena', pp. 321 and 325. See also Richard Hodgson, 'The Value of Evidence for Supernormal Phenomena in the Case of Eusapia Palladino', *JSPR*, vol. 7 (1895–6), pp. 36–55, p. 39.

[106] Lodge, 'Experience of Unusual Physical Phenomena', p. 325.

[107] Lodge, 'Experience of Unusual Physical Phenomena', p. 324.

Lodge's suggestion that Palladino more resembled an instrument that unconsciously mediated effects than a person who consciously 'demonstrated' or fabricated them would not have convinced those who suspected that she had more control over these effects than she claimed, and who therefore did not deserve to be humoured. But it was welcomed by Barrett and would have carried weight with those of Lodge's auditors who were aware that, in his recent researches on electromagnetic waves, he seemed to have demonstrated the virtues of this approach to instruments.[108] His development of a form of 'coherer' detector of these waves arose from carefully studying the capricious effect of bad electrical contacts, which an impatient physicist might have dismissed as an inherently deceptive feature of the apparatus. Lodge hinted at this methodological lesson in an 1894 lecture where he modestly suggested that "[p]erhaps some of the capriciousness of an anathematised bad contact was sometimes due to the fact that it was responding to stray electric radiation".[109] Similar to Crookes's view on residual effects, Lodge clearly believed that a patient study of mediums might trace capricious behaviour to genuinely novel psychological or physiological causes, whereas a hasty investigation would simply dismss the behaviour as conscious trickery.[110]

Lodge's rejection of a rigid distinction between physical and psychic instruments was shared by another physicist who investigated Palladino: J. J. Thomson. He had participated in the tests of her mediumship that Lodge, Rayleigh, Myers, the Sidgwicks and other SPR members had staged in Cambridge during July and August of 1895. He later recalled that the handful of sittings he had with her were "exciting and interesting" but on balance unconvincing.[111] On the one hand, he could not explain how heavy curtains had billowed out behind her, despite there being no draught and no evidence of a mechanism by which this could have been done. On the other hand, he was satisfied that the medium had cheated in a way that would become one aspect of Hodgson's sensational exposé of Palladino as a skilled fraudster. By insisting that the two people either side of her at the seance table – Thomson and Rayleigh – lightly touched rather than firmly held her hands, Palladino probably managed to wriggle her hands so that one of Thomson's hands ended up on one of Rayleigh's. This fooled the scientists

[108] Barrett, *Threshold of a New World of Thought*, p. 49.
[109] Lodge, 'Work of Hertz', p. 343.
[110] On Lodge's defence of patience as a scientific virtue see Michael Whitworth, 'Transformations of Knowledge in Oliver Lodge's *Ether and Reality*', in Navarro, *Ether and Modernity*, pp. 30–44.
[111] Thomson, *Recollections and Reflections*, p. 148. Thomson's participation in these seances is documented in notes taken by Eleanor Sidgwick's assistant Alice Johnson: Notes on Eusapia Palladino seances, SPR.MS 44/1/1/1, Society for Psychical Research Archive, Cambridge University Library.

into thinking contact had not been broken and allowed her to fraudulently produce "very lively" movements of furniture and other effects.[112]

Thomson's opinion of Palladino, however, was not as unfavourable as this might suggest. In his 1936 autobiography, he clearly felt that it was time to express a more equivocal view about her than he had felt acceptable in the late 1890s when the SPR officially agreed to dissociate itself from the 'tainted' medium. The grounds for his equivocation lay in considerations of psychical *and* physical research. He accepted that Lodge's evidence for the genuineness of some of Palladino's earlier telekinetic performances (revealed in *his* autobiography of 1931) arose from seances held under better conditions – where the medium had been more at ease with her sitters and less inclined to cheat.[113] But Thomson also revealed an attitude towards her erratic performances that derived from his experiences using tricky instruments of physics. It was no coincidence that in the midst of reflecting on Palladino he recalled a humorous observation made by one of his Cambridge colleagues, the physicist Coutts Trotter, in the 1890s: "the law of constancy of Nature was never learned in a physical laboratory".[114] Trotter's quip seemed especially appropriate to Thomson, whose research students often joked about his frustration with wayward instruments and who, in the 1890s, was often hindered by the capricious effects of apparatus built to probe the mechanisms by which electricity discharged through gases.[115] Thomson's experiences in the physical laboratory seem to have made him tolerant of Palladino's caprices. Given that a complex physical apparatus was "simplicity itself compared with a human being", it was even more presumptuous to expect a medium to produce consistent effects, especially in surroundings made unsettling by the presence of physical instruments.[116] Indeed, for Thomson, experimental physics was more important for engendering a tolerance of mediumistic performance than in providing the instrumental testing strategies discussed in Chapter 4.

The appeal to the tricky instruments of physics was mobilised by physical–psychical scientists to justify a patient approach to puzzling and capricious effects, and to defend themselves against accusations that their observational powers could not cope with the specific

---

[112] Thomson, *Recollections and Reflections*, p. 150.
[113] Thomson, *Recollections and Reflections*, p. 152; Lodge, *Past Years*, pp. 292–311.
[114] Thomson, *Recollections and Reflections*, p. 153.
[115] A. A. R[obb], 'J. J.', in *Post-Prandial Proceedings*, pp. 8–9; J. J. Thomson, 'On the Passage of Electricity Through Hot Gases', *Philosophical Magazine*, vol. 29 (5th Series) (1890), pp. 441–9; J. J. Thomson, 'On the Effect of Electrification and Chemical Action on a Steam-Jet, and of Water Vapour on the Discharge of Electricity Through Gases', *Philosophical Magazine*, vol. 36 (5th Series) (1893), pp. 313–27.
[116] Thomson, *Recollections and Reflections*, p. 153.

circumstances of seances. Few instruments proved rhetorically more useful here than the spectroscope, partly because physical–psychical scientists were aware that some their most redoubtable critics accepted this as an outstanding application of technical expertise.[117] In 1899, Benjamin Davies used the case of the spectroscope to challenge those, like Carpenter, who regarded the "common sense of educated mankind" as the ultimate court of appeal in assessing psychical phenomena. His argument was a response to a letter in a Liverpool newspaper insisting that until Crookes satisfactorily exhibited a materialised spirit before scientific witnesses, his notorious claims about manifestations that contradicted most people's experiences had to be dismissed as delusory. A similar charge, Davies explained, could be levelled at spectroscopy, a technique that he taught at Liverpool. Most people found it difficult to see the bright emission lines beyond the violet region, but this did not mean that others were deluding themselves: the "unanimous opinion of mankind", he urged, was "useless" here because the "experience of the majority is narrower than that of the individual who perceives the ultra-violet band".[118]

That vastly more experienced spectroscopist, Crookes, agreed and in 1909 told Davies's mentor that both psychical research *and* spectroscopy would be jeopardised if faint and transient effects only visible to people with "sharper eyes" had to be dismissed.[119] This was not the first time he had appealed to spectroscopy. In 1871, the astronomer William Huggins had suggested to him that they form a scientific committee to verify their claims about psychic force. Crookes was taken aback because he did not think their claims any more extraordinary than either his regarding the spectrochemical analysis of terrestrial materials or Huggins's regarding the spectrochemical study of celestial bodies: "You do not ask for a committee to examine spectra", he explained, "I did not ask for one to convince me that thallium was true".[120] For at least Crookes and Davies, Carpenter was right to expect deference to be paid to the more experienced or sharper-eyed participants, but that might involve accepting that those good at seeing spectra could also be trusted to see spectres.

---

[117] See, for example, William B. Carpenter, *Mesmerism, Spiritualism, & c, Historically & Scientifically Considered* (London: Longmans, Green, and Co., 1877), p. 70; John Nevil Maskelyne to Oliver Lodge, 13 September 1895, SPR.MS 35/1235, OJL-SPR.

[118] Benjamin Davies, 'Professor Richet's Address', *Liverpool Daily Post*, 11 August 1899, p. 3.

[119] William Crookes, typescript memorandum to Oliver Lodge, [circa 1909], SPR.MS 35/366, OJL-SPR.

[120] William Crookes to William Huggins, 16 June 1871, quoted in Fournier d'Albe, *Life of Sir William Crookes*, pp. 209–10, p. 210.

Psychical Researchers and Conjurors

When, in 1899, Schiller argued that scientists only became experts in psychical research after long and often bitter experiences of experimental error and trickery, he expressed a complex attitude towards trained scientific experts in psychical research that was echoed by other SPR members. By the early 1900s, the organisation was as keen as ever to exploit the scientific reputations of Crookes, Lodge, Richet and others in creating an image of intellectual credibility that many doubted. This is strikingly apparent from the choice of presidents and, more tellingly, the composition of the SPR's 'Committee of Reference'. Founded in 1885 and lasting well into the 1940s, this Committee's purpose was to referee papers intended for publication, and it was dominated by Council and other high-ranking SPR members whose scientific and medical qualifications evidently made them invaluable in giving the impression that the organisation had robust mechanisms of quality control.[121]

A long-serving member of this Committee, Lodge, was acutely aware of this symbolism. In October 1890, when his long and positive experiences of Leonora Piper's mediumship had persuaded him to get more closely involved in the SPR, he told Myers that while he felt near the "adult stage" of development as a psychical researcher, it was "by sticking definitely to physics that I shall be of most service, for only so shall I carry any weight".[122] To underline his point, he contrasted himself to Barrett, whom he judged to be someone whose "comparative neglect of Physics has made him less powerful than he ought to be".[123] There was some justification for Lodge's swipe at his fellow physicist because, despite being seven years older, Barrett's output of original physical research was meagre and he still lacked the fellowship of the Royal Society that Lodge had secured in 1887.[124]

Valuable as trained and active scientists were in bolstering the image that the SPR desired for itself, their value in executing psychical research had become, by the early 1900s, one of many sources of disagreement within the Society. Internal conflict over the proper way to prosecute psychical investigation seemed to be inevitable from the organisation's

---

[121] [Anon.], 'Council Meetings', *JSPR*, vol. 1 (1884–5), pp. 259–61.
[122] Lodge to Myers, 21 October 1890, SPR.MS 35/1309, OJL-SPR.
[123] Lodge to Myers, 21 October 1890. J. G. Piddington, the SPR's honorary secretary, deplored the idea of Barrett serving a second term as president because of his "constant fussiness & suspiciousness" and lack of a British and European scientific reputation: J. G. Piddington to Oliver Lodge, 24 September 1904, SPR.MS 35/1632, OJL-SPR.
[124] Barrett finally received his in 1899.

birth. When, in early 1882, the eminent spiritualist William Stainton Moses replied to Barrett's invitation to attend his conference on a possible psychical research organisation, he warned of the "disintegrating forces" threatening the future of the organisation, forces arising from the "marked social distinctions" and methodological differences between the spiritualists and intellectuals participating in the initiative.[125] One such intellectual, Henry Sidgwick, was more optimistic and later recalled that, as the SPR's first president, he tried hard to reconcile its "two heterogenous elements – persons convinced of the genuineness of the alleged effects of spiritual or occult agency, and persons like myself, who merely thought the evidence for their genuineness strong enough to justify serious inquiry".[126] The disintegrating forces were never stronger than in the wake of the SPR's 'exposure', in 1886, of William Eglinton, a medium dear to the heart of many spiritualists, many of whom reacted by resigning their SPR membership and intensifying their attack on the organisation's elitist and needlessly critical and painstaking approach to 'their' subject.[127]

Even after shedding many spiritualists, the SPR's membership was divided on a host of other questions, including the evidence for survival, the fruitfulness of studying the physical phenomena of spiritualism and the value of Myers's theory of the subliminal self. When it came to choosing a president, the diversity of opinion on these questions often became acute. In 1905, for example, Eleanor Sidgwick battled with Barrett to elect Frank Podmore because he represented the "hard-of-belief section", which did not go beyond accepting that telepathy had been proved and which had not been represented by the three previous presidents (Barrett, Lodge and Richet), who stood for the "forward" section of the society insofar as their attitudes to the physical phenomena of spiritualism and (in the case of Barrett and Lodge) survival were more positive.[128] For Eleanor and Henry Sidgwick and many other leading SPR members, the cautious section was ultimately of greater intellectual and symbolic importance because it was this section that was more likely to persuade the outside world, and especially the outside scientific world, that the SPR had to been taken seriously. It was precisely this view that underpinned the SPR's critical approach to the physical phenomena of spiritualism: as Eleanor Sidgwick explained in an 1886 critique of the

---

[125] William S. Moses to William F. Barrett, 12 January [1882], SPR.MS 3/A4/54, WFB-SPR.
[126] Henry Sidgwick, 'Remarks', *PSPR*, vol. 4 (1886–7), pp. 103–6, p. 103.
[127] Cerullo, *Secularisation of the Soul*, chapter 4; Williams, 'The Making of Victorian Psychical Research', chapter 8.
[128] Eleanor Sidgwick to William F. Barrett, 23 November 1905, SPR.MS 3/115.

evidence for such phenomena, it was because physical phenomena were "more probable than not" that they deserved to be more thoroughly tested than had been done before and to have their true evidential basis more firmly established.[129]

The desire of leading SPR members to persuade the world that it was subjecting the physical phenomena of spiritualism to the most searching tests possible explains why they increasingly valued conjuring as a form of expertise in psychical research. The organisation included prominent members such as Hodgson and Robert Strutt (4th Baron Rayleigh), who were amateur conjurors, and actively involved professional conjurors such as Maskelyne and Angelo Lewis ('Professor Hofmann') in its investigations (Figure 5.2).[130] One of the problems arising from this was that, as the cases of Eglinton, Palladino and Slade dramatised, conjuring expertise was considered most useful in precisely those kinds of psychical phenomena where physical–psychical scientists believed their expertise gave them an advantage. Disagreements over the relative importance of these forms of expertise were inevitable.

These disagreements certainly predated the SPR's foundation. In the wake of their psychical investigations in the 1870s, Crookes, Barrett and Zöllner found themselves attacked by conjurors, medical practitioners and fellow scientists for failing to take precautions against potential sources of deception well known to conjurors. In the heated discussion following Barrett's paper at the British Association meeting in 1876, Crookes opposed the argument that a "trained physical inquirer is no match for the professional conjuror" who might be masquerading as a medium.[131] By "close examination", Crookes insisted, the physical scientist was fully able to outwit the conjuror, whose performance depended strongly on the choice of venue, "apparatus" and the ability to control who participated and where they stood. Another physical–psychical scientist in attendance was Rayleigh, who departed from this view in maintaining that the "special accomplishments" of a conjuror were of "great value" in assessing rather than creating a psychical performance, and he revealed that one such person had accompanied his private investigation of Henry Slade, which resulted in the conjuror being unable to explain the medium's performance.[132]

---

[129] [Eleanor] Sidgwick, 'Results of a Personal Investigation into the Physical Phenomena of Spiritualism', *PSPR*, vol. 4 (1886–7), pp. 45–74, p. 74.

[130] On Strutt's conjuring see Strutt, 'Robert John Strutt', p. 1111.

[131] Crookes cited in [Anon.], 'British Association at Glasgow', p. 89. Crookes wrongly ascribed the argument to Barrett. Stewart does not appear to have attended the meeting but later agreed with Crookes: Stewart reported in [Anon.], 'Report of the General Meeting', *JSPR*, vol. 2 (1885–6), pp. 338–46, p. 346.

[132] Rayleigh cited in [Anon.], 'British Association at Glasgow', p. 89.

5.2 Handbill for one of John Nevil Maskelyne and George Cooke's late-Victorian stage entertainments at the Egyptian Hall, London. It shows many of the familiar phenomena of spiritualistic seances – including levitating tables and ghostly forms – that the performers claimed to reproduce by conjuring. Reproduced by permission of the Bill Douglas Centre, University of Exeter.

Rayleigh's concession to the expertise of a conjuror is particularly significant in the light of his negative experience of Crookes's behaviour during Annie Eva Fay's seances the previous year. He was temporarily "disgusted" with Crookes, apparently because Crookes had tried to justify an instance of Fay's fraudulence as the "only course" open to her after Rayleigh's disbelief upset the psychological conditions favouring manifestations.[133] Myers shared Rayleigh's misgivings about Crookes, whom Myers believed had been duped by Fay on one occasion and who seems to have felt his authority threatened by the scientifically weightier Rayleigh.[134] These problems would have persuaded Myers and Rayleigh that even a physicist with Crookes's extensive experience of seances needed a conjuror's help.

One of the problems that Myers and other leading SPR members had with Crookes was that despite these failings, they greatly admired his earlier investigations of Home and could not afford to lose the scientific lustre that he brought to the organisation through public attacks. This is apparent from the strategic way in which Eleanor Sidgwick, Barrett and Myers handled his work in SPR publications. In a footnote to her notorious paper of 1886, Sidgwick apologetically opined that it was "undesirable that even men of established scientific reputation like Mr. Crookes and Professor Zöllner should hold themselves exempt from mentioning even the simplest precautions".[135] Yet, partly because Zöllner was now dead, Sidgwick clearly felt it easier to criticise him than Crookes for specific failings in this direction. She attacked the astrophysicist for methodological lapses in the Slade tests that would have allowed even an amateur conjuror to persuade him of the existence of four-dimensional beings, but she merely referred to Crookes's experiments on Home's alteration of the weight of a board as one of the reasons why it was the duty of the SPR to make further investigations of the physical phenomena. Three years later, in reviewing a biography of Home by his widow, Barrett and Myers accepted the "drawback" that the other mediums through whom Crookes had "obtained striking results" had subsequently been exposed, but that this did not "necessarily affect the experiments with Home", which still amounted to unrivalled testimony of physical phenomena and whose main weakness was that they had not been confirmed by other scientists.[136]

---

[133] Lord Rayleigh to Frederic W. H. Myers, 15 January 1875, Myers 2⁴² (1), Frederic William Henry Myers Papers, Trinity College Library, Cambridge.

[134] Frederic W. H. Myers to Oliver Lodge, 15 March 1892, SPR.MS 35/1331, OJL-SPR; Lord Rayleigh to Evelyn, Lady Rayleigh, 10 January 1875, Rayleigh Family Papers, Terling Place, Terling, Essex.

[135] Sidgwick, 'Results of a Personal Investigation', p. 65n.

[136] William F. Barrett and Frederic W. H. Myers, 'D. D. Home, His Life and Mission', JSPR, vol. 4 (1889–90), pp. 101–16, p. 105.

A more telling assessment of Crookes appeared in one of the addresses that Lodge gave as SPR president, a position he agreed to fill to give the organisation the kind of intellectual leadership it sorely needed following the deaths of Henry Sidgwick and Myers in 1900 and 1901 respectively. Here, Lodge followed Eleanor Sidgwick's example of contrasting Crookes and Zöllner. As far as he knew, Crookes had never testified to an instance of the passage of matter through matter, while Zöllner's evidence for this phenomenon did not "leave a feeling of conviction on the unprejudiced mind".[137] But more significant was the fact that he judged Crookes's testimony for materialised spirits to be "detailed" and almost as hard to resist as accepting the "things testified".[138] This is not how Lodge had seen it in a draft of the address on which he had asked Crookes to comment. There, Lodge had implied that the "impositions of a conjuror" were the *veræ causæ* of materialisation phenomena, a claim that undoubtedly owed something to what Myers had told him in 1892 about the Fay seances.[139] Unsurprisingly, Crookes objected to this implication and Lodge was evidently persuaded to remove the offending sentences.

I want to suggest that Lodge's decision was helped by at least two related anxieties. First, he was clearly trying to protect Crookes's reputation in an organisation that appeared to be more equivocal about his work. Not long after Lodge's address, Frank Podmore, one of the SPR's most industrious workers, published *Modern Spiritualism*, a lengthy critical history of the subject that, while not presented as an official SPR publication, came with the blessing of such leading figures in the organisation as Eleanor Sidgwick, Hodgson and Myers. The problem for Lodge and, *à fortiori*, Crookes was that Podmore voiced in public what some SPR members had privately suspected for years: that Crookes's evidence for Cook and Fay was poor and left open the strong possibility of fraud. Podmore even questioned Crookes's hallowed evidence for Home and effectively reiterated the views of so many psychologist critics of spiritualism when he implied that Home was a "practised conjuror" who, by dictating the conditions of seances to suit his purposes, was able to fool such "untrained observers" as Crookes, Huggins and Cox.[140] In a Carpenteresque salvo, he emphasised that Crookes's "previous training did not necessarily render him better qualified to deal with problems differing widely from those presented in the laboratory" and, drawing on the warnings of Hodgson,

---

[137] Lodge, 'Address by the President', p. 46.
[138] Lodge, 'Address by the President', p. 45.
[139] William Crookes to Oliver Lodge, 28 January 1902, SPR.MS 35/346, OJL-SPR; Myers to Lodge, 15 March 1892.
[140] Frank Podmore, *Modern Spiritualism. A History and a Criticism*, 2 vols. (London: Methuen and Co., 1902), vol. 2, p. 240.

Davey and Sidgwick about the difficulties of continuously observing mediums, he charged that Crookes's attention may have been focussed so strongly on his apparatus that he failed to spot the "seemingly irrelevant movements" by which Home achieved his goals.[141]

Podmore's critique would have delighted Maskelyne, who had made a similar point to Lodge in 1895 by inverting the same psychical–spectroscopic analogy deployed by physical–psychical scientists to assert their authority. The "twitch of a muscle" of a medium, he warned, would be as telling to an expert on trickery as the "slight displacement of a line in the spectrum of an element would signify to a physicist".[142]

Lodge's second anxiety was that, as Podmore argued, critiques of Crookes could be turned into critiques of the capacity of *all* trained scientists to compete with the conjuror in psychical investigation, especially those physical scientists whose gaze was often focussed on apparatus. He had long found this difficult to accept. In an early paper on thought-transference, he strategically emphasised that had he "merely witnessed facts as a passive spectator" and accepted "imposed conditions", he could have been duped by a conjuror, but because he controlled the "circumstances" and arranged his own experiments, he had "acquired a belief in the phenomena observed quite comparable to that induced by the repetition of ordinary physical phenomena".[143] As critics of psychical research were at pains to stress, however, this confidence in experimental control was often deceptive and made the trained scientist an easy victim for a conjuror masquerading as a genuine medium.

Lodge's actions and writings suggest that he clearly understood that scientists and conjurors offered complementary forms of expertise to psychical investigation. In 1887, for example, he agreed to join an SPR committee for investigating spiritualistic phenomena whose members included a conjuror (Angelo Lewis), Gurney, Myers and two fellow physical scientists (Barrett and Crookes).[144] Moreover, he clearly had little difficulty with Maskelyne participating in the Palladino seances in Cambridge, and in the controversy following her 'exposure' in 1895, he

---

[141] Podmore, *Modern Spiritualism*, vol. 2, p. 240.
[142] John Nevil Maskelyne to Oliver Lodge, 13 September 1895, MS.SPR.35/1235, OJL-SPR. The "slight displacement" to which Maskelyne referred was the shift in the position of the dark lines in the absorption spectra of stars. The shift was relative to the position of the bright lines in the emission spectra of the vapours of the chemical elements corresponding to these lines and to the reference absorption spectrum for all celestial bodies: the solar spectrum. By analogy with the Doppler shift, the movement was explained in terms of the change in wavelength of light caused by the movement of the star towards or away from the observer.
[143] Lodge, 'Account of Some Experiments', p. 190.
[144] Balfour Stewart, 'Address', *PSPR*, vol. 4 (1886–7), pp. 262–7, p. 267.

was moved to praise Maskelyne's open-minded approach and said that his feats, as well as those of other conjurors, had been "constantly" in his mind when judging Palladino's genuineness.[145] But while accepting that Maskelyne and others had *some* authority in psychical research, he maintained that they did not have *supreme* authority. In a telling footnote to his Palladino report he denied that a conjuror's evidence was "necessarily of a kind such as to render other evidence superfluous" or "always superior to that of a person whose life-long study has been the pursuit of truth".[146]

N-rays and Psychical Expertise

Lodge's confidence in the capacity of trained scientists to produce reliable evidence of psychical effects became particularly visible in the early 1900s when two of the most powerful figures in the SPR published papers questioning this capacity. The figures were Eleanor Sidgwick and Alice Johnson. In the wake of the deaths of Gurney (in 1888), Myers and, above all, her husband Henry, Sidgwick played an enormously significant role in ensuring the continuity of the SPR's work.[147] In 1888, she took over from Gurney as the editor of the SPR's *Journal* and *Proceedings* and between 1907 and 1932 she served as honorary secretary, in which capacity she (and the relatives and colleagues whom she co-opted) dominated the SPR's development well into the interwar period. In these official roles and in her studies of telepathy, sensory and motor automatisms and the physical phenomena of spiritualism, she embodied what she had called the "hard-of-belief" section of the SPR.[148] When, in 1903, Alice Johnson took over as editor of the SPR's periodicals, there was no question about the continuity of Sidgwick's robust editorial line, since she had been so closely involved in Johnson's scientific and psychical education. Johnson had studied natural sciences at Sidgwick's Cambridge college (Newnham) and worked as a demonstrator in animal morphology in the college's Balfour Laboratory, and in the 1890s served as Sidgwick's private secretary and assistant.[149] In 1907, she became the SPR's first

---

[145] Oliver Lodge, 'The Exposure of Eusapia', *Daily Chronicle*, 5 November 1895, p. 3. Citation from Oliver Lodge, 'Unusual Physical Phenomena', *Liverpool Daily Post*, 2 January 1895, p. 7.

[146] Lodge, 'Experience of Unusual Phenomena', p. 315n.

[147] Alice Johnson, 'Mrs Henry Sidgwick's Work in Psychical Research', *PSPR*, vol. 44 (1936), pp. 53–93.

[148] Myers defined 'automatisms' as actions of the body taking place independently of the conscious will. The sensory variety included hallucinations and the motor variety of trance writing ascribed by spiritualists to spirits. See Myers, *Human Personality*, vol. 1, pp. xiv–xv.

[149] Helen de G. Salter and Isabel Newton, 'Obituary – Alice Johnson', *PSPR*, vol. 46 (1940–1), pp. 16–22.

salaried research officer, and was thus especially mindful of the problematic nature of psychical expertise.

Sidgwick would have surprised few of her auditors when, in her presidential address to the SPR in 1908, she contrasted the great strides that had been achieved in studies of hypnotism, phantasms of the living and telepathy with those in the physical phenomena of spiritualism, which, as far as she was concerned, stood "almost exactly where it did" when the SPR was founded.[150] The only progress had been in the negative direction, insofar as Hodgson and Davey had yielded "more definite experimental knowledge of the possibilities of mal-observation" and others had shown that unconscious muscular action played a bigger part in some phenomena than first assumed.[151] Sidgwick certainly did not think spiritualism should be abandoned given the quality of some evidence relating to physical phenomena, but progress towards stronger evidence required better policing of fraudulent mediums and a better understanding of the sensory hallucinations experienced by investigators. Many of Sidgwick's auditors would have known that she had recently argued that an underestimate of this latter problem had blighted the Italian neurologist Enrico Morselli's new evidence in favour of Palladino's genuineness: Morselli had been overly confident in his observational acumen and mistaken in his assumption that an absence of evidence of fraud was evidence of its absence.[152]

What may have surprised Sidgwick's auditors was that she chose to illustrate the experimental dangers of sensory hallucination by means of a potted history of a recent controversial topic in the physical sciences: N-rays. In 1903, the French physicist René Blondlot caused a scientific sensation when he claimed to have produced experimental evidence for a new form of invisible radiation that seemed to be emitted by electric sparks, incandescent lamps, sunlight and some solid bodies experiencing mechanical strain.[153] Later, both Blondlot and his colleagues at the University of Nancy reported that the rays emanated from organic sources such as animal muscles and brains and the human eye. Unsurprisingly, news of the discovery of invisible rays associated with the human body was hailed by many spiritualists, occultists and psychical researchers as new experimental evidence for Reichenbach's od and vital effluences, and a possible mechanism for brain waves.[154]

---

[150] Sidgwick, 'Presidential Address', p. 9.   [151] Sidgwick, 'Presidential Address', p. 10.

[152] E[leanor] M[ildred] S[idgwick] Review of Psicologia e 'Spiritismo', PSPR, vol. 21 (1908–9), pp. 516–25.

[153] See Mary Jo Nye, 'N-Rays: An Episode in the History and Psychology of Science', Historical Studies in the Physical Sciences, vol. 11 (1980), pp. 125–56.

[154] See, for example, Arthur Lovell, 'Reichenbach's Researches', Transactions of the Vril-ya Club, no. 2 (1904), pp. 1–34; Jules Regnault, 'Odic Phenomena and the New Radiations', Annals of Psychical Science, vol. 1 (1905), pp. 145–63.

One of the most controversial features of N-rays was that they were difficult to detect. The methods of detection included observing the marginal increase in the brightness of faint electric sparks or phosphorescent spots exposed to the rays. The difficulty that many had in replicating the effects reported by Blondlot and his colleagues engendered suspicions that the subtle visual changes, normally experienced in darkened rooms, were subjective. Among those who failed to repeat Blondlot's findings were Crookes, Rayleigh and one of Rayleigh's closest American colleagues, the optics expert Robert Wood.[155] Wood visited Blondlot's laboratory and eventually published damning evidence that Blondlot had merely imagined the effects of the alleged rays (in one test, Wood surreptitiously replaced a metal file believed to be a source of the rays for a piece of wood having a similar size and shape, and Blondlot still claimed that it made faint objects brighter). While Wood's denouement quickly destroyed scientific interest in N-rays, Sidgwick would have known from her brother-in-law's response that this did not settle the question of whether N-rays existed.[156] But the episode also gave Sidgwick an example of the possibility that, even in their physical researches, trained scientists could be victims of hallucinations caused by a strong expectation of particular effects. When physicists made forays into a domain where darkened rooms, faintly luminous effects and replication problems were even more common, they were at even greater risk of self-delusion and the scientific world had an even greater right to expect caution.

The next time that N-rays were mentioned in an SPR publication was in 1910, when Lodge used them to develop a counter-argument that trained physicists did have the appropriate skills for investigating certain classes of psychical phenomena. He was responding to a recent study published in the SPR's *Proceedings* in which Alice Johnson effectively extended her mentor's argument that perfectly well-educated individuals could experience hallucinations while witnessing the physical phenomena of spiritualism, although this did not signify mental imbalance, since hallucinations were common in everyday life.[157] Johnson's focus was primarily on witnesses to the physical phenomena occurring in Home's presence. More than any official SPR publication to date, however, she challenged Lindsay's and Crookes's accounts of the medium's levitations: Lindsay appears to have been the victim of an "excited imagination", probably caused by Home's powers of

---

[155] Crookes revealed this in the Minutes for 4 January 1905, Ghost Club Minutes, Volume 5, Add. 52262, GC-BL; Rayleigh, *Third Baron Rayleigh*, pp. 358–9; Robert W. Wood, 'The *n*-Rays', *Nature*, vol. 70 (1904), pp. 530–1.

[156] Rayleigh, *Third Baron Rayleigh*, p. 359.

[157] Alice Johnson, 'The Education of the Sitter', *PSPR*, vol. 21 (1907–9), pp. 483–511.

suggestion and Crookes may have mistaken a visual illusion for a genuine effect.[158]

What irked Lodge was that Johnson did not distinguish more sharply reports of psychical phenomena from novices in the "art of making and learning from experiments" and those who were skilled in these regards.[159] The Principal of Birmingham University and long-time promoter of scientific and technical instruction was bound to argue that these skills made the "experienced experimentalist" (including Crookes and Lindsay) trustworthy in studying all kinds of phenomena, and gave them an "undeniable advantage" over others in forming a "valid and secure opinion".[160] The trained scientist was, as Johnson had contended, a "novice" in the "particularly treacherous field" of psychical research, but their ability to "manipulate and deal critically with phenomena", and especially "faint" physical phenomena, made them less likely to mistrust their judgement and more likely to become an expert in the new field.[161]

The N-rays episode well served Lodge's argument. Novices invited to witness the rays "sometimes saw them" but rejected their observations as "anticipatory hallucination".[162] "Skilled physicists" from Britain did not see the rays, Lodge opined, and this was evidently because they were better at manipulating and dealing critically with sometimes "faint" phenomena.[163] Whether or not Lodge knew of Crookes's and Rayleigh's failures to see N-rays, he clearly felt that some physicists were simply more skilled than others (including Blondlot) in this regard, and that they embodied many of the elements of true psychical expertise.[164] Ultimately, Lodge accepted Johnson's argument that experts in a range of fields could contribute significantly to, and become experts in, psychical research, but it is unlikely that he ever completely shared her confidence in fields other than the sciences nurturing the scientific "habit of mind" required in the

---

[158] Johnson, 'Education of the Sitter', p. 493.
[159] Oliver Lodge, 'The Education of an Observer', *JSPR*, vol. 14 (1909–10), pp. 253–8, p. 254.
[160] Lodge, 'Education of an Observer', pp. 254–5.
[161] Lodge, 'Education of an Observer', pp. 254 and 257.
[162] Lodge, 'Education of an Observer', pp. 257–8.
[163] Lodge, 'Education of an Observer', pp. 257–8.
[164] Rayleigh's son Robert (later the Fourth Baron Rayleigh), who witnessed his father's N-ray experiments, agreed that this episode showed the lessons that experimental physics had for psychical research. In 1932, he argued that just as N-rays were rejected and cathode rays accepted on grounds other than such "objective records" as photographs, so disbelief in physical aspects of psychical phenomena could not hinge on the absence or presence of similar records: Lord Rayleigh to Theodore Bestermann, 19 September 1932, in Theodore Besterman, notebook, Add. MS 57729, British Library.

psychical researcher.[165] Lodge's problem was that Johnson's psychical researcher was not the kind of person that he believed could be trusted to successfully revive an area of investigation that, in his and Barrett's opinion, had been neglected by the SPR for too long and for which physicists and physiologists had "direct and repeated experience": the physical phenomena of spiritualism.[166] It is not at all surprising that Lodge's staunch defence of the capacities of British physicists to tackle telekinesis and N-rays took place in a period when he thought it high time that physicists or physicists collaborating with other kinds of trained scientist should spearhead this initiative.

***

This chapter has shown that the proposed solutions to the problems of expertise in psychical investigations were closely bound up with particular interpretations of psychical phenomena. The conflicts that Carpenter had with Crookes and Barrett, Wundt with Zöllner, and Tuckett and Johnson with Lodge suggest that those who believed that psychologists and medical practitioners wielded supreme scientific authority tended to agree that psychical phenomena were primarily subjective and therefore best studied by those with experience of abnormal mental conditions; and those who believed that physicists had an important part to play in psychical research tended to agree that some psychical phenomena had an objective physical existence outside the mind and body of the psychic subject, and could therefore be studied by the tools and techniques of the physical sciences. Likewise, those who privileged spiritualists' more sympathetic approaches to mediumship were more likely to agree that the spiritual interpretation of manifestations was the most plausible; and those who privileged conjurors and detectives as spiritualistic investigators were more likely to accept that fraudulence explained much of what happened in seances.

The need to speak of tendencies rather than definite correlations is required by the complexity of the positions encountered in this chapter. Rayleigh, for example, did not think all the physical phenomena of spiritualism were the result of mediumistic trickery or self-deception, but he was prepared to defer to the expertise of conjurors in this question more than were Crookes and Lodge. Eleanor Sidgwick and Alice Johnson accepted that subjective impressions accounted for a significant amount of evidence for the physical phenomena of spiritualism, but they were as

---

[165] Alice Johnson, 'Note on the Above Paper', *JSPR*, vol. 14 (1909–10), pp. 259–60, p. 260.

[166] Lodge, 'Scientific Attitude to Marvels', p. 471; Barrett remarks in [Anon.], 'General Meeting', pp. 182–4. Citation from Lodge, 'Education of an Observer', p. 257.

likely to defer to somebody with extensive experience in psychical investigation as a trained psychologist.

The entanglement of questions of expertise and interpretation in psychical research was not limited to SPR circles. Well into the twentieth century, contributors to general periodicals and newspapers frequently sided with psychologists and conjurors in declaring physical scientists to be out of their depth in studying phenomena that clearly seemed to be the result of either mental aberration or fraudulence.[167] Undoubtedly of greater consequence to physical–psychical scientists were the views of close professional colleagues, which indicate the limited success of physical–psychical scientists in persuading others that there was, to adopt a phrase of Gurney's, a 'definite corner' of psychical research where physical scientists could lead. A telling example is George F. FitzGerald, who, as we saw in Chapter 2, argued in 1893 that since psychical research occupied a "borderland" in "close proximity to hysteria, lunacy &c" then its "proper students" were "physicians, not physicists", and certainly not people "without scientific scepticism, like Theosophists".[168] Another of Lodge's Maxwellian colleagues, Heaviside, proved only marginally more hopeful for physicists. The "physical basis" of spiritualism would "some day be attacked by physicists", he suggested to Lodge in 1895, but in its present state spiritualism was "more fit for the hypnotic doctors to study".[169] The example of Crookes, whom he thought had been humbugged by Home and "got taken in" by 'Katie King', evidently confirmed his doubts about physicists' fitness for spiritualistic investigation.[170]

Lodge's response to criticism from such close quarters is unknown, but his subsequent writings do not suggest any *immediate* loss of confidence in the contributions that physical scientists could make to psychical research. But in the medium and long term, his confidence and that of other physical–psychical scientists ebbed or least changed form. By the early 1900s, they were finding it difficult to achieve this goal, partly because of the want of strong psychical cases, and because of heavy professional commitments that prompted fears that leading SPR members were too busy to devote the necessary time and energy to psychical research.[171] For many physical–psychical scientists inside and outside the SPR, however, a pragmatic solution to these difficulties lay in exploiting

---

[167] See, for example, [Anon], 'Transcendental Physics', *Saturday Review*, 11 September 1880, pp. 327–8; [Anon.], 'The Professors and the Conjuror', *Speaker*, vol. 12 (1895), pp. 467–8; Robert Hughes, 'Seeing Things: The Scientists and Spiritualism', *Pearson's Magazine*, vol. 21 (1909), pp. 188–97.

[168] FitzGerald quoted in [Anon.], 'Response to the Appeal', p. 19.

[169] Oliver Heaviside to Oliver Lodge, 11 January 1895, MS Add. 89/50(ii), OJL-UCL.

[170] Oliver Heaviside to Oliver Lodge, 28 January 1895, MS Add. 89/50(ii), OJL-UCL.

[171] Ferdinand C. S. Schiller, 'The Future of the S.P.R.', *JSPR*, vol. 10 (1901–2), pp. 74–7.

forms of expertise besides the experimental, and which most of them had long used in the context of the physical sciences: scientific writing and lecturing. How they used popular and semi-popular scientific texts to safeguard and invigorate the possible connections between physics and psychics in the early twentieth century is the task of Chapter 6.

When, in 1900, Lodge accepted an invitation to become the first Principal of the University of Birmingham, he did so partly on condition that he was given a laboratory and assistants for maintaining his profile as an original physical researcher and that he be allowed to continue his work in psychical research, a field of inquiry whose results and legitimacy continued to divide the scientific and intellectual world.[1] For the SPR, Lodge's prestigious new role, in addition to a knighthood in 1902 and a successful scientific career at Liverpool behind him, made him an obvious choice of president following the deaths of Sidgwick and Myers, and a failed attempt to persuade the scientifically more eminent Rayleigh to assume the helm.[2]

The presidential address that Lodge gave to the SPR in 1903 revealed the extent to which he saw similarities between his SPR and Birmingham roles. The solution to the SPR's ongoing problem of becoming a scientific organisation with a "sound and permanent basis" converged with the solution to the problem of establishing a modern university.[3] Both problems required financial investment, albeit on vastly different scales. As SPR president, Lodge announced the initial contributions to the SPR's endowment fund and sketched out his ideas for its possible uses. When it reached £8,000, it could support a "Research Scholarship in Psychical Science" and thus give "young people of genius" a monetary incentive to contribute to the subject; and if it grew significantly larger, the fund could support a "much-needed laboratory" for staging experimental inquiries into phenomena that were "improperly accepted or improperly rejected".[4] Similarly, in a manifesto published only a month earlier, the

---

[1] Lodge, *Past Years*, pp. 315–16.

[2] Arthur J. Balfour et al., 'The Memorial, of Which a Copy Is Printed Below, Was Forwarded to Lord Rayleigh on 18th February', *JSPR*, vol. 10 (1901–2), pp. 58–60.

[3] Lodge, 'Presidential Address', p. 6.

[4] Lodge, 'Presidential Address', pp. 8–9. By the time Lodge spoke, the fund had already reached £2,000: [Anon.], 'Endowment Fund for Psychical Research, *JSPR*, vol. 11 (1903–4), pp. 44–5.

Birmingham principal argued that an investment of at least £5 million in university buildings (including laboratories), equipment, salaries and other items would help alleviate a tendency of the world to be "wasteful of genius" and to allow such talented individuals to push back the frontiers of existing departments of knowledge.[5]

To attract investment, however, the practical benefits of the Midlands varsity and the SPR needed to be more strongly impressed upon the public. Echoing decades of arguments for technical and scientific instruction, Lodge emphasised that investment in Birmingham's areas of academic study would have significant cultural, economic, moral and other benefits and an altogether "extraordinary influence on the progress of the country".[6] Investing in the SPR promised to lead to such important "practical results" as new understandings of the "obscure" workings and treatment of the criminal mind and steps towards the scientific basis of religious beliefs.[7]

Lodge's strategy for the SPR was mainly driven by a genuine fear that the organisation faced a "long period of danger and difficulty".[8] His fears were partly justified. Although its membership had grown five times since 1882 and was in its strongest-ever financial position, the SPR had lost its two greatest intellectual champions in close succession, the scientific lustre of its highest ranks had diminished with the deaths of such individuals as Adams, Stewart and Stone, and of the prestigious scientific members still living, Crookes, Ramsay, Rayleigh and Thomson no longer had the time or enthusiasm for new investigations with which to bolster the organisation's profile.[9]

One of SPR's greatest difficulties, however, remained confronting a world of "official science" that was ambivalent at best about the fruitfulness of psychical research per se, let alone the arguments for the existence of specific types of psychical phenomenon.[10] Much critical opinion held that the SPR had produced strong but not convincing evidence for the existence of telepathy but that its biggest weaknesses lay in its claims regarding survival and theories of the unconscious.[11] Lodge encountered

---

[5] Oliver Lodge, 'The University in the Modern State', *Nature*, vol. 67 (1902–3), pp. 193–6, p. 196.

[6] Lodge, 'University in the Modern State', p. 193.

[7] Lodge, 'Presidential Address', p. 15.   [8] Lodge, 'Presidential Address', p. 7.

[9] Between 1882 and 1902 the SPR's balance rose from over £200 to nearly £3,000. See [Anon.], 'Anniversary Meeting', *PSPR*, vol. 1 (1882–3), pp. 158–60; [Anon.], 'Report of the Council to Members and Associates of the Society for Psychical Research for the Year 1902', *JSPR*, vol. 11 (1903–4), pp. 38–43.

[10] This is well captured in [Anon.], 'La Métapsychique', *Lancet*, 18 February 1905, pp. 449–50.

[11] [Anon.], 'Modern Spiritualism', *Edinburgh Review*, vol. 198 (1903), pp. 304–29; W. H. Mallock, 'The Gospel of Mr. F. W. H. Myers', *Nineteenth Century and After*, vol. 53 (1903), pp. 628–44.

an especially stark reminder of psychical research's troubles in 1909 when he read the declaration of Simon Newcomb, the veteran American astronomer and one-time president of the American branch of the SPR, that the outcome of 25 years of painstaking investigation into telepathy had been "[s]cientifically, nothing at all".[12] The evidence for this alleged mental faculty only comprised isolated facts that could not be reduced to laws, and the conditions under which it could be reproduced remained unknown. Branding psychical research a form of "modern occultism", a pejorative label that the SPR had spent decades trying unsuccessfully to eschew, Newcomb sided with many critics in warning of the extent to which the evidence of psychical effects was gravely weakened by observational error, lapses in memory and fraudulence.[13]

Lodge's response to Newcomb was a now-familiar critique of authority, insofar as the astronomer's judgements seemed to be based on a woefully limited understanding of what psychical research had achieved – notably the amassing of evidence for telepathy – and the robust methods it had employed to do so. What particularly surprised Lodge was that Newcomb had relied on poor sources of information, including a recent popular book by someone Lodge excoriated as a "not specially competent and quite irresponsible journalistic writer".[14] Given the difficult situation in which the SPR found itself, such ill-informed popular works were particularly unwelcome. An obvious solution to this problem was for those who did consider themselves competent to offer their own articles, books and lectures, for similar audiences, that would rebut and displace such misrepresentations.

Leading physical–psychical scientists were, of course, no strangers to these popularisation strategies, especially when they felt the need to defend their authority in psychical research. In this chapter we shall see that from the early 1900s, physical–psychical scientists were more likely to contribute to psychical research by these literary means than by experimental investigation. Their texts gave them liberties that they were not permitted in other works. In official SPR publications, they were obliged to avoid theological controversy to safeguard the claimed objectivity of the

---

[12] Newcomb, 'Modern Occultism', p. 131. Cf. Newcomb's compatriot, the Harvard physicist John Trowbridge, who, in 1903, argued that "there is no science in the subject of telepathy" because the evidence was weak and the phenomena could neither be repeated nor measured: John Trowbridge, 'Telepathy', *Nation*, vol. 76 (1903), pp. 308–9, p. 308.

[13] Long into the twentieth century, psychical research was deemed to be a form of occultism by both sympathisers and critics. See, for example, the contents of the generally sympathetic *Occult Review* and Edward Clodd's critical perspective in his 'Occultism', *Fortnightly Review*, vol. 107 (1920), pp. 757–68.

[14] Lodge, 'Attitude of Science to the Unusual', p. 207. The book was Beckles Willson, *Occultism and Common Sense* (London: T. Werner Laurie, 1908).

organisation; but in popular scientific, philosophical and religious books and journals, they could debate the religious and moral uses of psychical research that they believed might ultimately attract financial support for the SPR.[15] These were the very publications where, as Peter Bowler has shown, scientists, liberal theologians, philosophers and others discussed ways of reconciling scientific or naturalistic and religious interpretations of the cosmos, and where scientists could explore ideas that were increasingly inappropriate in the secularised discourses of professional scientific research and teaching.[16] As Fournier d'Albe perceptively explained in his book on a physical theory of the soul, the tome did "not work under the limitations of a college text-book" or just "give only that which is generally accepted", but took controversial "facts" as the basis for illuminating the "obscure" psychical problems sidestepped by "official science".[17] Like so many of the texts explored in this chapter, Fournier d'Albe's book cut across the distinctions between the physical and what were increasingly upheld as psychical 'sciences', and between scientific and religious ways of doing and knowing – distinctions that were reinforced in such modern scientific institutions as Birmingham University, where Fournier d'Albe had taught for several years.

One of the reasons why writing on psychical research and spiritualism became attractive to Fournier d'Albe and other physical–psychical scientists was because the market for books, articles and other texts on these subjects boomed in the early twentieth century, and especially during and after the First World War, when many of the millions who had lost loved ones in conflict sought solace in evidence of life after death and mind independent of matter.[18] By creating and taking advantage of opportunities to raise the profile of psychical research, physical–psychical scientists were, of course, opening themselves up to scrutiny from a large readership, among whom were many who felt that they had pushed the connections between physics and psychics too far or not far enough.

Busy Men

"Our officers are able men, self-sacrificing men, but they are also busy men, in some cases extremely busy men."[19] Thus, in May 1901, Ferdinand

---

[15] The SPR's policy was expressed by Eleanor Sidgwick in [Eleanor Sidgwick], 'Correspondence', *JSPR*, vol. 6 (1893–4), p. 214.

[16] Bowler, *Reconciling Science and Religion*.

[17] Fournier d'Albe, *New Light on Immortality*, pp. 204–5.

[18] This trend is strikingly illustrated by an analysis of the titles of books and other texts listed in *JISC Library Hub Discover* and the titles and contents of articles in the ProQuest digital library, *British Periodicals Collections I–III*, and the *British Newspaper Archive*.

[19] Schiller, 'Future of the S.P.R.', p. 76.

C. S. Schiller doubted whether the SPR's leading figures had the "time and energy" to compensate for the deaths of the organisation's most industrious founders, and whether it was "rank and file" or new members who would continue where Gurney, Sidgwick and Myers had left off.[20] Schiller's description would not have been inappropriate for the SPR's leading scientific officers who survived into the twentieth century.

The oldest was Crookes. Now in his 70s and boasting a knighthood (in 1897) for his scientific work, this elder statesman of British chemistry and physics no longer had as much energy for the kinds of work he had pursued in earlier decades. Much of the energy he did have was channelled into new researches on radioactivity and spectroscopy, as well as government consultancy and business ventures.[21] His interest in psychical phenomena was unabated: he continued discussing such matters in private, he religiously attended the SPR's London meetings, he had occasional sittings with spiritualist mediums and, above all, he joined the London-based private dining society, the Ghost Club, where he regaled fellow 'ghosts' with recollections of his old investigations of Home, Cook and other mediums, and more recent forays into spirit photography.[22] Yet Crookes was left with little time and energy for anything more in the psychical direction and did not even have time to write a book that was to amplify his old articles on the subject.[23] Inevitably, his reticence prompted much speculation on his attitude towards spiritualism, and particularly his ambiguous views on the existence of disembodied spirits.[24]

Lack of time was no less of a problem for Crookes's younger and more energetic colleagues. Rayleigh's heavy commitments to government advisory and other official scientific duties, as well as ongoing private researches into acoustics, optics and hydrodynamics, were evidently the main reasons why he declined the offer of the SPR presidency in 1901. But as a close relative of Eleanor Sidgwick, Arthur and Gerald Balfour, he could hardly have been unaware of the organisation's activities, and it was doubtless their encouragement that persuaded him to deliver a presidential address in 1919, only six weeks before he died. The address suggested that Rayleigh had had few psychical experiences since the Palladino investigations of 1895 and was as undecided as ever about telepathy, survival and the physical phenomena of spiritualism. Yet

[20] Schiller, 'Future of the S.P.R.', p. 76.
[21] On Crookes's later career see Brock, *William Crookes*, chapters 21–4.
[22] See, for example, W. W. Baggally, 'Some Sittings with Carancini', *JSPR*, vol. 14 (1909–10), pp. 193–211; Crookes's letters to Lodge in SPR.MS 3/344–67, OJL-SPR; Minute Books vols. 4–8, Add. MS 52261–52265, GC-BL.
[23] William Crookes to Oliver Lodge, 15 March 1913, SPR.MS 35/359, OJL-SPR.
[24] See, for example, 'Perplexed', 'Is Sir William Crookes a Spiritualist?', *Light*, vol. 25 (1905), p. 35.

there was no doubt in Rayleigh's mind that the need to build convincing evidence for telepathy remained strong and that scientists still needed to be warned of their unscientific habit of dismissing psychical effects simply because they conflicted with "ordinary experience".[25]

The institutional burden of Rayleigh's successor at Cambridge's Cavendish Laboratory, J. J. Thomson, and his leadership of research into positive rays and other aspects of the electrical discharge through gases meant that he, like Rayleigh, could only sustain armchair interests in psychical research. With Eleanor Sidgwick on campus, there was no need for Thomson to worry that psychical research lacked powerful local advocates who could encourage students and staff to get involved.[26] But Thomson was not too busy to inspire students of physics to at least contemplate the possible relevance of physics to psychical research. In the early decades of the twentieth century, many physics undergraduates read the multi-volume *Textbook of Physics* that he co-authored with Lodge's Birmingham colleague, John Henry Poynting. Significantly, in explaining the "universal rule" that the disappearance of one form of physical energy was always followed by the appearance of another, one volume of the work tantalisingly suggested that if telepathy were "placed beyond question" then the energy associated with it was probably converted into another, unknown form as it transited from one person to another.[27]

The most committed of all SPR physical scientists, Barrett and Lodge, were no less busy than Rayleigh and Thomson in the early twentieth century. Barrett was greatly preoccupied by the relocation and expansion of the Royal College of Science for Ireland, but still managed a modest output of physical research in the areas of metal alloys and physiological optics, a demanding schedule of public lecturing activities, and the local charitable work that he had pursued in Dublin for decades.[28] Lodge's administrative duties as Birmingham principal, and especially the task of expanding an institution in a parlous financial state, proved so onerous that he was forced to delegate much of his new physical research in wireless and cable telegraphy to assistants such as Benjamin Davies.[29]

[25] Rayleigh, 'Presidential Address', p. 285.
[26] Distinguished Newnham College members who were involved with the SPR during Sidgwick's time there as college principal included the geologist Gertrude Elles, the ornithologist Agnes Elliot, the plant geneticist Edith Saunders, the zoologist Alice Johnson, and the classicists Jane Harrison, Margaret de Gaudrion Verrall (née Merrifield), Helen Woolgar de Gaudrion Salter (née Verrall), and Florence Stawell.
[27] J. H. Poynting and J. J. Thomson, *A Textbook of Physics. Heat* (London: Charles Griffin, 1904), p. 115.
[28] On the transformation of the college see *Royal College of Science for Ireland: Its Origin and Development* (Dublin University Press, 1923).
[29] On Davies's work see Clow, 'Laboratory of Victorian Culture', chapters 4–5; Richard Noakes, 'Industrial Research at the Eastern Telegraph Company, 1872–1929',

Lodge's research publications certainly declined after 1900, and this is consistent with his later private complaint that the principalship hampered his ability to keep abreast of modern developments in physics.[30]

Barrett and Lodge were genuinely busy, but this had more of an effect on *what* they chose to contribute to psychical research than *how much* they contributed. Barrett's investigative work tended to focus on that which could be done from Ireland, including local cases of poltergeists and mediumship, and his renowned large-scale study of water-divining, which combined experimental tests of dowsers in the Irish countryside and critical analyses of historical and modern testimony.[31] Much of this latter research, which Barret believed yielded strong evidence of an obscure sensory capacity (akin to clairvoyance) by which dowsers perceived hidden springs and metallic ores, was presented to the Dublin branch of the SPR, which Barrett had founded in 1908 and which successfully catered to local interest in psychical research.[32] The Dublin branch gave Barrett a more accessible forum for psychical research, but it also symbolised his increasing misgivings about the main organisation, whose leadership's "autocratic treatment" of some members' investigations and overly cautious approach to the physical phenomena of spiritualism exasperated him almost to the point of resignation.[33] To express what he believed psychical research had been invented to accomplish, Barrett also turned increasingly to publications and venues outside the SPR's direct control.

Lodge did not share Barrett's misgivings about the SPR leadership's cautious approach to phenomena (especially for 'public' purposes), but was increasingly disappointed that it had not pursued physical phenomena with the seriousness of European scientific practitioners with psychical interests.[34] These were the individuals who seemed to be trying harder

---

British Journal for the History of Science, vol. 47 (2014), pp. 119–46; Roberts, 'Training of an Industrial Physicist'. On the University of Birmingham's financial difficulties see Ives, Drummond and Schwarz, *First Civic University*, chapter 8.

[30] Oliver Lodge to Andrew C. Bradley, 30 April 1923, typescript copy, OJL/1/55/6, Oliver Lodge Papers, Cadbury Research Library, University of Birmingham. See also Theodore Besterman, *A Bibliography of Sir Oliver Lodge F.R.S.* (Oxford University Press, 1935) and the declining number of entries in Lodge's research notebooks: Lodge research notebook no. 9, MS.3.18, Oliver Lodge Papers, University of Liverpool Library.

[31] William F. Barrett, 'Poltergeists: Old and New', *PSPR*, vol. 25 (1911), pp. 377–412; William F. Barrett, 'On the So-Called Divining Rod or *Virgula Divina*', *PSPR*, vol. 13 (1897–8), pp. 2–282; William F. Barrett, 'On the So-Called Divining Rod', *PSPR*, vol. 15 (1900–1), pp. 130–383.

[32] [Anon.], 'Society for Psychical Research: Dublin Section', *JSPR*, vol. 14 (1909–10), pp. 63–4.

[33] William F. Barrett to Oliver Lodge, 21 October 1912, SPR.MS 35/73, OJL-SPR.

[34] Oliver Lodge, 'Introduction to the Earl of Dunraven's Record of Experiences with D. D. Home', *PSPR*, vol. 35 (1925), pp. 1–20, esp. pp. 4–5.

than anybody else to build on the examples of Crookes, Varley and others in employing the resources of experimental physics in psychical research. Given his 1894 vision of a 'psychical laboratory', Lodge must have envied the foundation, in the early 1900s, of laboratories at the Institut Général Psychologique in Paris and at Baron Albert von Schrenck-Notzing's palatial residence in Munich, both of which were stocked with instruments for detecting, measuring and controlling the physical phenomena associated with mediumship.[35] It was at the Insitut that such fellow scientists as Éduoard Branly, Marie and Pierre Curie and Paul Langevin participated in physiological, mechanical and electrical tests of Palladino, which effectively supported his, Richet's and Myers's conclusion that, in the midst of fraudulent activity, there were some genuinely novel phenomena. Of particular interest to Lodge would have been the evidence that Palladino could discharge electroscopes without touching them, thus suggesting that the strange vital prolongations from her body had ionising properties.[36]

In many ways, however, Lodge's decisions about how he was going to contribute to psychical research may have been partly responsible for the SPR's failure to follow the European examples. He was certainly not too busy for fresh investigative work, as demonstrated by further seances with Leonora Piper, whom he brought over from Boston in 1906, and the more notorious sittings (in 1915–16) with the British medium Gladys Osborne Leonard, through whom he claimed to have gained proof of the survival of his son Raymond, who had been killed in the First World War.

Participating in investigations of physical mediumship where the resources of experimental physics were most likely to matter, however, seems to have caused Lodge more difficulties. Part of the problem was that most of the mediums who offered the best opportunities for this resided overseas, and Lodge's academic commitments would have made such trips difficult to schedule. He even found it difficult to personally investigate cases of physical phenomena much closer to home. In late 1910, he had to ask Davies to visit the London clinic of the medical electrician Walter J. Kilner, who, inspired by the Reichenbach investigations of his former mentor William H. Stone, claimed to have constructed a glass screen filled with a special dye through which the human 'aura' could be

---

[35] On the Institut see Lachapelle, *Investigating the Supernatural*, pp. 75–85. On Schrenck-Notzing's laboratory see Wolffram, *Stepchildren of Science*, chapter 3.

[36] Jules Courtier, *Documents sur Eusapia Palladino. Rapport sur les séances d'Eusapia Palladino à l'Institut général psychologique en 1905, 1906, 1907, 1908* (Paris: Institut général psychologique, 1908), pp. 508–11.

seen.[37] Davies found himself acting as Lodge's psychical deputy again in 1916, when he witnessed the mechanical tests devised by the Belfast academic engineer William J. Crawford that appeared to yield powerful new evidence of mediumistic levitation.[38] Equally telling was Lodge's response to Harry Price's request, in 1925, for support in the foundation of Britain's first psychical laboratory, an organisation designed specifically to implement some of the investigations into the physical phenomena of spiritualism that Lodge cherished. The physicist disappointed this rising star of British psychical research on the grounds that he had too many other commitments in "writing and calculating", which he now regarded as his "special function".[39]

Lodge's later view of himself as primarily a writer and calculator highlights the extent to which the sharp decline in his output of original psychical investigations owed more to a conscious career choice than dwindling opportunities. After accepting the Birmingham position, he saw himself as much more of a public figure than ever before, and as someone who needed to engage with a variety of audiences on a far wider range of topics than he had previously published or lectured on.[40] The marked increase in his literary output and the frequency of his public lectures testifies to the seriousness with which he took this goal after 1900.[41] There were intellectual and moral motivations for this. Lodge clearly felt passionate about many topics – notably ether physics, psychical research, the liberalisation of Christian theology, and social reform – and

---

[37] Davies concluded from his visit that it was hard to tell whether the aura he saw surrounding his hands was not a "fancy": copy of a letter from Benjamin Davies to Oliver Lodge, 20 February 1911, letter 564, File 5, Box 3, BD-NLW. His conclusion supported Lodge's earlier expectation that the effect was as dubious as N-rays: Oliver Lodge to Benjamin Davies, 15 December 1910, letter 560, File 5, Box 3, BD-NLW. See also Kilner, *Human Atmosphere*.

[38] Oliver Lodge to Benjamin Davies, 23 February 1916, letter 618, File 5, Box 3, BD-NLW. Davies was reasonably impressed with Crawford's experiments, but Lodge's interest owed more to Barrett, whose attendance at seances with Crawford and his medium (Kathleen Goligher) in 1915 had persuaded Barrett that table movements could not be due to physical contact with participants: William F. Barrett, 'Report of Physical Phenomena Taking Place at Belfast with Dr. Crawford's Medium [1916]', *PSPR*, vol. 30 (1918–19), pp. 334–7. Lodge also emphasised the importance of Barrett's testimony to Fournier d'Albe, who, in 1921, concluded from tests of Goligher that her telekinetic and ectoplasmic effects were fraudulently produced. See Oliver Lodge to E. E. Fournier d'Albe, 17 August 1921, E. E. Fournier d'Albe Papers, Private Collection, Christine Fournier d'Albe; E. E. Fournier d'Albe, *The Goligher Circle, May to August 1921* (London: John M. Watkins, 1922). For Crawford see Allan Barham, 'Dr. W. J. Crawford, His Work and Legacy in Psychokinesis', *JSPR*, vol. 55 (1988), pp. 113–38.

[39] Oliver Lodge to Harry Price, 9 May 1925, HPC/4B/150, Harry Price Collection, Senate House Library, University of London.

[40] Peter J. Bowler, *Science for All: The Popularisation of Science in Early Twentieth-Century Britain* (Chicago University Press, 2009), pp. 219–20.

[41] Besterman, *Bibliography of Sir Oliver Lodge*.

sought to exploit his academic authority in shaping public understanding of them. But there were financial motivations too. Like Crookes, Lodge saw scientific work as a diverse source of income and by plying skills in teaching, public lecturing, writing, commercial consultancy, patenting and other activities he could ensure that his large family had a respectable quality of life.[42] Lodge had long experience negotiating financially advantageous agreements with publishers and journal editors and by 1900, now boasting scientific and intellectual credentials that made him a valuable commodity in the literary marketplace, he could make writing even more profitable.[43] Many of Lodge's books largely comprised reprints or revised versions of published articles for which he had already been remunerated, and from which he clearly believed more monetary value could be extracted.

Even before he arrived at Birmingham, Lodge was aware that he could make a bigger difference to the profile of psychical research by writing and lecturing than by any other means. In 1897, his former Liverpool colleague Arthur Chattock had opined that it was mainly because of his writings that "centres of interest" in psychical research had emerged in various colleges, including those represented by Chattock's and Lodge's own students.[44] But neither Lodge nor any other physical–psychical scientist seems to have been able to turn such "centres of interest" into anything institutionally more concrete. Lodge, who wielded more power within his academic institution than most other physical–psychical scientists, had contemplated something along these lines in 1905 – a professorship or a permanent endowment – but seems to have been persuaded against taking it any further.[45] He clearly believed that a more effective way of bringing psychical research within the University of Birmingham's walls was via lectures to students.[46]

---

[42] On Lodge's consultancy and commercial ventures see Anna Guagnini, 'Ivory Towers? The Commercial Activity of British Professors of Engineering and Physics, 1880–1914', *History and Technology*, vol. 33 (2017), pp. 70–108, esp. pp. 78–93. On his patenting see Stathis Arapostathis and Graeme Gooday, *Patently Contestable: Electrical Technologies and Inventor Identities on Trial in Britain* (Cambridge, MA: MIT Press, 2013), esp. chapter 6.

[43] See Lodge's correspondence with Macmillan's regarding various editions of such volumes as *Modern Views of Electricity* (1889) and *Pioneers of Science* (1893): ff. 49–173, Add. 55220, Macmillan and Company Ltd Archive, British Library.

[44] Arthur P. Chattock to Oliver Lodge, 11 April 1897, MS ADD 89/23, OJL-UCL.

[45] Arthur Rücker to Oliver Lodge, 31 December 1905, MS Add. 89/91, OJL-UCL. Rücker, a physicist, university administrator and SPR member, questioned Lodge's proposals on the grounds that psychical "facts" were not yet sufficiently reproducible or testable to constitute the "solid" basis on which an academic chair or endowment fund could be justified. There is no mention of Lodge's proposal in the University Council's Minute Index Books from 1903 to 1906: UB/COU/3/3–4, University of Birmingham Special Collections.

[46] See, for example, Oliver Lodge, 'University of Birmingham. Lectures to Medical Practitioners on Physics Applied to Medicine', Summary of Lectures V and VI in Lodge, *University of Birmingham. Lectures to Medical Practitioners on Physics Applied to*

Anyone who missed Lodge's lectures could easily have apprehended his views on psychical research via the astonishing number of publications he managed to produce from the early 1900s. Lodge's literary output on psychical and other subjects certainly helped him become one of the best-known scientific personalities of the day, and the sheer volume of his writings on what he often described as "physics and psychics" means that he necessarily dominates this chapter.[47] Lodge's output certainly dwarfs that of most other physical–psychical scientists who survived into the twentieth century and who contributed very little to the literature on psychical research, or what, from the early twentieth century onwards, was increasingly referred to as 'psychic research', 'psychic science' or simply 'psychics'. However, Lodge's work draws attention away from some physical–psychical scientists, both inside and outside the SPR, who contributed important other perspectives on 'physics and psychics'. As we shall see, what they managed to contribute repays close attention and demonstrates that Lodge's views did not represent a consensus.

Applied Psychical Research

In *Survival of Man* (1909), his first and commercially most successful book on psychical research, Lodge explained that while religion encroached too closely on the "region of emotion to be altogether suitable for consideration" by such a "scientific Society" as the SPR, it was one of the "practical applications" of psychical research by which the value of the fledgling science to humanity would be judged.[48] Lodge thus captured an important place that religious questions occupied in the literary activities of physical–psychical scientists in the early twentieth century. The religious significances that he and others gave to psychical research would also give value to a welter of new proposals regarding the possible connections between 'physics and psychics'. Yet as a book primarily focussed on analysing the SPR's evidence for telepathy and survival (including his own studies of Piper's mediumship), *Survival of Man* was typical of many early popular and semi-popular books on psychical research in sidestepping detailed discussion of religious applications. In concluding his volume on the subject for a new series of popular science books, Barrett

---

Medicine (University of Birmingham, 1904), pp. 11–13. Copy at OJL3/3/103, Oliver Lodge Papers, Cadbury Research Library, University of Birmingham.

[47] See, for example, Lodge, *Beyond Physics*, p. 7; Oliver Lodge, *Phantom Walls* (London: Hodder and Stoughton, 1929), p. 99.

[48] Lodge, *Survival of Man*, p. 36. By 1926, this book had reached its 10th and final edition, and French, German and Italian editions were published.

warned that psychical research, "though it may strengthen the foundations, cannot take the place of religion" because it only concerned itself
with the "*external*, though it be in an unseen world" and at best provided
a "stepping stone in the ascent of the soul to its own self-apprehension".[49]

In texts that were *not* primarily designed to persuade readers of the
scientific credibility of psychical research's evidence, Barrett and Lodge
were far less restrained in their religious discussion, and on many fundamental questions agreed with other writers on the religious significances
of psychical research and spiritualism.[50] Moreover, as Peter Bowler has
shown, their writings were among the intellectually weightier contributions to the vigorous early-twentieth-century debate in Britain on the
relationships between the sciences and religions.[51] They joined company and entered dialogues with a host of professional scientists, science
popularisers, philosophers and representatives of liberalising and modernising trends in Christian theology, who devised new reconciliation
strategies based largely on relatively recent idealistic, vitalistic and
otherwise non-mechanistic trends in physical, biological and psychological thought.

Barrett's and Lodge's output of religious writing was far greater than
that of most other physical–psychical scientists of the early twentieth
century, who, even in forms of publication where there was ample room
for religious discussion, tended to repeat the common trope that psychical
research led to a more satisfactory understanding of the future state than
either Christian dogma or materialistic science. For example, in the early
1920s, the leading British forum of religious and philosophical debate the
Hibbert Journal published articles on telepathy and survival by the physical
chemist and barrister William R. Bousfield. He criticised those whose
early religious education made them blind to new evidence of "miraculous" phenomena indicating the operation of "higher" natural laws and
whose scientific education nurtured a "materialistic complex" that made
them abandon all ideas about God and the future life and reject the
religious implications of psychical research.[52] Despite its brevity,
Bousfield's religious application of the results of psychical research was
still more extensive than that of most other physical–psychical scientists,

---

[49] Barrett, *Psychical Research*, p. 246. The series was the Home University Library. On this
and similar book series see Bowler, *Science for All*, chapter 7.
[50] Examples of such writers include Arthur Conan Doyle, *The New Revelation* (London:
Hodder and Stoughton, 1918) and James Hervey Hyslop, *Psychical Research and the
Resurrection* (Boston: Small, Maynard and Company, 1908).
[51] Bowler, *Reconciling Science and Religion*.
[52] Bousfield, 'Telepathy', pp. 498–9. See also William R. Bousfield, 'Human Survival',
Hibbert Journal, vol. 22 (1923–44), pp. 501–14.

who, like an increasing number of professional scientists, do not seem to have been willing or able to engage publicly in religious debates at all.[53]

From its launch in 1902 until the 1910s, the *Hibbert Journal* was strongly associated with some of Lodge's most provocative interventions on religious and philosophical questions of the day, although much of what he wrote had been rehearsed since the late 1890s in discussions at the Synthetic Society, whose surviving members would also contribute to the periodical. Lodge's earliest contribution – a 1902 essay on the "outstanding controversy" between science and Christian faith – marked a significant change from his earlier reticence on the religious applications of psychical research.[54] Nine years earlier, he had hinted that he was aware of such applications but was clearly unable or unwilling to go further: he merely suggested that studies of both psychical phenomena and the ether would eventually provide a "glimpse into a region of the universe which Science has never entered yet" but which had been "perhaps blindly apprehended, by painter or poet, by philosopher or saint".[55]

The confidence that Lodge showed in his later arguments for connections between psychical research, ether physics and religion owed much to the "tuition" that he received from the person he would later venerate as "among the chief influences in my life": Myers.[56] From the 1880s, the two men corresponded at least once a week and frequently met at each other's homes or at meetings of the SPR and other learned societies.[57] Among the major consequences of such encounters were Lodge's abandonment of his belief, encouraged by such early teachers as Tyndall and Huxley, that mind and body were inseparably connected, and his appreciation of Myers's interpretations of the evidence amassed by psychical research. One of the most important of these interpretations was Myers's idea of a "metetherial" environment.[58] This environment, which Myers believed was continuous with but ultimately distinct from the ether of

[53] Bowler, *Reconciling Science and Religion*, p. 20.
[54] Lodge, 'Outstanding Controversy'. Complementing the analysis here are John D. Root, 'Science, Religion and Psychical Research: The Monistic Thought of Sir Oliver Lodge', *Harvard Theological Review*, vol. 71 (1978), pp. 245–63; David B. Wilson, 'On the Importance of Eliminating *Science* and *Religion* from the History of Science and Religion: The Cases of Oliver Lodge, J. H. Jeans and A. S. Eddington', in Jitse M. van der Meer (ed.), *Facets of Faith and Science Volume 1* (Lanham, MA: University Press of America, 1996), pp. 27–47.
[55] Lodge, 'Interstellar Ether', p. 862.
[56] Lodge, *Past Years*, p. 220. See also Lodge, 'The Life Work of My Friend F. W. H. Myers', *Nature*, vol. 144 (1939), pp. 1027–8.
[57] Their friendship is evident from the correspondence at SPR.MS 35/1298–1572, OJL-SPR.
[58] Myers, *Human Personality*, vol. 1, p. 216.

space, was the permanent location of our "spiritual life", which psychical research had "proved by actual evidence" did not depend on the "material world" or our bodily existence.[59]

Lodge's debt to Myers is particularly apparent in an incestuous review that he wrote of the book from which the above quotes have been taken: Myers's posthumous magnum opus *Human Personality and its Survival of Bodily Death* (1903). He explained that his late friend's "admission of telepathy as the *vera causae* in psychical science" opened the "flood-gates to a torrent of new ideas", including the possibility of a "channel of communication" between humans and between both incarnate and discarnate "intelligences in the universe".[60] The communication channel probably extended to the divine, which would provide the much-sought-after "rational interpretation" of the efficacy of prayer and the Christian doctrine of the communion of saints.[61] Myers's book had also rendered the afterlife less abstract and altogether less terrifying a prospect. According to Lodge, Myers had suggested that the conditions in the future life were not "revolutionarily different": our personalities were, for a long time after death at least, essentially the same, even though they were constantly being enlarged by the "etherial" environment in which they now resided.[62] This environment may not have been "material" but, reassuringly, it still gave us a "kind of semi-bodily existence", which Christian writers called the "spiritual body" but which could also be regarded as a "sort of etherial" one.[63]

Lodge's debt to Myers is also obvious from his earliest *Hibbert Journal* articles, the most substantial of which he adapted for re-publication in his first and commercially most successful theological book, *Man and the Universe* (1908).[64] Here, Lodge invoked the well-established metaphor of the conflict between science and religion. The kernel of the "controversy" was that "orthodox modern science" regarded the universe as "self-contained and self-sufficient" and denied that any supposed "transcendent beings" could influence it, while religion (by which he meant Christianity) required humanity to be in close contact with "a power, a mind, a being or beings, entirely out of our sphere, entirely beyond our

---

[59] Myers, *Human Personality*, vol. 1, p. 215.

[60] Oliver Lodge, 'The Survival of Personality', *Quarterly Review*, vol. 198 (1903), pp. 211–29, p. 220. On *Human Personality* see Carlos Alvarado, 'On the Centenary of Frederic W. H. Myers's *Human Personality and Its Survival of Bodily Death*', *Journal of Parapsychology*, vol. 68 (2003), pp. 3–43; Gauld, *Founders of Psychical Research*, chapters 12 and 13; Hamilton, *Immortal Longings*, chapters 5 and 9.

[61] Lodge, 'Survival of Personality', p. 220.   [62] Lodge, 'Survival of Personality', p. 226.

[63] Lodge, 'Survival of Personality', p. 226.

[64] Lodge, *Man and the Universe*. By 1928, the book had reached its 21st and final edition, and Czech and Danish editions were also published.

scientific ken", agencies on which the origin, maintenance and progress of the universe depended.[65] The reconciliation between these extreme positions hinged critically on one question – the efficacy of prayer – because this represented the most obvious example of external guidance of the apparently self-contained universe. Telepathy provided the most important step towards answering this question because if

> we are open to influence from each other by non-corporeal methods, may we not be open to influence from beings in another region or of another order? And if so, may we not be aided, inspired, guided, by a cloud of witnesses – not witnesses only, but helpers, agents like ourselves of the immanent God?[66]

The evidence for telepathy, which Lodge and many others now believed to be conclusive, suggested that prayer was an aspect of a cosmic law "hitherto unimagined by science" and through which transcendent beings lovingly guided humanity on a path of moral and spiritual progress.[67] By the 1920s, Lodge would turn evidence of immaterial guidance into contributions to a growing literature attempting to reconcile biological evolution with the idea of a beneficent creative power.[68]

Myers's "tuition" featured especially strongly in Lodge's critical approach to Christian theology and his particular argument that the Christian teachings on the soul were strengthened by the evidence of telepathy and post-mortem communications, which suggested that mind existed independently of the material body and that the essential constituent of human personality was "permanent".[69] This argument was one of many ways in which Lodge ultimately hoped to free "official" Christianity from outdated and unintelligible doctrines and to tackle what he perceived to be the corresponding popular indifference to the "outward and visible" forms of the church.[70] Christianity would be less depressing if Christ's Resurrection was seen as the "pattern" for the resurrection of all Christian souls and, moreover, as a form of survival, or continuity in the existence of an immaterial 'body' or soul carrying the "undying essence or spirit" rather than the "infantile" and unproven idea

---

[65] Lodge, 'Outstanding Controversy', p. 49.

[66] Lodge, 'Reconciliation Between Science and Faith', p. 223.

[67] Lodge, 'Outstanding Controversy', p. 60. See also Oliver Lodge, 'Religion, Science and Miracle', *Contemporary Review*, vol. 86 (1904), pp. 798–807.

[68] See, for example, Oliver Lodge, *Evolution and Creation* (London: Hodder and Stoughton, 1926). For Lodge's evolutionary writings see Bernard Lightman, 'Lodge and the New Physics, 1919–1933', in Graeme Gooday and James Mussell (eds.), *Oliver Lodge: Continuity and Communication, 1875–1940* (Pittsburgh, PA: University of Pittsburgh Press, in press).

[69] Oliver Lodge, 'The Immortality of the Soul. Part II. The Permanence of Personality', *Hibbert Journal*, vol. 6 (1907–8), pp. 563–85, p. 563.

[70] Oliver Lodge, 'The Alleged Indifference of Laymen to Religion', *Hibbert Journal*, vol. 2 (1903–4), pp. 235–41, pp. 84 and 90.

of the particles of the material body being resuscitated.[71] Both Christian doctrines of the resurrection of the 'body' and the immortality of the soul would be more appealing if the soul was regarded as the immaterial yet still physical vehicle of spirit, rather than something closely tied to the material body that dissolved at death. For Lodge, the pursuit of the ether problem promised to illuminate these and other profound theological questions: its apparently perfectly continuous, robust, immaterial but physical nature made it a possible candidate for the substance of the soul or spiritual body and the means by which divine spirit interacted with the material world.

Lodge's ideas about the ways that Christian doctrine could be reinterpreted in light of psychical research changed very little from the early 1900s to his death in 1940. His high intellectual profile ensured that they provoked much debate. His ideas appealed to many fellow Christian-minded scientists and to proponents of liberalising and modernising trends in Christianity, but alarmed plenty of agnostics and Christian traditionalists alike.[72] Critics saw his attacks on such doctrines as the Atonement, the Resurrection and the Virgin Birth as hopelessly amateur, his claims for divine immanence as essentially pantheistic, his appeal to the evidence for survival as morally and theologically perilous, and his preoccupation with the etherial or other physical vehicles of spirit as too materialistic.[73] By the 1930s, the important role that Lodge claimed for the ether in his modernising efforts regarding the soul was also proving to be a major weakness for many Christian thinkers. For these individuals, the ether's existence had been cast into doubt by the theory of relativity, which, as interpreted by Arthur Stanley Eddington, James Jeans and other popularisers of the new physics, represented a more promising scientific vindication of the Christian idea that the physical universe was an expression of the Divine Mind.[74]

---

[71] Lodge, *Man and the Universe*, pp. 284–5, 292.

[72] Sympathisers included the Scottish biologist Patrick Geddes, the German physicist Phillip Lenard and the major British exponent of 'new' and 'liberal' Christian theology Reginald Campbell: Geddes cited in A. H. Tabrum, *Religious Beliefs of Scientists* (London: North London Christian Evidence League, 2nd ed., 1913), pp. 70–2; Phillip Lenard to Oliver Lodge, undated letter, MS Add. 89/66, OJL-UCL; Reginald J. Campbell, *A Spiritual Pilgrimage* (London: Williams and Norgate, 1917), esp. p. 258.

[73] Examples of criticism include: [Anon.], 'Professor Lodge's Theology', *Church Times*, vol. 50 (1908), p. 767; [Anon.], 'Science', *Athenaeum*, 28 November 1908, pp. 686–9; Charles Gore, *The New Theology and the Old Religion* (London: John Murray, 1907); Joseph McCabe, *The Religion of Sir Oliver Lodge* (London: Watts and Co., 1914); E. S. Talbot, 'Sir Oliver Lodge on "The Re-interpretation of Christian Doctrine"', *Hibbert Journal*, vol. 2 (1903–4), pp. 649–61.

[74] See, for example, E. O. James, 'The Return of Materialism', *Church Times*, vol. 104 (1930), pp. 161–2. On Eddington, Jeans and religion see Bowler, *Reconciling Science and Religion*, pp. 101–21; Matthew Stanley, *Practical Mystic: Religion, Science and*

Barrett had been writing about the relationship between one of the focal points of psychical research – spiritualism – and religion long before Lodge. Many of the fundamental hopes and fears that he had for the religious uses of spiritualism were unchanged by his later experiences in psychical research. His first book on spiritualism, *On the Threshold of a New World of Thought* (1908), reiterated old arguments that it showed the "immanence of the spiritual world" that could undermine a "paralysing materialistic philosophy" and provide a welcome "hand-maid" to Christian faith.[75] It also repeated his oft-expressed warnings to those who treated spiritualism as a religion in its own right, and failed to distinguish between the often gross and deceptive phenomena of the seance and the true spirituality to which they faintly pointed.

Barrett's hopes for the Christian uses of spiritualism were strengthened and complemented by the religious interpretations that he and others made of telepathy. Like Lodge, he believed it gave credence to Christian ideas of prayer, divine inspiration, and of a Divine Mind that transcended, unified and gave meaning to the material cosmos, and he later claimed for such an immaterial influence a role in biological evolution.[76] Yet this Congregationalist minister's son proved he was more orthodox than Lodge in eschewing radical reinterpretations of fundamental Christian doctrines, notably the immortality of the soul. While Lodge believed that we were born with a "permanent human element" and that psychical research could point to this, Barrett insisted that there was no Christian warrant for the natural immortality of the soul and that this was accordingly beyond the scope of psychical research.[77]

Lodge's deviation from Christian orthodoxy was, however, moderate compared to that of William Kingsland. By the early 1900s, he had largely turned from electrical engineering to scientific and philosophical writing, and was using the results of psychical research and the new understandings of the nature of matter and ether to develop what he upheld as a "truly *Scientific Idealism*".[78] While Kingsland was rare among British writers on idealism in his engagement with the sciences, his appeal to psychical research and physics was hardly original given that at least one

A. S. Eddington (Chicago University Press, 2007); Wilson, 'On the Importance of Eliminating *Science* and *Religion*'.

[75] Barrett, *Threshold of a New World of Thought*, pp. 5 and 100.

[76] Barrett, 'Psychic Factor'; William F. Barrett, 'The Spiritual Significance of Nature', *Contemporary Review*, vol. 105 (1914), pp. 791–9.

[77] Oliver Lodge, *Why I Believe in Personal Immortality* (London: Cassell and Company, 1928), p. 148; Barrett, *Threshold of a New World of Thought*, p. 78.

[78] Kingsland, *Scientific Idealism*, p. vii. For Kingsland's strong allegiance to Blavatsky and his attack on the SPR's exposure of her see William Kingsland, *The Real H. P. Blavatsky. A Study in Theosophy, and the Memoir of a Great Soul* (London: John M. Watkins, 1928).

of the scientific authorities on whom he drew for his understanding of the ether – Lodge – had already been discussing the congruent tendencies of physics to "elevate matter and all existence to the level of mind and spirit" and of psychical research to elevate mind over body.[79] But Kingsland's idealism differed markedly from Lodge's in being far less indebted to psychical research and concerned with the religious benefits of this enterprise. Its principal debt was to modern Theosophy, a form of occultism with which Kingsland had been strongly associated since the 1880s, but which Lodge seems to have disliked and which certainly divided opinion among physical–psychical scientists.[80]

Kingsland essentially updated modern Theosophy's approach of challenging what was perceived to be materialistic science and Christian theological dogma with a synthesis of 'secret' knowledge in world religions and philosophies (including esoteric forms of Christianity, Buddhism and Hinduism), and modern scientific revelations about the hidden aspects of human psychology and matter. Psychical research and physical science, he insisted, confronted the same problem of tracing things to the same higher plane of existence. Just as psychical research showed that our normal consciousness, personality and memory were fragments of a far larger self operating on an "astral" or "etheric" plane, so physical science was resolving matter into the "substance" of that same, higher plane.[81]

Scientific evidence of this higher plane lent support to the idea of still higher and subtler planes of consciousness and reality that modern Theosophy had been teaching for decades. Indeed, the key to the reconciliation between materialism and idealism lay in the fact that the highest of the seven macrocosmic planes taught in modern Theosophy was the infinite, eternal "Absolute Primordial Substance" or "Noumenon".[82] As the ultimate embodiment of cosmic unity, the Absolute sustained "inherent" motions that gave rise on the one hand to consciousness and subjectivity, and on the other to matter and objectivity.[83] On this basis the entire universe, including matter, was conscious and spiritual, and distinctions between material and spiritual, and natural and supernatural were illegitimate. The claimed unity between the inner self and what Kingsland called

[79] Lodge, 'Scope and Tendencies of Physics', p. 355. On British idealism and the sciences see W. J. Mander, *British Idealism: A History* (Oxford University Press, 2011), esp. 547–51.

[80] Oliver Lodge, untitled paper dated 26 January 1904, in *Papers Read Before the Synthetic Society 1896–1908*, pp. 460–2, p. 460. Barrett, however, was interested in the modern Theosophical interpretations of karma and reincarnation: Barrett, *On the Threshold of a New World of Thought*, pp. 42– 3n. For Crookes's Theosophical interests see Brock, *William Crookes*, pp. 337–43.

[81] Kingsland, *Scientific Idealism*, p. 361.   [82] Kingsland, *Scientific Idealism*, p. 286.

[83] Kingsland, *Scientific Idealism*, p. 163.

the "Self of the Universe" reflected modern Theosophy's teaching of the divinity of all humans, their capacity to apprehend the higher planes of reality, and their evolutionary development towards truly divine beings, of whom Christ was an exemplar.[84]

There were some overlaps between Kingsland's form of scientific idealism and Lodge's writings on Christian theology and new conceptions of matter and ether. Both sought to use modern science to rid Christian doctrine of superstitious accretions; both challenged the distinctions that orthodox Christianity made between God and humanity; both appealed to the ether of space as the connecting link between material and spiritual domains; and both saw the moral and spiritual evolution of humanity as part of the upward progress of the whole universe. But for all its unorthodoxy, Lodge's theology remained Christian because it accepted the "unique degree" to which God had revealed himself to humanity via the Incarnation of Jesus Christ, a teaching whose "hold" upon the "race" would strengthen if incarnation was seen as something happening to "ordinary humanity".[85] Kingsland's idealism and, a fortiori, the Theosophical teachings on which it built, however, had no interest in privileging Christianity. On the contrary, by replacing the God of Christian theology with an abstract Absolute Principle and by denying that Jesus Christ was more heavily endowed with the divine spirit than Gautama Buddha and other "higher types of men", Kingsland ought not have been surprised to report that Christians were modern Theosophy's "bitterest opponents".[86]

Lodge's Etherial Body

In 1909, after reading Lodge's first popular book on the ether, Crookes praised his friend for "giving this elusive substance a solid place in our thoughts" but confessed that the "apparent contradictions are more appalling than ever!"[87] Crookes merely echoed the frustrations that so many physicists of the early twentieth century had with the hypothetical space-filling medium. The astonishing developments in wireless or 'etherial' telegraphy since the 1890s certainly seemed to provide indirect experimental evidence of the reality of an electromagnetic ether that could be manipulated for long-distance communication purposes. Yet even the ether's staunchest advocates agreed that the entity's constitution

---

[84] Kingsland, Scientific Idealism, p. 217.   [85] Lodge, Man and the Universe, p. 293.
[86] Kingsland, Scientific Idealism, p. 322; William Kingsland, 'Theosophy and Christianity', Lucifer, vol. 14 (1894), pp. 335–40, p. 336.
[87] William Crookes to Oliver Lodge, 22 May 1909, SPR.MS 35/356, OJL-SPR. Crookes was referring to Oliver Lodge, The Ether of Space (London: Harper & Brothers, 1909). The following section develops my 'Making Space for the Soul'.

remained a problem.[88] Attempts to produce direct experimental evidence of the ether's existence had failed and there remained numerous conceptual and theoretical problems. It was still not clear how a medium fluid enough to allow matter to pass through it without any resistance could be rigid enough to transmit the transverse vibrations constituting light; how a medium filling all space transmitted the cohesive electrical, magnetic and gravitational forces between material bodies as well as electromagnetic radiation; and how a medium that was supposed to be perfectly uniform developed the localised singularities that Larmor, Lodge and other physicists regarded as the origins of the electrical or 'electronic' constituents of matter. This latter theory led Lodge to one of the contradictions that probably appalled Crookes: if electrons arose from etherial singularities then the ether's density needed to be about 10^{12} times denser than water, which challenged a long-held assumption that it was supremely rarefied.[89]

Physicists' lack of success in devising satisfactory mechanical explanations of the ether's puzzling physical properties fuelled existing arguments that it could no longer be described in terms of the properties of gross matter and prompted suggestions that it necessitated a new form of mechanics that was more fundamental than Newtonian. While Lodge, Larmor, J. J. Thomson and other British ether theorists had, by 1900, abandoned the old mechanical ethers of William Thomson and Stokes with which they had grown up, they never regarded the ongoing ether problem as a good reason to give up hope for some kind of mechanical explanation or of abandoning the medium per se. This latter path was taken by a younger generation of physicists who admired the conceptual simplicity and mathematical power of Albert Einstein's Special Theory of Relativity, which notoriously rendered the ether superfluous to the analysis of moving electrical and magnetic bodies.

Several years after the publication, in 1905, of Einstein's Special Theory of Relativity, Larmor, Lodge and other elder statesmen of British physics upheld numerous physical and philosophical arguments against abandoning the ether.[90] In his entry on the entity for the 11th edition of the Encyclopaedia Britannica, Larmor noted the narrow "modern" trend of

[88] This was expressed in Joseph Larmor, 'Aether', Encyclopaedia Britannica, 29 vols. (Cambridge University Press, 11th ed., 1910–11), vol. 1, pp. 292–7, esp. p. 294; Lodge, Ether of Space, p. xv; Thomas Preston, The Theory of Light (London: Macmillan and Co., 4th ed., 1912), p. 33.

[89] Lodge, Ether of Space, p. 82.

[90] See Stanley Goldberg, 'In Defense of the Ether: The British Response to the Special Theory of Relativity, 1905–1911', Historical Studies in the Physical Sciences, vol. 2 (1970), pp. 89–125. Contra Goldberg, however, Warwick has argued that the lukewarm reaction of Cambridge mathematical physicists to relativity owed less to an adherence to the ether

merely resting "content" with mathematical laws describing mechanical interactions between bodies at a distance before turning to what he clearly preferred as the "wider view", which filled space with a "dynamical process" or a continuous "aethereal transmitting medium".[91] This "view" was a "predilection" nurtured partly by the successes of continuum mechanics and by a fundamental human sense of exerting mechanical effects on outside bodies via "limbs and sinews".[92] As someone who had upheld the continuous ether as the "foundation" of the human conviction of an "orderly" cosmos, he was bound to declare that the problem of the entity's constitution could be "attacked and continually approximated to" and possibly "definitely resolved".[93]

Lodge's numerous defences of the ether strongly echoed Larmor's arguments. In his widely discussed presidential address to the British Association meeting of 1913, he urged a "conservative attitude" towards new scientific theories that seemed to be undermining the foundations of long-cherished views of the physical world, notably the idea of the "*ultimate* Continuity" of the cosmos.[94] Unable to imagine the "exertion of mechanical force across empty space", he saw the ether as "at least" the

great engine of continuity. It may be much more, for without it there could hardly be a material universe at all. Certainly, however, it is essential to continuity; it is the one all-permeating substance that binds the whole of the particles of matter together. It is the uniting and binding medium without which, if matter could exist at all, it could exist only as chaotic and isolated fragments.[95]

The prime threats to continuity were relativity's assault on the ether and the new quantum theory's revolutionary proposal that energy, including that associated with light, was not continuous but concentrated in discontinuous packets or 'quanta', and that this applied to energy in its absorption and emission by the discrete constituents of matter and in its transit through the allegedly continuous ether. As someone who strongly believed that direct experimental evidence of the engine of continuity

than to Einstein's theoretical techniques, which were alien to those practised in the ancient English varsity: Warwick, *Masters of Theory*, chapter 8.

[91] Larmor, 'Aether', p. 293.   [92] Larmor, 'Aether', p. 293.

[93] Joseph Larmor, 'Physical Aspects of the Atomic Theory [1908]', in Joseph Larmor, *Mathematical and Physical Papers*, 2 vols. (Cambridge University Press, 1929), vol. 2, pp. 344–72, p. 372; Larmor, 'Aether', p. 294. For analysis of British physicists' defences of the 'continuity' of the ether and of the traditions of dynamical physics see Imogen Clarke, 'Ether at the Crossroads of Classical and Modern Physics', in Navarro, *Ether and Modernity*, pp. 14–29.

[94] Oliver Lodge, 'Continuity', in *Report of the Eighty-Third Meeting of the British Association for the Advancement of Science. Birmingham: 1913* (London: John Murray, 1914), pp. 3–42, pp. 15 and 19.

[95] Lodge, 'Continuity', pp. 19 and 27.

would one day be found, Lodge was clearly delighted to help himself to J. J. Thomson's plea of 1909 that, as an ether was required to explain how the earth's inhabitants benefited from the sun's "gifts", its study represented "the most fascinating and important duty of the physicist".[96] The ongoing search for ether's drift relative to the Earth and the continued debate on the ether's existence well into the 1930s suggests that Lodge's and Thomson's views were much more widely shared.[97]

Thomson's reference to the "gifts" conferred by the ether suggests a providentialist interpretation of the ether that was congruent with Lodge's belief in the possibility of the ether's "mental and spiritual functions" and harked back to the religious significances of the ether upheld by Thomson's teachers Stewart and Stokes.[98] Yet Thomson and Lodge were far from being the only physicists of their generation for whom the ether had a religious significance, and that this was relevant to questions about its reality. In 1908, George F. C. Searle, a devout Christian and one of the leading teachers of experimental physics in Thomson's Cavendish, told fellow Anglicans that the ether constituted "evidence of the unity of the universe" and that this evidence led to the "conviction that the whole universe, the ether included, is the work of a single Creator".[99] Searle's interpretation would have compounded the scientific reasons that he had at this time for being indifferent to and baffled by Einstein's Special Theory of Relativity.[100] Religious considerations weighed equally heavily with one of Lodge's old Liverpool colleagues, the electrical engineer William Thornton. As late as 1930, he warned fellow Christians that while relativity was useful "as far as the equations go", the existence of an ether that could potentially store vastly more energy than just the "material universe" rendered more intelligible the idea that the whole universe was the physical immanence of an infinitely energetic God.[101]

Thornton's balancing of the ether's wider significance against relativity's technical power owed much to Lodge. When, in 1919, sensational astronomical evidence was produced in support of Einstein's General Theory of Relativity, Lodge's resolve to broadcast the philosophical virtues of the

[96] Lodge, 'Continuity', p. 27. Lodge was quoting from Thomson's presidential address to the British Association: J. J. Thomson, 'Address', in *Report of the Seventy-Ninth Meeting of the British Association for the Advancement of Science Winnipeg, 1909* (London: John Murray, 1910), pp. 3–29, p. 15.
[97] Swenson, *Ethereal Aether*, chapters 9–12.   [98] Lodge, 'Continuity', p. 27.
[99] George F. C. Searle, 'The Modern Conception of the Universe', in *Pan-Anglican Papers. Being Problems for Consideration at the Pan-Anglican Congress, 1908. Religion and Science* (London: Society for the Promotion of Christian Knowledge, 1908), pp. 1–8, p. 4.
[100] Warwick, *Masters of Theory*, pp. 399–404.
[101] William M. Thornton, *The Scientific Background of the Christian Creeds* (Newcastle-on-Tyne: Andrew Reid and Company, 1930), p. 17.

ether only strengthened. As he insisted in a popular article of 1921, the Special and General Theories of Relativity represented an impressive but ultimately "blindfold mathematical method of arriving at results" relating to constant and accelerated motion, because the method sidestepped the question of physical cause, which an ether, embodying the "solid ground of inductive dynamical physics" and the fundamental human conception of force, answered.[102]

Yet Lodge's response to General Relativity was more equivocal than we might expect from this self-confessed "fervent believer in the Ether of Space".[103] Einstein's "real achievement", he contended, was in extending the work of Faraday, Maxwell and others in raising the importance of 'empty' space and in bringing the elusive force of gravity within the ether's remit. Lodge's argument ignored the fact that the 'ether' that Einstein had recently accepted as the necessary physical basis for the spacetime continuum of General Relativity differed significantly from even the abstract dynamical ether of Larmor and Lodge in being entirely devoid of mechanical qualities.[104] But for Lodge, such differences over the ether's nature were outweighed by the usefulness of Einstein's work in illustrating a welcome tendency in physics towards multiplying the ether's functions. Writing in the same periodical where he had first hinted at the ether's possible psychical significance, he explained that relativity encouraged the idea that the ether's functions might one day include "other forms of existence which for simplicity Science feels it convenient at present to ignore".[105]

Some of Lodge's readers may have known that a few years earlier, in 1919, he had published his first detailed but tentative hypothesis about the ether's relationship to one such form of "existence".[106] This concerned the nature of the "etherial body", which was essentially Lodge's most elaborate attempt to render the Christian idea of the soul or spiritual body more intelligible and appealing than either the vulgar materialistic view that the soul was merely a rarefied form of matter that disappeared at death, or the equally discomforting idealistic view that it was completely disembodied, outside space and time and otherwise beyond human apprehension.[107] Over the remaining decades of his life, it came to

---

[102] Oliver Lodge, 'Einstein's Real Achievement', *Fortnightly Review*, vol. 110 (1921), pp. 353–73, pp. 358 and 366.

[103] Lodge, 'Continuity', p. 15.

[104] Albert Einstein, 'Ether and the Theory of Relativity [1920]', in *Sidelights on Relativity* (New York: E. P. Dutton, 1922), pp. 3–24.

[105] Lodge, 'Einstein's Real Achievement', p. 372.

[106] Oliver Lodge, 'Ether, Matter and the Soul', *Hibbert Journal*, vol. 17 (1918–19), pp. 252–60.

[107] Lodge, *Raymond*, p. 391; Lodge, *My Philosophy*, p. 256.

represent the core of a 'philosophy' dominated by the known and anticipated properties of the ether and provided the key to two related problems
he had pondered for decades: how mind or spirit interacted with matter,
and how 'physics and psychics' were linked.

In Chapter 3 we saw that in 1902 Lodge had briefly discussed the idea
of an etherial body as a possible means by which disembodied spirits
materialised themselves to the living.[108] At this stage, Lodge's hypothesis
was no more sophisticated than the ethereal, etherial or 'astral' body that
spiritualist, Theosophical and occult writers had envisioned as the subtler
physical vehicle of the spirit or instrument by which spirits manifested
themselves on the material plane.[109] It owed something to Myers's notion
of a 'metetherial' world and to the Unseen Universe, which had, like many
texts explored in this study, sidestepped questions about the ultimate
nature of spirit but turned to physics (and particularly ether physics) to
illuminate the nature of the spiritual body.[110]

By 1919, Lodge had at least two reasons for believing that a new
argument for the etherial body was both necessary and possible, and
both related to the First World War. First, the etherial body symbolised
the values that he and others ascribed to the ether and which he believed
were being threatened by modern German culture. In his fiercely anti-
German War and After (1915), he interpreted German physicists' apparent denial of an ether as an "allegory" of the "larger scheme" in which the
focus of Britain's chief enemy on the material side of things and the
military power of the German state had destroyed its "spiritual
sense".[111] His revulsion towards German barbarism turned to anger
months later, in 1915, when his son Raymond was killed in action near
Ypres. Raymond's death also informed the second reason why the idea of
an etherial body needed articulating. Lodge channelled much of his grief
into a campaign to reassure those who had lost loved ones in the conflict,
and who may already have turned to spiritualism for consolation, that
death was not final.[112]

---

[108] Lodge, 'Address by the President', p. 47.

[109] See, for example, Helena Petrovna Blavatsky, Isis Unveiled: A Master Key to the Mysteries
of Ancient and Modern Science and Theology, 2 vols. (New York: J. W. Bouton, 1877), vol.
1, pp. 280–1; Sargent, Scientific Basis of Spiritualism, pp. 54–6 and 196–213.

[110] Lodge was aware of the Unseen Universe by at least the 1890s: William F. Barrett to
Oliver Lodge, 18 October 1890. Oliver Lodge, 'The Ether of Space', Contemporary
Review, vol. 93 (1908), pp. 536–46, p. 540.

[111] Oliver Lodge, The War and After: Short Chapters on Subjects of Serious Practical Import for
the Average Citizen in A.D. 1915 Onwards (London: Methuen and Co., 1915), p. 18.

[112] George M. Johnson, Mourning and Mysticism in First World War Literature and Beyond
(Basingstoke: Palgrave, 2015), chapter 2; Jay Winter, Sites of Memory, Sites of Mourning:
The Great War in European Cultural History (Cambridge University Press, 1995), chap-
ter 3.

It is no coincidence that it was in his account of how he had coped with Raymond's death – via establishing evidence for his post-mortem manifestation in seances – that Lodge sought to lend his scientific authority to the comforting ideas that death was "not a word to fear" and that those whose material bodies had been blown apart in battle continued to live, whole in an etherial continuum.[113] Death probably involved the complete transit of our spirit to the ether, which, though immaterial, intangible and elusive, was nonetheless physical and "substantial enough" to give spirit some kind of a body that could, under certain circumstances, manifest itself to the living.[114]

The idea of an etherial body was mentioned by the 'disembodied' spirit 'Feda' who controlled Gladys Leonard's voice during the seances she gave to Lodge, and which was the primary source of information about Raymond.[115] Lodge was gratified that the denizens of the other world gave their "general approval" to the idea that they had some kind of body, but for public purposes the argument for the etherial body's possibility needed a more this-worldly basis.[116] A stronger argument for the etherial body was possible because, as far as Lodge was concerned, the ether had become a more significant physical entity than it had been in the late nineteenth century. The extraordinary physical properties it seemed to possess – notably its enormous elasticity and inertia, its apparently flawless continuity and transparency, and freedom from friction and disintegration – underpinned Lodge's frequent references to it as the most "substantial" and "perfect" entity in the universe, and his argument that it was especially promising as an abode of life and mind.[117] Since life and mind were associated with gross matter, then it was not unlikely that they were also associated with a physical entity (the ether) whose elasticity and inertia made it vastly better than gross matter in carrying other immaterial phenomena such as light, electricity and magnetism. Since ether was also believed by many (notably older British physicists) to be intimately related to gross matter – whether as the medium out of which the constituents of matter were formed or as the agent of cohesion and other forms of physical influence between all material bodies – then it promised to be the perfect "intermediary third" class of entity that could explain how immaterial mind, whether human or divine, interacted with the material world.[118]

---

[113] Lodge, *Raymond*, p. 298.   [114] Lodge, *Raymond*, p. 319.
[115] Lodge, *Raymond*, p. 195.   [116] Lodge, *Beyond Physics*, p. 7.
[117] Oliver Lodge, *Ether and Reality: A Series of Discourses on the Many Functions of the Ether of Space* (London: Hodder and Stoughton, 1925), pp. 154–5.
[118] Lodge, *Beyond Physics*, p. 22.

The significance of the ether in constituting and welding together matter was critical to Lodge's stronger argument for the etherial body. It led him to propose that it was probable that all sensible objects had material and etherial aspects, even though the latter aspect was invisible and intangible.[119] The psychical significance of *animate* objects arguably applied to both their material and etherial aspects and this had a major implication for human beings. Our etherial aspect or body was bereft of the "temporal disabilities" blighting gross matter, such as "fatigue, imperfect elasticity, friction, dissolution", and when freed from such matter it could lead a "less abstracted and livelier existence".[120] The possibility that our etherial body had a psychical significance, and that this survived the dissolution of the material body, was, Lodge conceded, a question to be settled by evidence from psychical research rather than by "dogmatism".[121] Difficult as this task had been and would continue to be, it was worth pursuing because it promised to render "vaguely explicable" and physically more intelligible Christian conceptions of the soul or spiritual body and how incorporeal intelligences appeared to interact with terrestrial matter as manifested in the communications and movements of spiritualist seances.[122]

Lodge's preoccupation with the ether's possible properties only intensified after his retirement, in 1919, from the University of Birmingham. This transition gave him more time for writing, lecturing and, later, radio broadcasting, and permitted him more freedom in contemplating the controversial physical, psychical, religious and philosophical topics that had interested him for decades. Much of this material combined surveys of established knowledge of the ether's known physical properties with articulations of the vaguer grounds he had for believing in the medium's wider significances. In *Ether and Reality* (1925), his most commercially successful book on the ether, he moved from a popular exposition of the ether's "ascertainable physical properties" to the "instinct" he had for the idea that since we were "more in direct touch with the ether than with matter", then this made the physical medium "our real primary and permanent instrument".[123] "Instinct" led him to push the religious significance of etherial contact further than he had done before. In one of the most audacious of all his books' conclusions, this volume declared that

---

[119] Lodge, 'Ether, Matter and the Soul', p. 258. For a complementary analysis of Lodge's hypothesis see Wilson, 'The Thought of Late-Victorian Physicists'.
[120] Lodge, 'Ether, Matter and the Soul', p. 258.
[121] Lodge, 'Ether, Matter and the Soul', p. 259.
[122] Lodge, 'Ether, Matter and the Soul', p. 259.
[123] Lodge, *Ether and Reality*, pp. viii, 173 and 176–7. The book was still being reprinted in 1930. An incisive study of the volume is Whitworth, 'Transformations of Knowledge'.

the ether was the "primary instrument of Mind, the vehicle of Soul, the habitation of Spirit. Truly it may be called the living garment of God".[124]

Like all of Lodge's later writings on the ether, *Ether and Reality* balanced a sense of the significant uncertainties about the ether (notably the precise means by which matter and spirit interacted with it) with a sincere belief that these uncertainties were among the greatest scientific problems for the future. In the 1920s and '30s, Lodge often spoke of the ether as one of the ways he felt that "physics and psychics" were "interlocked", and this reflected his hopes for fruitful convergences of physical, psychical and theological forms of enquiry as well as his belief in the capacity of the ether to illuminate the interactions between phenomena of the physical and psychical domains.[125] As he explained in his last book:

The bringing in of the ether into the scheme of psychics, as it has already been partially brought into the scheme of physics, is the work which I feel sure is lying ahead for generations of men. Then – when a serious beginning in this direction has been made – the term "soul" will acquire a definite and clear connotation; no longer will the idea of a spiritual body seem vague and indefinite and difficult of apprehension – there is nothing indefinite about future existence – soul will no longer be regarded as a term to be avoided, but will become as real and recognisable, as concrete and tractable, as are the corpuscles of electricity.[126]

Like Larmor and Thomson, Lodge could not forget the tradition of ether-based physics that had been most successful in the late nineteenth century, and this was clearest in his hopes for the discovery of a hydrodynamical or other "perfect mechanism" for the medium.[127] But, unlike those other physicists, he believed that the success of rendering the problems of physics more intelligible via the ether could be repeated in the domains of psychical research and Christian theology. What strengthened the "hold" that he admitted the quasi-mechanical ether had on him was not just its potential psychical and theological significances, but the way that it fitted into his fundamental belief that everything in the cosmos – matter, life, mind and spirit included – were ultimately parts of a cosmic "chain of causation", many parts of which were beyond human grasp.[128] So closely tied was the etherial mechanism and cosmic orderliness in his mind that he judged "preposterous" the argument, associated closely with such younger rivals in popular physics as Arthur Stanley Eddington and James Jeans, that the interdeterminacy

---

[124] Lodge, *Ether and Reality*, p. 179.   [125] Lodge, *Beyond Physics*, p. 114.

[126] Lodge, *My Philosophy*, p. 238. Chapter 21 of this book expands Lodge, 'Ether, Matter and Soul'.

[127] Oliver Lodge, 'The Ether and Relativity', *Nature*, vol. 126 (1930), pp. 804–5, p. 804.

[128] Lodge, 'Ether and Relativity', p. 805; Oliver Lodge, *Modern Problems* (London: Methuen and Co., 1912), p. 4.

implied by quantum mechanics made the idea of a Divine Mind scientifically more acceptable.[129] It also meant that he could be confident that while in the 1930s it was profoundly difficult to conceive of the interaction between mind and ether – an interaction between something so unphysical and immaterial with something with ascertainable physical and mechanical properties – it was a problem that would eventually be solved by the ceaseless human quest for intelligibility.

Interpreting Lodge's Physics and Psychics

In late 1926, the British comic periodical *Punch* featured Lodge in its long-running series of cartoons of leading personalities of the day (Figure 6.1). Showing the ageing wireless pioneer sitting near a radio receiver, the image was accompanied by a short poem parodying verses from William Cowper's anti-slavery poem of 1784, 'The Task'. Claiming that Lodge had found the "mundane" world too "cramping for his style", it described how he had strayed into the "psychic sphere", which seemed to the "average F.R.S. / A Lodge in some vast wilderness".[130] Lodge would not have been surprised, and may have been amused, to find his apparent scientific marginality portrayed in this way. He had long perceived, with some justification, that his "psychic utterances" had done "harm" to his scientific reputation, and the numerous attacks on *Raymond* and similar works would have reminded him that doubts about his capabilities as a psychical researcher, as well as the scientific credibility of psychical research per se, were as severe as ever.[131] His forays into theology compounded perceptions of his professional marginality: most scientists remained in some sense religious but were increasingly reluctant to speak out on matters lying beyond their scientific expertise, thus undermining the efforts of the older generation of scientists, liberal Christians and others to find common ground between science and religion.[132]

Yet the *Punch* cartoon alluded to one reason why Lodge had grounds for believing that perceptions of his scientific marginality were not necessarily shared beyond circles of elite scientists. As a prominent figure in

---

[129] Lodge, *Beyond Physics*, p. 73.
[130] Bernard Partridge, 'Mr. Punch's Personalities. XXX – Sir Oliver Lodge, F.R.S.', *Punch*, 24 November 1926, p. 585.
[131] Oliver Lodge to J. Arthur. Hill, 23 December 1914, in Hill, *Letters from Sir Oliver Lodge*, p. 49. Examples of hostile responses to Lodge's psychical works include Viscount Halifax, 'Raymond; Or, Life and Death', *Church Times*, vol. 72 (1917), pp. 181–2; J. W. A. Hickson, 'Sir Oliver Lodge and the Beclouding of Reason', *University Magazine*, vol. 16 (1917), pp. 379–97; Charles Mercier, *Spiritualism and Sir Oliver Lodge* (London: Watts and Co., 1919); Tuckett, 'Psychical Researchers'.
[132] Bowler, *Reconciling Science and Religion*, pp. 20–4.

MR. PUNCH'S PERSONALITIES.

XXX.—SIR OLIVER LODGE, F.R.S.

AS ALEXANDER found erstwhile
One world too cramping for his style,
Our OLIVER still asks for more
Than can be gleaned from mundane lore;

Until the wireless pioneer,
Straying into the psychic sphere,
Seems to the average F.R.S.
"A LODGE in some vast wilderness."

6.1 Bernard Partridge, 'Mr. Punch's Personalities. XXX – Sir Oliver Lodge, F.R.S.', *Punch*, 24 November 1926, p. 585.

early radio broadcasting, Lodge reinforced his profile as one of the most familiar figures in British science who was often called upon to interpret the ideas of 'modern' physics and a host of other subjects to a wide range of scientific and non-scientific audiences.[133]

Few of these ideas divided Lodge's different audiences more than those relating to the ether. In a lecture to fellow electrical engineers in 1926, Alexander Pelham Trotter opined that "most of those of us" who knew Lodge's broadcasts and writings agreed with him that relativity's champions could not dismiss the ether "by a summary negation".[134] The British physicist Edward Andrade was more equivocal. Reviewing *Ether and Reality* in *Nature*, he echoed the praise that Lodge frequently received for his literary "charm and simplicity" but deemed the author's ideas about the ether confusing and antiquated.[135] He was puzzled by Lodge's insistence on an ether that was not composed of gross matter but had mechanical properties (for example, inertia and elasticity) commonly associated with such matter, and by his appeal to Einstein – a physicist who, unlike Lodge, had rejected the idea of the ether as mechanical and as embodying the absolute reference frame for space and time. It was for these reasons, and Lodge's apparent appeal to "mystic inspiration and psychic experiences" for some of his etherial arguments, that Andrade believed that the book was better suited to those seeking a "wider outlook" than to professional physicists such as himself.[136] In his review of the same book, the leading British science journalist J. W. N. Sullivan denied that many physicists adhered to Lodge's "modified form of the old ether theory" but, unlike Andrade, questioned whether he had also served those seeking a wider outlook.[137] The fundamental problem of understanding the interaction of mind and matter was not simplified by appealing to something as substantial as Lodge's ether.

[133] On Lodge as a populariser of modern physics see Clarke, 'Ether at the Crossroads'; Lightman, 'Lodge and the New Physics'; Whitworth, 'Transformations of Knowledge'.

[134] Alexander P. Trotter, 'Illumination and Light', *Journal of the Institution of Electrical Engineers*, vol. 64 (1926), pp. 367–71, p. 370.

[135] E[dward] N. [da] C[osta] A[ndrade], 'Ether and Erdgeist', *Nature*, vol. 116 (1925), pp. 305–6, p. 305. Cf. Andrade's review of Lodge's earlier book, *Atoms and Rays* (1924): E[dward] N. [da] C[osta] A[ndrade], 'A Veteran's View of Modern Physics', *Nature*, vol. 114 (1924), pp. 599–601. Some reviews of *Ether and Reality* agreed that its faults were balanced by its capacity to stimulate thought: see, for example, E. E. Free, 'The Ether and the Soul', *Popular Radio*, vol. 8 (1925), p. 280.

[136] Andrade, 'Ether and Erdgeist', p. 305.

[137] [J. W. N. Sullivan], 'Ether and Reality', *Times Literary Supplement*, No. 1217 (14 May 1925), p. 325.

Sullivan was hardly the first to doubt whether Lodge's ether had the appropriate qualities for the psychical and spiritual purposes that the physicist sought for it. In 1906, the English writer W. H. Mallock had demurred to Lodge's claim that the ether made it easier to comprehend a domain where free will and immortality were possible because, despite being different from gross matter, it was still "determined" and had structures that were temporary.[138] When, in 1918, Eleanor Sidgwick replied to Lodge's proposal to have his paper on the etherial body published by the SPR, she warned him that a "good many physicists" would question his ideas about the ether and offered the personal estimate that the entity seemed "too materialistic a conception" to illuminate "psychics", which she, like so many SPR members, had long believed was primarily a psychological question.[139]

The substantial, objective and quasi-mechanical ether cherished by Lodge fared rather better among spiritualists and modern Theosophists. By the 1920s, spiritualists had come to regard Lodge as one of their greatest scientific allies and they typically welcomed the intellectual gravitas that his books, lectures and radio broadcasts brought to questions relating to survival and telepathy. His speculations on the etherial body found plenty of admirers among contributors to the leading spiritualist weekly *Light* and were appropriated in one of the bestselling British spiritualist texts of the interwar period: Arthur Findlay's *On the Edge of the Etheric* (1931).[140]

Lodge's writings directly or indirectly helped modern Theosophists develop their old argument that modern science was catching up with, and helping to support, occult wisdom. Lodge's theory of a super-dense ether was used by Alfred Percy Sinnett in a 1919 edition of *Occult Chemistry* – Annie Besant and Charles Leadbeater's notorious account of their clairvoyant "observations" of the atomic, subatomic and "ultra-physical" constituents of the chemical elements.[141] For Sinnett, Lodge's

---

[138] W. H. Mallock, 'Sir Oliver Lodge on Life and Matter', *Fortnightly Review*, vol. 80 (1906), pp. 33–47, p. 37.

[139] Eleanor Sidgwick to Oliver Lodge, 4 October 1918, SPR.MS 35/2255, OJL-SPR. In 1930, Lodge speculated that the new wave mechanics, which represented subatomic particles as vibrations, provided a possible etherial mechanism by which mind could interact with matter. But the young British physicist and SPR member Guy Burniston Brown warned that because this mechanism could be described in terms of the Schrödinger wave equation, it could "hardly be said to bristle with spontaneity and free will!": Guy B. Brown, Review of Oliver Lodge's *Beyond Physics*, *Philosophy*, vol. 5 (1930), pp. 624–6, p. 626.

[140] [Anon.], 'The Ether and Human Survival', *Light*, vol. 45 (1925), p. 198; H. A. Dallas, 'The Etherial Body: Its Nature and Scope', *Light*, vol. 44 (1924), p. 116; Arthur Findlay, *On the Edge of the Etheric* (London: Psychic Press, 1931), chapter 2.

[141] Besant and Leadbeater, *Occult Chemistry*, p. 4.

theory converged with clairvoyant perception that space was filled with an infinitely dense fluid ('koilon') and that structures within the fluid comprised the building blocks of matter. Similarly, in *Rational Mysticism* (1924), William Kingsland revealed his debt to Lodge when he wrote that the "real substantiality, that which is more real in the sense of being more permanent, more enduring, more *inner*, is not physical matter, but something which lies quite beyond the reach of our senses".[142] This "something" was the ether of space, which Kingsland was pleased to insist had not been destroyed by relativity theory. Although this surviving ether remained a purely physical and "*dead* substance", it was a stepping-stone to Theosophical and mystical ideas of higher planes of substance, including the ultimate "Primordial or Root Substance" of the cosmos where mind and matter were unified.[143]

Kingsland's book illustrates the diverse ways in which modern Theosophists and spiritualists tried to cope with the contradictions between that part of 'modern' physics upholding the existence of an ether and the more recent part denying its existence. His approach challenged the latter position by emphasising Einstein's and Eddington's arguments for space possessing physical qualities and by upholding the "deeper", non-empirical evidence of hidden levels of reality, of which the physical ether was one.[144] The British physical chemist and modern Theosophical writer, William Coode-Adams, was altogether more equivocal. Like many Theosophical texts of the period, his *Primer of Occult Physics* (1927) used Lodge's view of cosmic significance of the immaterial ether to support the general occult principle that "reality is always behind the appearances".[145] Elsewhere, it characterised relativity as a powerful scientific endorsement of occult perceptions of a reality transcending ordinary and rigid notions of space, time and matter, and as a theory whose conception of time as a fourth dimension of spacetime lent credence to clairvoyant travels into the past and future.

Quantum theory was no less useful to Coode-Adams. The idea that energy existed in an atomic form provided an analogy for the Theosophical idea that matter was energy on a lower plane of reality, and seemed to be confirmed by clairvoyant observation. Yet Coode-Adams's attempt to show the congruence of modern and occult physics forced him to make a puzzling compromise. Having accepted the challenge that relativity and quantum theories posed to the idea of an objective ether, he had

---

[142] William Kingsland, *Rational Mysticism: A Development of Scientific Idealism* (London: George Allen and Unwin, 1924), p. 81.
[143] Kingsland, *Rational Mysticism*, p. 83.   [144] Kingsland, *Rational Mysticism*, p. 71.
[145] W. R. C. Coode-Adams, *A Primer of Occult Physics* (London: Theosophical Publishing House, 1927), p. 53.

to conclude that this did not affect the status of the three higher "grades" of ether taught in Theosophy or of Lodge's view of the ether as our primary instrument.[146]

Spiritualists were certainly aware of the challenge that relativity theory posed to their cherished idea, partly encouraged by Lodge's writings, of an etherial body surviving the death of the material body. Some followed Findlay in ignoring relativity; some explicitly challenged the philosophical implications of the theory; and some welcomed it as another argument against the concepts of matter, time and space propping up materialism, and for the existence of higher spatial dimensions where spiritualists had long located spirits.[147] Some tried to forge compromises between relativity and ether physics that, owing partly to Lodge, had proven such a useful source of scientific speculation. One compromise involved focussing on how both kinds of physics involved a welcome challenge to nineteenth-century materialism, even though this was reached by the ostensibly divergent routes of abandoning traditional concepts of space, time and matter and embracing Lodge's idea of an immaterial and objective ether.[148]

In their engagements with Lodge's writings, spiritualists and psychical researchers revealed a variety of interpretations of Lodge's attitude towards the connection between the ether and telepathy or other forms of psychical transmission. In 1921, the civil engineer Stanley De Brath alluded to Lodge's warning of 1903 that spiritual and psychical events might occur in realms beyond the physical and so may not involve the ether of physics.[149] Lodge's warning would have seemed appropriate because in the 33 years between the SPR's first major work on telepathy – *Phantasms of the Living* – and De Brath's article, psychical researchers had amassed further evidence for experimental and spontaneous cases of the obscure human faculty, but many were mindful that the evidence was still inadequate to convince the scientific world and to support etherial, wireless or other physical explanations of telepathy.[150] One of the chief contributors to the more recent studies of telepathy, Eleanor Sidgwick, spoke

---

[146] Coode-Adams, *Primer of Occult Physics*, p. 48. Other popular expositors of relativity were inconsistent in their views on the ether: see Jaume Navarro, 'Ether and Wireless: An Old Medium into New Media', in Jaume Navarro, *Ether and Modernity*, pp. 130–54.

[147] Tudor A. Morgan, 'The Ether and Spiritual Science', *Occult Review*, vol. 50 (1929), pp. 83–9; George Lindsay Johnson, *The Great Problem and the Evidence for Its Solution* (London: Hutchinson and Co., 1927), pp. 35–6; Herbert S. Redgrove, 'Mathematics and Psychical Research', *Psychic Research Quarterly*, vol. 1 (1920–1), pp. 220–34.

[148] Frederick Stephens, 'Science and the Unseen World', *Light*, vol. 49 (1929), pp. 338–9, 356–7.

[149] Stanley De Brath, 'Relativity', *Light*, vol. 41 (1921), pp. 520–1.

[150] [Eleanor] Sidgwick, 'Phantasms of the Living', *PSPR*, vol. 33 (1923), pp. 23–429; Rudolf Tischner, *Telepathy and Clairvoyance*, translated by W. D. Hutchinson

for many psychical researchers in the early 1920s when she called for more experiments that would address such questions as why telepathy was so capricious, what were the psychological conditions of success, whether it depended on the energy of the transmitting agent, and what exactly was transmitted from one mind to another.[151] Sidgwick's reservations about the employment of physical analogies in understanding this transmission were moderate compared to those of Barrett and Lodge, who agreed that the development of wireless telegraphy – the iconic technology of transmitting intelligence through 'empty' space – had encouraged many etherial analogies for and theories of telepathy that they still questioned.[152]

Yet Barrett and, *a fortiori*, Lodge were also partly responsible for the very analogising and theorising that they wanted to curb because so many of their writings emphasised the psychical and spiritual significances of the ether.[153] Someone who read Lodge in this way was Cyril 'Jack' Frost, a former British army officer who by the mid-1920s had emerged as an authority on wireless telegraphy and as a writer and lecturer on spiritualism. In 1926, he told one spiritualist audience that Lodge's *Ether and Reality* aroused "deep interest" in the possibility of a connection between the medium of radio transmission and the medium of communication between discarnate and incarnate souls.[154] As a former official at the British Broadcasting Corporation, he had good reason to think that the organisation had, by making it "commonplace" for people to hear voices of distant souls, also made it easier to contemplate the reality and wider, possibly psychic possibilities of the etherial carrier of those voices.[155]

Frost is one of many individuals from the 1920s and '30s who demonstrate that the connections between the cultures of psychical research and of wireless telegraphy are stronger and more complex than we might assume from the obvious, but actually complicated case of Lodge. In Britain and America, the 'psychical' and 'occult' content of magazines

(London: Kegan Paul, Trench, Trübner and Co., 1925); René Warcollier, *La télépathie recherches expérimentales* (Paris: Librairie Félix Alcan, 1921).

[151] Eleanor Sidgwick, 'Experimental Telepathy: The Need of Further Experiments', in *L'état actuel des recherches psychique d'aprés le travaux du IIme Congrés International tenu à Varsovie en 1923* (Paris: Les Presses Universitaires de France, 1924), 174–80.

[152] Barrett, *Psychical Research*, p. 107; Lodge, *Survival of Man*, pp. 125–6. Examples of such analogies are [Anon.], 'Wireless and Telepathy', *Light*, vol. 47 (1927), p. 78; J. C. F Grumbine, *Telepathy: Or, The Science of Thought Transference* (London: L. N. Fowler and Co., 1915), esp. chapters 4–5.

[153] Examples of Barrett's continued psychical uses of the ether are Barrett, 'Spiritual Significance of Nature', pp. 795–6; William F. Barrett, *On the Threshold of the Unseen: An Examination of the Phenomena of Spiritualism and of the Evidence for Survival After Death* (London: Kegan Paul, Trench, Trübner and Co., 1920), p. 118.

[154] Jack Frost, 'Radio and Psychic Science', *Light*, vol. 46 (1926), p. 221.

[155] [Anon.], 'Wireless and the Next World', *Light*, vol. 46 (1926), pp. 544–5, p. 544.

serving the burgeoning professional and amateur interests in wireless, as well as those catering to less specialist scientific readerships, overlapped more with the 'technical' content of spiritualist, psychical and other occult periodicals than we might expect. Like so much popular scientific and technical literature, this material speculated to a degree that would not have been permitted in the publications of those elite scientific organisations that seemed to marginalise Lodge. Susan Douglas has argued that in the United States it was precisely the hunger for "otherworldly contact, for communion with disembodied spirits, for imaginative escapades that affirmed there was still wonder in the world" that helps explain the astonishing growth of professional and amateur interest in wireless.[156] The hunger for such wireless possibilities was present on both sides of the Atlantic. Articles on 'wireless' theories of telepathy, electrical devices that seemed to capture thoughts and discarnate spirits, experimental evidence of brain waves, and reflections on the classic scientific studies of spiritualism were common to popular wireless and occult magazines in Britain and America[157] (Figure 6.2).

Much of this material demonstrates the fact that the concept of an objective ether was used by professional and amateur wireless practitioners long after it was deemed to have been killed off by relativity theory and become of only marginal interest to professional physicists.[158] For many in these former technical constituencies, the ether's alleged demise was a legitimate topic of debate, not least because the concept still made the quotidian practices of wireless telegraphy intelligible and was indispensable in teaching fundamental principles to novices.[159] Given their acknowledgement of Lodge as a major authority on wireless, it is hardly

---

[156] Susan J. Douglas, *Listening In: Radio and the American Imagination* (Minneapolis, MN: University of Minnesota Press, 2004), p. 52.

[157] See, for example, [Anon.], 'Notes of the Month', *Occult Review*, vol. 28 (1918), pp. 187–203; [Anon.], 'Wireless and Telepathy'; Hereward Carrington, 'Sir William Crookes's Psychical Researches', *Electrical Experimenter*, vol. 7 (1919), pp. 407, 440, 442 and 444; C. H. Collings, 'Wireless: Some Facts and Speculations', *Light*, vol. 42 (1922), p. 461; E. E. Free, 'Have "Brain Waves" Been Discovered?', *Popular Radio*, vol. 9 (1926), pp. 366–7; Hugo Gernsback, 'The Thought Recorder', *Electrical Experimenter*, vol. 7 (1919), p. 12, pp. 84–5; Philip J. Risdon, 'Psychic Phenomena and Wireless', *Popular Wireless Weekly*, vol. 1 (1922), pp. 237–8; Cesar de Vesme, 'Human Brain Waves', *Light*, vol. 54 (1934), p. 115. For analysis of this material see Richard Noakes, 'Thoughts and Spirits by Wireless: Imagining and Building Psychic Telegraphs in America and Britain, circa 1900–1930', *History of Technology*, vol. 32 (2016), pp. 137–58.

[158] For a fuller discussion of this point see Navarro, 'Ether and Wireless'.

[159] See, for example, Philip R. Coursey, 'Aether the Substratum of the Universe', *Wireless World*, vol. 8 (1920), pp. 37–40; Philip J. Risdon, 'There Is No Ether?', *Popular Wireless Weekly*, vol. 1 (1922), p. 145; Charles P. Steinmetz, 'There Are No Ether Waves', *Popular Radio*, vol. 1 (1922), pp. 161–6.

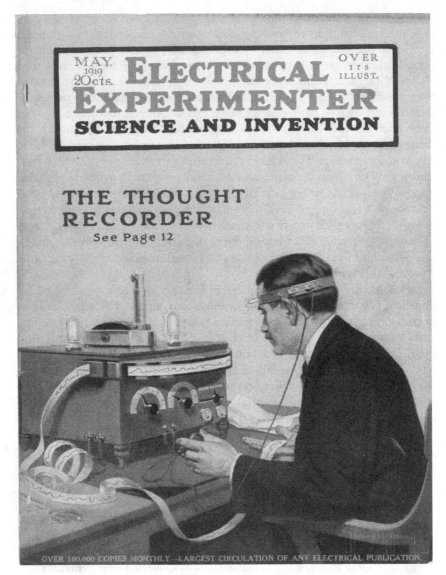

6.2 A cover of *Electrical Experimenter*, one of many popular scientific magazines launched by the American inventor and publisher Hugo Gernsback. In this issue, Gernsback discussed a machine that might be able record the etherial waves transmitted by the brain: Hugo Gernsback, 'The Thought Recorder', *Electrical Experimenter*, vol. 7 (1919).

surprising that wireless practitioners should have helped themselves so readily to the physicist's views on the ether. However, some wireless practitioners were frustrated by Lodge's refusal to extend his etherial speculations in the directions of new theories and experimental investigations connecting psychical and etherial forms of communication.

This is suggested by the comments of Quentin Craufurd, a former Royal Navy wireless officer who, like Frost, started contributing to the spiritualist press in the 1920s. Writing in *Light* in 1927, he deviated from Barrett, Lodge and many other physical–psychical scientists in denying that spiritualists' use of scientific terms was necessarily misguided. Spiritualists' talk of "magnetism" associated with the body or of "invisible light" perceived by clairvoyants was no longer risible given scientific evidence of electrical motion in the body and of wireless waves respectively.[160] For this reason, spiritualists deserved to be respected as individuals who had the potential to make important scientific discoveries, in much the same way that other 'amateurs' were. Indeed, Craufurd insisted that it was the

free-lance and the amateur who, in the most recent discoveries, have far outstripped the orthodox teachings of science in wireless practice of the present day. The bold experimenter has got ahead of the most cautious professor in many cases, and in nothing is this more obvious than in communication by "magnetism" (if you like to call it by that name) to distant regions".[161]

As we shall see, Craufurd would pursue one such "bold" experimental project with Frost, and this clearly challenged the "cautious" approach of Lodge and others towards magnetic communication with spiritual "regions".

The actions of Craufurd and others with combined interests in wireless telegraphy and psychic communication were at least as telling as their words. In many ways their actions and words represented some of the most creative new attempts of the interwar period to explore connections between physics and psychics. In 1922, the American psychical researcher and science populariser Hereward Carrington fuelled speculation about the occult significances of wireless signals by explaining in the American-based *Popular Radio* magazine that he had recently established a Psychical Institute in New York where he and fellow investigators were exploring, among other questions, the possible enhancement of telepathic impressions by superposing them onto carrier waves generated by a powerful electric field.[162] Although it failed to produce any conclusive

---

[160] Q. C. A. Craufurd, 'The Unknown Force: An Electrician's View', *Light*, vol. 47 (1927), p. 339.

[161] Craufurd, 'Unknown Force'.

[162] Hereward Carrington, 'Will We Talk to the Dead by Radio?', *Popular Radio*, vol. 1 (1922), pp. 92–7. See also Hereward Carrington, 'New York's Laboratory of the Mysterious', *New York Tribune*, 9 April 1922, p. 3.

results, Carrington's experiment represented a double challenge to Lodge's authority: it not only assumed that thought might create detectable etherial vibrations but drew upon French psychologist Hippolyte Baraduc's claim to have produced 'thought photographs' – images to which Lodge and other leading SPR members paid little attention. Indeed, Carrington's Psychical Institute represented one of many organisations founded in the 1920s partly to explore lines of enquiry (notably into the physical phenomena of spiritualism) that the SPR was reluctant to follow because such enquiries seemed to involve compromising the robustness of the investigative protocols that it had developed over previous decades.[163]

One of the most striking of Carrington's lines of enquiry was the construction of instruments that could detect psychical effects, including the human will and disembodied spirits, without the need for mediums, and which accordingly reduced the risk of fraudulence. Carrington's ideas owed something to the examples of Crookes and, moreover, to Thomas Alva Edison. In 1920, the doyen of American electrical inventors had caused a sensation by proposing to build a machine, based on the principle of the thermionic valve used in wireless detectors, to register the subtle "effort" exerted by spirits of the dead.[164] Unlike Crookes, Carrington believed that the problem of building such an instrument could be solved by amateur scientific and technical practitioners rather than by professional scientists. It was partly owing to the post-war explosion in the availability of thermionic valves that he could invite readers of *Popular Radio* to "undertake experiments of a similar character" to those in his Psychical Institute focussing on the construction of a valve that might pick up "subtle etheric waves" from beyond the grave.[165]

Quentin Craufurd shared more with Carrington than a sense that the most exciting new connections between physics and psychics were being forged by the amateur rather than the professional. He would also have reason to think that the SPR was too blinkered in its approach to psychical research and he too was preoccupied by distinguishing between the "prejudices" that mediums subconsciously contributed to spirit messages

---

[163] On this point see Inglis, *Science and Parascience*, pp. 214–20. The SPR revealed its position following the dramatic resignation of Arthur Conan Doyle over the organisation's hypercritical attitude towards spiritualism: [Anon.], 'Sir Arthur Conan Doyle's Resignation', *JSPR*, vol. 26 (1930), pp. 45–52.

[164] Austin C. Lescaraboura, 'Edison's Views on Life and Death', *Scientific American*, 30 October 1920, pp. 446, 458–60, p. 446. Carrington also reported on two Dutch inventors, J. L. W. P. Matla and G. J. Zaalberg van Zelst, who had built an electromechanical apparatus reputedly enabling 'spirits' to print words: Hereward Carrington, 'Are the Dead Trying to Reach Us by Radio?', *Popular Radio*, vol. 1 (1922), pp. 188–93.

[165] Carrington, 'Are the Dead Trying to Reach Us by Radio?', p. 191.

and the ostensibly "pure tone" of vibrations from the other world.[166] Like Carrington, Edison and others, he also believed that the most promising solution to this old problem of filtering psychical signal from noise lay in the construction of 'delicate' instruments. By 1928, he and fellow spiritualist–wireless engineer Frost were collaborating on an undisclosed wireless invention that promised to function as a direct spirit communication device.[167] Despite disappointing initial results, Craufurd was by 1933 reputedly enjoying "some" success with the device, although, like the spirit apparatus with which Edison seems to have achieved no results, nothing more was heard of it.[168]

In 1931, Carrington had linked the inconclusive outcomes of attempts to detect spirits by purely instrumental means to the possibility that the only kind of energy connecting the spiritual and physical worlds was that associated with life, thus affirming the need for mediums and other living subjects in psychical research.[169] Craufurd's limited success with his device, however, may well have helped dispel Carrington's doubts. By 1939, Carrington had published an entire book on laboratory- and instrument-based psychical investigations, whose purpose was partly to showcase the "minute fraction" of the "planned and partially completed" instrumental strategies for detecting, independently of mediums, the subtle energies associated with spirits and a host of other psychical effects.[170] The contents of popular technical and occult periodicals from the 1930s suggest that Carrington had not wildly over-estimated the extent of this activity and that some of the most elaborate attempts to make the instruments of physics relevant to the solution of psychical puzzles were continuing to flourish, albeit far from the SPR.[171]

---

[166] Q. C. A. Craufurd, 'Vibrations', *Light*, vol. 47 (1927), p. 632. See also Q. C. A. Craufurd, 'The Crisis in the SPR', *International Psychic Gazette*, vol. 18 (1930), p. 135.

[167] Q. C. A. Craufurd and Jack Frost, 'Psychic Communication and Wireless: A New Instrument', *Light*, vol. 48 (1928), p. 305.

[168] [Anon.], 'Wireless and Mediumship Problems', *Light*, vol. 53 (1933), p. 150. In 1933, one popular American technical magazine reported the failure of one of Edison's inventions to register spirits: [Anon.], 'Edison's Own Secret Spirit Experiments', *Modern Mechanix and Inventions*, October 1933, pp. 34–6. I owe this reference to Phillipe Baudouin.

[169] Hereward Carrington, *The Story of Psychic Science* (London: Rider and Co., 1930), p. 234.

[170] Hereward Carrington, *Laboratory Investigations into Psychic Phenomena* (London: Rider and Co., 1939), p. 23.

[171] [Anon.], 'Radio Psychometry', *Radio Craft*, vol. 5 (1933), p. 264; [Anon.], 'Model Engineers. Record Number at Exhibition – The Electric Psychometer', *Electrician*, 9 September 1932, p. 324; T. B. Franklin and V. J. Vickers, 'A Radio Divining Rod', *Modern Wireless*, vol. 8 (1933), p. 278; Vesme, 'Human Brain Waves'. See also [Anon.], 'Notes by the Way', *Quarterly Transactions of the British College of Psychic Science*, vol. 15 (1937), pp. 342–3. This article reported on the dispute between the American physicist R. A. Watters and British physicist Bernard Hopper over Watters's claim to have

Interwar Transitions

When, in 1932, the SPR celebrated its 50th anniversary, Lodge and Eleanor Sidgwick accepted invitations to take stock of the organisation's achievements. Both in their 80s, they were now among the SPR's longest-serving members and two of only a handful of individuals listed in Table 2.1 who were still actively contributing to psychical research. Since 1900, many physical scientists in the organisation had either died or resigned, the deaths of Crookes and Barrett in 1919 and 1925 respectively robbing a fast-shrinking network of physical–psychical scientists of two of its most important nodes.

The anniversary addresses that Lodge and Sidgwick gave revealed much agreement between these icons of British psychical research. They both insisted that the SPR's cautious methods and goals were as justified as ever: in one of many quotes from Myers's *Human Personality*, Lodge emphasised the continued need for an organisation to study obscure psycho-physical phenomena with the scientific methods and intellectual virtues that had now become necessary for "salvation".[172] Yet both Lodge and Sidgwick feared for the capacity of the SPR to continue its intellectual mission. Its membership had peaked at 1,305 in 1920, a rise owing much to the obvious wartime and post-war interest in the question of survival that had also fuelled a growth of interest in spiritualism.[173] But by 1932, membership had fallen to 809. The decline was attributed to the global economic depression, which made it difficult for some to maintain their subscriptions, but it also owed something to the emergence of rival British psychical and 'psychic' research organisations offering enquiries into spiritualism, to which many believed the SPR devoted too little attention or was methodologically incapable of testing fairly.[174]

---

photographed the souls of dying animals in a cloud chamber, a device normally used to show the paths of invisible radioactive emanations via their condensation of droplets in supersaturated water vapour. Watters believed that souls were 'intra-atomic' forms of energy on which the cloud droplets formed. Hopper claimed to have replicated Watters's experiments but not produced any results attributable to non-physical causes.

[172] Oliver Lodge, 'The Past and the Future', *PSPR*, vol. 41 (1932–3), pp. 61–74, p. 69.

[173] Sidgwick, 'Society for Psychical Research', p. 2.

[174] [Anon.], 'Annual Report of the Council for 1932', *JSPR*, vol. 28 (1933–4), pp. 19–29, p. 29. On interwar psychical research see Jenny Hazelgrove, *Spiritualism and British Society Between the Wars* (Manchester University Press, 2000), chapter 7; Inglis, *Science and Parascience*, chapters 6–7; Mauskopf and McVeigh, *Elusive Science*, chapters 2 and 8; Robertson, *Science and the Seance*; Joanna Timms, 'Ghost-Hunters and Psychical Research in Interwar England', *History Workshop Journal*, vol. 74 (2012), pp. 88–104; Valentine, 'Spooks and Spoofs'. Rival organisations included the pro-spiritualist British College of Psychic Science (founded 1920) and Harry Price's National Laboratory of Psychical Research (founded 1925). The fortunes of the American SPR were no better, with its membership halving in size in the 1930s.

The SPR's membership was not only shrinking but increasingly short of those whose professional scientific credentials had always lent much intellectual lustre to the organisation. Lodge believed, with some justification, that the scientific world had become less hostile to psychical research than it had been decades earlier, even though James Jeans would in 1930 provide a reminder of the way professional colleagues continued to attack the subject without having adequate knowledge of it.[175] Marginally less discomforting to him would have been a recent survey of the religious beliefs of Royal Society Fellows, which revealed that slightly more believed in than rejected the idea of the afterlife, but that many considered it a question beyond the scope of the sciences because it concerned personalities lacking the physical form to which scientific methods could be applied.[176] On this question, in other words, psychical research was misguided.

More regrettable to Lodge was the fact that many of the "younger men" of science felt that their talents were better occupied in more "remunerative" areas of scientific enquiry.[177] Lodge may well have been reflecting on the difficulty that he and other SPR scientists with university careers had in encouraging students to follow their examples: Barrett, Chattock, Lodge, Rayleigh, Stewart and J. J. Thomson had certainly interested some of their charges in psychical research, but these younger individuals did not display a significant commitment to the subject. The younger men who did not come under their direct influence were even less likely to participate and, as the cases of Eddington and Jeans illustrate, were either indifferent or actively hostile towards the subject.[178] For many scientists, psychical research simply did not seem as professionally rewarding as the more established areas of enquiry they had been trained to pursue.

The area of psychical enquiry that Lodge had long hoped would prove most "remunerative" to those with expertise in physical sciences – the

---

[175] Oliver Lodge to J. A. Hill, 6 March 1930, in Hill (ed.), *Letters from Sir Oliver Lodge*, pp. 234–6. Lodge was reacting to Jeans's comments in a recent BBC radio broadcast. See also note 178.

[176] C. L. Drawbridge (ed.), *The Religion of Scientists: Being Recent Opinions Expressed by Two Hundred Fellows of the Royal Society on the Subject of Religion and Theology* (New York: Macmillan Company, 1932), pp. 94–110. In 1933, *Nature* expressed unusual confidence in psychical research's methods and possible future achievements: [Anon.], 'Science and Psychical Research', *Nature*, vol. 132 (1933), pp. 945–6.

[177] Lodge, 'Past and the Future', p. 71.

[178] Eddington noted his disbelief in spiritualism in Arthur Stanley Eddington, 'The Domain of Physical Science', in Joseph Needham (ed.), *Science, Religion and Reality* (London: Sheldon Press, 1925), pp. 189–218, p. 214. His attitude probably owed something to a rejection of the spiritualist belief in natural law extending to the spiritual domain: Arthur Stanley Eddington, *Science and the Unseen World* (London: George Allen & Unwin, 1929), pp. 32–3. See also James Jeans, untitled essay, in *More Points of View: A Second Series of Broadcast Addresses* (London: George Allen & Unwin, 1930), pp. 55–71. This was the text of a BBC radio broadcast.

physical phenomena of spiritualism – was one where his verdict differed significantly from Sidgwick's. He believed that the physical effects in seances were better evidenced than they had been in the early 1900s and, despite his own lack of experimental work in this area, still promised to enrich scientific knowledge of "semi-physiological phenomena".[179] Sidgwick was much more pessimistic. The SPR's involvement in testing physical mediumship had actually surged during the 1920s, but the encouraging results of its tests of the telekinetic powers of the Austrian mediumistic brothers Rudi and Willi Schneider were outweighed by disappointing outcomes of its investigations into the French ectoplasmic medium 'Eva Carrière' (Marthe Béraud), the British spirit photographer William Hope, and 'Margery' (Mina Crandon), the American medium who produced a range of telekinetic and ectoplasmic effects.[180] Sidgwick's conclusion that the question of physical phenomena was in the "same position" as in 1882 reflected the fact that researches into the question led by the SPR – and, moreover, by Richet, Schrenck-Notzing and others independently of the SPR – had not impressed her as much as it had Lodge.[181] Accordingly, her view of where the SPR's future lay – in enhancing the evidence for the psychological "departments" of telepathy, survival and clairvoyance – had changed little since the 1880s.[182]

The professional scientists who were members of the SPR in the 1920s and '30s seem to have agreed that psychical research was much more accurately regarded as a branch of psychology than of physics or even physiology. Compared to the 1880s, proportionately more hailed from the mental and life sciences than from the physical sciences. There is no doubt that given the investigative priorities of the organisation, even those physicists, chemists and astronomers who were sympathetic to the wider ambitions of psychical research would have echoed George F. FitzGerald and Augustus Trowbridge, who in earlier decades had argued that they lacked the training and interests to contribute to the subject.[183] As far as

---

[179] Lodge, 'Past and the Future', p. 68.
[180] On the Schneiders see Wolffram, *Stepchildren of Science*, chapter 3. For Hope see Martyn Jolly, *Faces of the Living Dead: The Belief in Spirit Photography* (London: British Library, 2006), chapter 5 and Andreas Schmidt, 'The Most Disrespectful Camera in the World', in Chéroux et al., *Perfect Medium*, pp. 72–91. On Eva Carrière see Lachapelle, *Investigating the Supernatural*, chapter 5. On Crandon see Robertson, *Science and the Seance*, esp. chapter 6.
[181] Sidgwick, 'Society for Psychical Research', p. 22. Barrett shared Lodge's enthusiasm for this work: see William F. Barrett, 'Ectoplasms', *Light*, vol. 41 (1921), p. 347.
[182] Sidgwick, 'Society for Psychical Research', p. 26.
[183] FitzGerald quoted in Stead, 'Response to the Appeal', p. 19; Augustus Trowbridge to Hereward Carrington, 23 June 1921, Folder 50, Box 2, Hereward Carrington Papers (C1159), Manuscripts Division, Department of Rare Books and Special Collections, Princeton University Library.

the SPR was concerned, the connections between physics and psychics were more fruitfully explored in popular and semi-popular scientific writings than in laboratories and other sites of psychical experimentation.

The differences between Lodge and Sidgwick regarding the physical phenomena of spiritualism reflected wider disagreements among psychical researchers over the most fruitful areas of study and the role of "co-workers" in the troubled field of enquiry.[184] When, in her anniversary address, Sidgwick praised Lodge for helping to keep the SPR together over the past few decades, she implicitly accepted that he had been more effective than her in working with spiritualists, who were of course the main sources of the physical phenomena.[185] Under the aegis of Sidgwick and her closest allies, the SPR's relationship with spiritualists had been at least as turbulent as it had been in the 1880s, not least because the SPR remained frustrated by mediums' trickery and spiritualists' credulity, and spiritualists still questioned the legitimacy of the SPR's ultra-cautious and critical approaches.

William Hope's spirit photography dramatised the widening rifts between the SPR and spiritualists inside and outside the organisation. Hope had come to the attention of spiritualists and psychical researchers in the early 1900s, but his claims sparked numerous controversies. An important, though predictably reticent, supporter was Crookes, who in 1916 strengthened his belief in survival on the basis of Hope's images of the spirit of his recently deceased wife.[186] Six years later, Harry Price, the amateur conjuror and relative newcomer to the SPR, tested Hope and concluded that he had employed fraudulent methods to produce images of spirit 'extras' on photographic plates. Among the many spiritualists who defended Hope's credibility was Arthur Conan Doyle. Undoubtedly one of the most energetic and conspicuous champions of spiritualism of the 1920s, the creator of Sherlock Holmes led arguments that Price had himself resorted to underhand methods to expose Hope.[187] For Doyle, the Hope controversy would be one of many episodes highlighting the "essentially unscientific and biased work" of the SPR, which eventually forced him and many other spiritualists to resign their membership.[188] Although Lodge sympathised with the SPR's damning verdict on Hope

---

[184] Lodge, 'Past and the Future', p. 71. See Mauskopf and McVaugh, *Elusive Science*, chapter 1.

[185] Sidgwick, 'Society for Psychical Research', p. 15.

[186] [Anon.], 'Important Interview with Sir William Crookes'. Crookes had revealed his conviction to Lodge the previous year: William Crookes to Oliver Lodge, 22 December 1916, SPR.MS 35/363, OJL-SPR.

[187] On Doyle see Inglis, *Science and Parascience*, pp. 83–92; Johnson, *Mourning and Mysticism*, chapter 3.

[188] Doyle quoted in [Anon.], 'Sir Arthur Conan Doyle's Resignation', p. 46.

and drew unfavourable comparisons between spiritualists' and the SPR's methods, he had long represented a side of the SPR most sympathetic to spiritualism.[189] He never accepted the spiritualist label, but the views he expressed on survival and mediumship outside formal SPR channels persuaded many spiritualists to share his hopes for a courteous and harmonious relationship between different psychical workers.[190]

Lodge's ability to engage with spiritualists depended partly, as we have seen, on the scientific credibility that his writings lent to the idea of an etherial body that carried the mind and spirit after bodily death. Yet the firm adherence of this most prominent physicist and psychical researcher to the increasingly antiquated idea of an objective ether not only marginalised him from the younger generations of physicists but ultimately hindered his ability to consider other possible theoretical connections between physics and psychics that were being explored elsewhere, typically by non-physicists.

By 1930, Coode-Adams's *Primer of Occult Physics* had been joined by many other texts that turned to both relativity and quantum theory for new ways of rendering psychical effects and powers intelligible. In 1928, for example, J. Malcom Bird, the American mathematician and science journalist who had led the *Scientific American*'s inconclusive investigations of Mina Crandon, insisted that the alleged ability of mediums to experience the future in the present "lined up exactly" with the hyperdimensionality of relativity, while telekinetic powers were easier to comprehend on the "reasonable" basis that the spacetime continuum around a medium was somehow distorted more than that constituting a gravitational field and that this caused untouched objects to defy gravity.[191]

Two years later, the French psychical researcher René Sudre highlighted quantum theory as an example of how "official science" was "clearing the way" for a "rational theory" of psychical effects.[192] The breakdown of strict determinism and causality at the atomic level opened the door to the possibility that mind was an integral part of physical reality. For Eddington, this was a critical part of an idealist worldview; for Sudre, it was much more. It promised a firmer scientific basis for a speculation that FitzGerald, Lodge, Myers and others had started in the

---

[189] Lodge expressed his scepticism of Hope in Oliver Lodge to William Crookes, 23 December 1916, typescript copy, SPR.MS 35/364, OJL-SPR and Oliver Lodge to Arthur Conan Doyle, 2 January 1925, typescript copy, SPR.MS 35/442, OJL-SPR.

[190] See Lodge's remarks in [Anon], 'A Symposium: Why I Became a Spiritualist', *Quarterly Journal of the British College of Psychic Science*, vol. 14 (1935–6), pp. 48–54, p. 48.

[191] J. Malcom Bird, 'Some Theoretical Aspects of Psychical Research', *British Journal of Psychical Research*, vol. 1 (1928), pp. 331–9, 335–6.

[192] René Sudre, 'Psychical Research and the New Physics', *Psychic Research*, vol. 24 (1930), pp. 117–21, p. 117.

1890s but largely abandoned: that a medium's mind could, by controlling individual molecules, move bodies at a distance, create ectoplasm and materialised spirits, and dissolve objects into thin air. The British psychical researcher Whately Carington agreed with Sudre that Eddington's writings created space for psychical effects. Having earlier regarded relativity as a reason to take seriously his four-dimensional 'mechanism' of survival and clairvoyance, he now linked Eddington's argument for subjectivity and mind in descriptions of physical reality to the idea of universal consciousness mediated by telepathy.[193]

One of the most conspicuous commentators on the possible relationships between what was often called the 'new physics' and psychical research was the leading American science journalist Waldemar Kaempffert. In the late 1930s, he championed the work of Joseph Banks Rhine, the American psychologist whose new approaches to clairvoyance and telepathy yielded powerful evidence for a general psychological faculty, christened 'Extra Sensory Perception' (ESP), that Rhine believed could not be put down to chance coincidence or fraud.[194] In its methodology and setting – notably the statistical analyses of a vast number of laboratory-based trials – Rhine's new science of psychical effects (christened 'parapsychology') sought to position itself far closer to the young but secure academic discipline of experimental psychology than to psychical research. Although the claims of Rhine and his followers were subject to plenty of scientific criticism (especially from psychologists), their methods helped parapsychology gain an academic credibility that psychical research had never enjoyed.

For Kaempffert it was not at all surprising that physicists who formerly scoffed at psychical investigations were now paying attention to Rhine. They had lost their "old cocksureness" about a "mechanistic universe" in favour of new conceptions of space, time, matter and causality that seemed to make psychical effects less implausible.[195] Relativity dissolved the distinction between space and time and made the idea of experiencing the future in the present more intelligible. Quantum theory encouraged a statistical view of physical reality which made Rhine's statistical arguments for the reality of ESP particularly appealing. And both theories represented a radically new way of looking at physical reality – as one

---

[193] Whately Carington, *The Death of Materialism* (London: George Allen and Unwin, 1933). See also Carington's work published under his original name: W. Whatley Smith, *A Theory of the Mechanism of Survival* (London: Kegan Paul, Trübner, Trench and Co., 1920).

[194] On Rhine see Asprem, *Problem of Disenchantment*, pp. 398–412; Mauskopf and McVaugh, *Elusive Science*.

[195] Waldemar Kaempffert, 'The Duke Experiments in Extra-Sensory Perception', *New York Times*, 10 October 1937, pp. 2 and 21, p. 2.

created by mind – that encouraged a more general change in conceptions of mind, of which ESP was one aspect.[196]

***

Lodge never quite lost the "old cocksureness" of nineteenth-century physics. He could not accept that indeterminacy was an ultimate feature of the physical world and that this was a plausible route to explaining psychical effects. In words echoing Einstein's famous quip that God "is not playing at dice", he remarked in 1930 that the "Divine Artificer does not work with the calculus of probability", and insisted that perceptions of indeterminacy at the atomic level merely reflected the limitations of human knowledge.[197] Having rejected ultimate physical indeterminacy, however, Lodge was forced to argue that free will (which he readily accepted) had to relate to the non-physical domain, even though this was interlocked with the physical domain via the ether. So strongly was Lodge attached to traditional ideas of space, time and determinism in the *physical* world that he believed that it was the less "orthodox" facts about a non-physical domain – those relating to clairvoyance, spiritualist mediumship and precognition – rather than relativity and quantum theories that would revolutionise fundamental understanding of the nature of time and space.[198]

By the late 1930s, Lodge, the long-reigning champion of physics and psychics, had little energy for the literary activities that had preoccupied him for decades. Virtually all the books on which his reputation depended were out of print and no longer judged commercially viable publishing strategies, and only a few articles now came from his pen. The handful of publications of his still in circulation embodied some of the profoundest speculations on the connections between physics and psychics, even if they were hampered by an ether hypothesis that only seemed to push back the question of how mind and matter interacted.[199]

For the younger generation of physicists, wireless engineers and other scientific practitioners, the explanatory difficulties of Lodge's ether hypotheses were overshadowed by the notorious problems of directly detecting the entity and investigating psychical effects, and for this reason

[196] Waldemar Kaempffert, 'Searching Out the Mind's Mysteries', *New York Times Magazine*, 17 October 1937, pp. 8 and 24.
[197] Albert Einstein to Max Born, 4 December 1926 in Max Born (ed.), *The Born-Einstein Letters: Correspondence Between Albert Einstein and Max and Hedwig Born from 1916 to 1955, with Commentaries by Max Born*, translated by Irene Born (London: Macmillan, 1971), pp. 90–1, p. 91. Lodge, 'Ether and Relativity', p. 805.
[198] Lodge, 'Past and the Future', p. 73.
[199] See, for example, Oliver Lodge, *Making of Man* (London: Hodder and Stoughton, 12th ed., 1938).

the ageing physicist's ideas were appreciated more as food for thought than as fruitful lines of research. This generation of scientific practitioners had grown up with conceptions of the physical world that were very different from the largely mechanistic ones on which Lodge, Crookes, Rayleigh, Thomson and other physical–psychical scientists were nurtured. But just as the puzzles created by mechanistic conceptions had made psychical phenomena especially alluring for the older scientists, so, suggested Kaempffert six months before Lodge's death, the implications of the newer conceptions of physics seemed to be drawing a younger generation of physical scientists into psychical investigation and to reverse a decline that had begun decades earlier.[200]

[200] Waldemar Kaempffert, 'Science in the News', *New York Times*, 25 February 1940, p. 51.

Conclusion

If Waldemar Kaempffert was right in thinking that physicists of the late 1930s were more sympathetic to psychical investigation than ever before then they certainly kept quiet about it.[1] By this period there remained only a handful of physical scientists, let alone physicists, who were open about their psychical interests, and even then, it was not at all clear how far such interests were informed by an embrace of new physical under-standings of space, time, matter and causality.

One such physicist was Robert Strutt, who succeeded his father as Baron Rayleigh in 1919. Born into the Strutt–Balfour–Sidgwick clans, it was inevitable that he should develop scientific interests in physics and psychical research and, in 1937–8, followed in the footsteps of his father, his aunt Eleanor and three of his uncles as an SPR president. What was not so inevitable was the critical spirit in which this ostensibly most sympathetic of all the younger physical–psychical scientists assessed many of the major episodes in the heyday of 'physics and psychics' that this book has studied.

In his presidential address to the SPR, the Fourth Baron Rayleigh questioned whether the investigations of Crookes and others had achieved enough control over the physical phenomena of mediumship to allow satisfactory repetition of effects and significantly reduce oppor-tunities for mediumistic fraudulence.[2] These were among the principal reasons why many scientists remained at best ambivalent about psychical research and believed that distinguished fellow workers from the past

---

[1] Among those physicists who appear to have been quietly sympathetic in this period were two Americans: Robert A. McConnell (1914–2006) and Joseph H. Rush (1911–2006). See Arthur S. Berger, *Lives and Letters in American Parapsychology: A Biographical History, 1850–1987* (Jefferson, NC: McFarland & Company, 1988), pp. 313–14, 320–1. Predictably, the earliest publications of McConnell and Rush included papers on the question of the physical basis of psychic effects: Robert A. McConnell, 'Physical or Non-Physical', *Journal of Parapsychology*, vol. 11 (1947), pp. 111–17; J. H. Rush, 'Some Considerations as to a Physical Basis of ESP', *Journal of Parapsychology*, vol. 7 (1943), pp. 44–9.

[2] Rayleigh, 'Presidential Address: The Problem of Physical Phenomena'.

such as Crookes and Lodge had probably been duped.[3] While praising Crookes's mechanical tests of psychic force as a "definite advance" towards "systematic investigation", Rayleigh concluded that they had not gone "far enough" because the "marvels" had not been repeated enough times.[4] Similarly, he dismissed the positive evidence that Lodge and others had claimed for Palladino's mediumship because the effects that they attributed to genuine telekinetic powers were too similar to those that she had been found to produce fraudulently. The electrical tests used by Varley and Crookes, which he would have known about from his father, were silently ignored.[5]

Elsewhere, Rayleigh balanced praise for Lodge's courage in publicly endorsing psychical evidence with criticism of his credulity regarding the often "absurd" utterances of trance mediums, and deemed the original Reichenbach investigations so admired by Barrett, Crookes and Varley to be "quite uncritical" and unworthy of the SPR's attention.[6] These latter criticisms derived mainly from the negative results of a fresh experimental investigation which he had evidently believed Reichenbach's claims still deserved: he exposed panchromatic photographic plates to a powerful permanent magnet for five months but, despite the plates being 600 times more sensitive than human vision, he reported no effect.[7]

Rayleigh's assessment of the experimental work of fellow physical scientists in psychical research was not unusual for psychical researchers and the first generation of parapsychologists of the 1930s and '40s. Countering those who upheld Crookes's work on Home and Cook, Lodge's on Palladino, and Barrett's on Reichenbach were many who chose to remember these physicists solely for their pioneering contributions to telepathy and the general scientific gravitas that they brought to the subject.[8] Neither was Rayleigh alone in levelling criticism at the more

---

[3] See, for example, [Anon.], 'Our Point of View', *Scientific American*, vol. 151 (1934), p. 121; [Anon.], 'Science and Psychical Research'.

[4] Rayleigh, 'Presidential Address: The Problem of Physical Phenomena', p. 5.

[5] Rayleigh's awareness of such tests is suggested in [Lord Rayleigh], 'MM. Osty's Investigations of Rudi Schneider', *Nature*, vol. 133 (1934), pp. 747–9.

[6] Rayleigh quoted in [Anon.], 'Sir Oliver Lodge. Further Tributes', *Times*, 27 August 1940, p. 7; Lord Rayleigh, 'The Question of Lights Supposed to Have Been Observed Near the Poles of a Magnet', *PSPR*, vol. 45 (1939), pp. 19–24, p. 19n.

[7] Rayleigh was not the only early-twentieth-century physicist to revisit Reichenbach's claims. In 1914, the Austrian physicist Eduard Hascheck agreed with some of Reichenbach's claims regarding the luminosity surrounding magnets and human bodies but ascribed them to photochemical processes rather than od: Eduard Haschek, 'Über Leuchterscheinungen des menschlichen Körpers', *Sitzungsberichte der Kaiserlichen Akademie der Wissenschaften. Mathematisch-Naturwissenschaften Klasse*, vol. 123 (1914), pp. 523–32.

[8] Those upholding their physical investigations include Carrington, *Story of Psychic Science* and Harry Price, *Fifty Years of Psychical Research: A Critical Survey* (London: Longmans, Green and Co., 1939). Works that only discussed their telepathic investigations include

theoretical and speculative aspects of physics and psychics. In an otherwise adulatory contribution to an obituary notice of Lodge for the SPR, he echoed earlier critics in questioning the physicist's "rather materialistic point of view about the ether".[9] It was puzzling because it contrasted with the "anti-materialistic" view underpinning Lodge's interest in psychical research and caused problems for the "analogy" he drew between the way material bodies received "inspiration" from the "empty space" around them and the way humans received impressions from a non-material or spiritual world.[10] As far as Rayleigh was concerned, the analogy worked equally well if, following Einstein, space was not defined in terms of a quasi-mechanical and quasi-material ether.

Rayleigh's comment bears comparison to Joseph Banks Rhine's roughly contemporary view that the brain wave theory of telepathy proposed by Crookes and others was not only unable to account for so many properties of ESP (including the capacity of one person to correctly select the waves of a particular person from myriad others in their neighbourhood) but represented altogether too narrow a "concept of physics".[11] There was clearly an "energetic link" between individuals in telepathic communion, and between an inanimate object and the mind of the individual clairvoyantly perceiving it, but this required "another order of energy, not radiant".[12] The concepts and theories of nineteenth-century physics, in other words, hampered rather than helped the interpretation of psychical effects.

The views of Rayleigh, Rhine and others illustrate the complex legacy of the late-nineteenth- and early-twentieth-century heyday of physics and psychics. For most proponents and critics of psychical research, the major representatives of physics and psychics had not succeeded in forging satisfactory connections between these enterprises: the theories and concepts of physics had proven of limited use in interpreting psychical effects, experimental physics had not yielded particularly strong evidence for such effects and, *pace* Lodge's vision of 1891, it was not clear that training in physics gave an individual the right to lead psychical investigation.

Yet even for critics such as Rayleigh there were reasons to think physics could still make decisive contributions to psychical research. Some rays of hope came from various attempts over the preceding decades to achieve

Joseph Banks Rhine, *Extra-Sensory Perception* (Boston: Boston Society for Psychical Research, 1934) and George N. M. Tyrrell, *Science and Psychical Phenomena* (London: Methuen and Co., 1939).

[9] Rayleigh quoted in [Anon.], 'Obituary. Sir Oliver Lodge, F.R.S., Sir J. J. Thomson, O. M., F.R.S.', *PSPR*, vol. 46 (1940–1), pp. 209–23, p. 216.

[10] Rayleigh, 'Obituary', p. 216.   [11] Rhine, *Extra-Sensory Perception*, p. 163.

[12] Rhine, *Extra-Sensory Perception*, p. 163.

that critically important goal of greater control in seances. One such attempt was William J. Crawford's investigations of the 1910s into the telekinetic and materialisation phenomena of the Irish medium Kathleen Goligher – investigations that had impressed Barrett and Lodge.[13] Another was French psychical researchers Eugene and Marcel Osty's study in the early 1930s of Rudi Schneider, a study to which Rayleigh himself had contributed. To determine whether Schneider consciously or unconsciously used his hands to achieve telekinetic effects, the Ostys guarded the objects to be moved with a beam of infra-red light. This kind of light preserved the conditions of darkness required by Schneider but revealed new insights into his abilities. Although the beam was cut during Schneider's seances, flashlight photography taken when the beam was interrupted revealed that the medium had not moved and nothing visible was obstructing the beam. Moreover, precision measurements of the beam's changing intensity revealed a close correlation with Schneider's breathing, and this suggested that something invisible but physiologically connected to his body (possibly an ectoplasmic extrusion) had absorbed part of the beam.

For Rayleigh, the Ostys' result was "one of the most valuable contributions ever made" to psychical research and the level of control achieved made their approach a model for others to follow in the quest for better studies of the physical phenomena of mediumship.[14] The example of the Ostys was doubly useful to Rayleigh. It demonstrated the capacity of experimental physics to contribute sophisticated instruments of control, detection and measurement to psychical research, and to furnish the investigator with important skills in distinguishing known from novel causes. Responding to allegations that Schneider, by means of resonant effects in the instruments induced by well-timed breaths, caused the partial absorption of the beam, Rayleigh counselled that the potential effect of vibration was "the sort of thing that constantly haunts the night thoughts of the careful experimentalist" and prompted them to make the appropriate checks the following morning.[15] Careful experimental physicists could still contribute to the robustness of psychical tests in ways that psychologists, conjurors and medical doctors could not.

Rayleigh's reference, in his obituary of Lodge, to relativity was one of several instances when he hinted that hopes for the future contributions of physics to psychical research also lay in the domain of revolutionary theoretical developments. In his presidential address to the SPR, he

[13] Barrett, 'Report of Physical Phenomena'; Oliver Lodge, 'The Reality of Psychic Phenomena', *Light*, vol. 37 (1917), pp. 66–7.
[14] Rayleigh, 'Presidential Address: The Problem of Physical Phenomena', p. 14.
[15] Rayleigh, 'Presidential Address: The Problem of Physical Phenomena', p. 14.

suggested that the "adjustments" made in the physical sciences over the previous decades were a warning to those who, whether in physics or psychics, dogmatically upheld notions of what nature's constitution ought to be.[16] However, Rayleigh went no further with this allusion and in this sense he resembled most physical–psychical scientists of his generation, who were largely uninterested in further explorations of the psychical significances of relativity and quantum theory.

Indeed, by the 1930s, many physical–psychical scientists were still interested in exploring the older physical theories of psychical effects, including electromagnetic theories of telepathy. A common argument against such theories – that the strength of telepathic impressions did not decline with distance – was challenged by at least one radio engineer of the 1930s with evidence that long-distance radio transmission did not always exhibit this common property of radiation.[17] More tellingly, by this period Soviet radio engineers and physicists associated with the emergent field of biophysics had long been participating in experimental tests of the radio wave theory of telepathy.[18] Much of this work was inspired by recent studies of the electric field surrounding the human body, Hans Berger's ground-breaking method of recording electrical activity in the human brain, and physiological theories of the electromagnetic wave activity of the nervous system. Most importantly, it drew on the Italian psychiatrist Ferdinand Cazzamalli's controversial experimental evidence of radio waves emitted by the brain during intense cerebral activity. The Soviet researchers failed in their bid to verify Cazzamalli's result and, moreover, found that telepathic impressions were unaffected by the presence of metallic screens that would have blocked such impressions had they been radio waves. Although they did not rule out the faint possibility that telepathy might be electromagnetic vibrations of very short or long wavelength or some other form of physical energy, they ultimately abandoned attempts to test these hypotheses experimentally.

The verdict on Cazzamalli, however, was not decisive, since in the 1950s there were many American radio engineers who, inspired by Berger's work and evidence that the 'psychic' faculty of water dowsing had an electromagnetic basis, debated the possibility of using microwaves, electronic amplifiers and other resources to pick up faint vibrations

---

[16] Rayleigh, 'Presidential Address: The Problem of Physical Phenomena', p..17.

[17] G. W. Fisk, 'Criticism of a Radiation Theory as an Explanation of Telepathy or Clairvoyance', *JSPR*, vol. 29 (1935–6), pp. 35–6.

[18] Leonid L. Vasiliev, *Experiments in Distant Influence: Discoveries by Russia's Foremost Parapsychologist* (London: Wildwood House, 1976), chapter 2. Vasiliev's book refers to Soviet researches started over fifty years before it was published.

associated with thought.[19] None of these proposals appear to have borne fruit, but other telecommunications engineers from this period seem to have gone further in applying their technical skills to psychical questions. The British radio expert Alec Reeves, for example, devoted much of his spare time during and after the Second World War to the construction of a host of instruments for detecting and measuring psychical effects. These included a delicate apparatus for exploring 'psychokinesis', Rhine's alternative term for 'telekinesis' describing the capacity of mind to directly influence matter independently of known energies or forces.[20]

The methods developed in the late 1930s by Rhine and his followers to produce evidence for psychokinesis mirrored those that they had developed for ESP. The emphasis was on subtle effects produced by normal and psychic experimental subjects rather than just the more startling phenomena associated with spiritualist mediums; on laboratory- rather than seance-based tests; and on statistical analyses of a large number of events rather than more qualitative approaches to a smaller quantity of evidence. In some ways these features of psychokinesis research created new possibilities for experimental connections between physics and psychics that were explored from the late 1960s onwards. These connections were distinguished by highly sophisticated instruments used to detect, control and process subtle physical effects, and which have since strengthened arguments for the scientific status of parapsychology and psychical research. They included German engineer Helmut Schmidt's invention of a machine that measured the extent to which psychokinesis skewed a normally random pattern of electric lights activated by electrons emitted during radioactive decay and the American engineer Robert Jahn's investigations into the effect of psychokinesis on other types of random event generators and on such sensitive phenomena as interferometric fringe patterns and the luminosity of gas discharge tubes.[21]

The period within which these new instrumental approaches to psychical effects were first developed represented the most promising new phase in the relationship between physics and psychics for decades, albeit

[19] See correspondence of Robert J. Bibbero, Henri Busginies, Arthur Hammond, Hans Hollmann, Ted Powell and Harry Stockman in the *Proceedings of the Institute of Radio Engineers*, vol. 38 (1950), pp. 979 and 1097; vol. 39 (1951), pp. 290, 841, 969 and 1571; vol. 40 (1952), pp. 605 and 995.

[20] On Reeves see David Robertson, 'The Radical Who Shaped the Future', *IEE Review*, May 2002, pp. 31–6.

[21] See Helmut Schmidt, 'Psychokinesis', in John White (ed.), *Psychic Exploration: A Challenge for Science* (New York: Perigee Books, 1974), pp. 179–93; Robert G. Jahn, 'The Persistent Paradox of Psychic Phenomena: An Engineering Perspective', in Charles T. Tart, Harold E. Puthoff and Russell Targ (eds.), *Mind at Large: Institution of Electrical and Electronic Engineers Symposia on the Nature of Extra Sensory Perception* (Charlottesville, VA: Hampton Roads, 2002), pp. 238–316.

one that never regained its late-nineteenth-century strength. In 1974, the American aerospace engineer James Beal hailed the arrival of the new science of "paraphysics", which he defined as the "study of the physics of paranormal processes – i.e., phenomena that resemble physical phenomena but are without recognizable physical cause".[22] It was a field whose emergence was fuelled by the growth of interest in the more general question of the relationship between the sciences (and especially physics) and consciousness.[23]

One of the most conspicuous indicators of the new field was the surge in the number of physicists who busied themselves with theories of psychical effects. This was particularly striking among west coast American physicists, who, as David Kaiser has argued, were more interested in the philosophical implications of quantum mechanics than the computational and other practical uses that had dominated this branch of physics for decades.[24] They spearheaded numerous explorations of the psychical implications of quantum mechanics that only a few physicists had explored before.[25] Some proposed that the 'Copenhagen interpretation' of the theory – one view of which proposes that the act of observing and measuring the physical state of a quantum mechanical system determines this state – rendered psychokinesis more plausible. Others interpreted quantum mechanics to suggest the capacity of two separated quantum-level objects to exchange information with each other instantaneously – an 'entanglement' violating causality and the relativistic prohibition on superluminal velocities but also creating a space within physics for psychokinesis, telepathy and the psychical effect posing one of the greatest

---

[22] James B. Beal, 'The Emergence of Paraphysics: Research and Applications', in White, *Psychic Exploration*, pp. 426–46, p. 426.

[23] This is evident from Fritjof Capra, *The Tao of Physics: An Exploration of the Parallels between Modern Physics and Eastern Mysticism* (London: Wildwood House, 1975); Jeffrey Mishlove, *The Roots of Consciousness: Psychic Liberation Through History, Science and Experience* (New York: Random House, 1975); Andrija Puharich (ed.), *The Iceland Papers: Select Papers on Theoretical and Experimental Research on the Physics of Consciousness* (Amherst, WI: Essentia Research Associates, 1979).

[24] David Kaiser, *How the Hippies Saved Physics: Science, Counterculture, and the Quantum Revival* (New York: W. W. Norton, 2011), chapter 1.

[25] My discussion draws on James E. Beichler, 'To Be or Not to Be! A "Paraphysics" for the New Millennium', *Journal of Scientific Exploration*, vol. 15 (2001), pp. 33–56; Bernard Carr, 'Worlds Apart? Can Psychical Research Bridge the Gulf Between Matter and Mind?', *PSPR*, vol. 59 (2008), pp. 1–96; Collins and Pinch, *Frames of Meaning*, chapter 4 and Douglas M. Stokes, *The Nature of Mind: Parapsychology and the Role of Consciousness in the Physical World* (Jefferson, NC: McFarland and Co., 1997), chapter 7. Older physicists who connected quantum mechanics and psychic effects included Henry Margenau and Pascual Jordan. See Henry Margenau, 'Physics and Psychic Research', *Newsletter of the Parapsychology Foundation*, 1956, pp. 14–15 and Pascual Jordan, 'Reflections on Parapsychology, Psychoanalysis and Atomic Physics', *Journal of Parapsychology*, vol. 15 (1951), pp. 278–81.

challenges to scientific intelligibility: precognition. Elsewhere, paraphysicists added new levels of mathematical sophistication to older speculations about psychical interactions involving four or more dimensions of spacetime and types of physical signalling via a particle, field or some other emanation. Amongst the latter category were theories explaining ESP in terms of neutrinos and precognition in terms of waves travelling backwards in time and superluminal particles (tachyons).

Much of this work, however, failed to meet the demands of parapsychologists, psychical researchers and their scientific critics for a comprehensive theory of psychical efforts. The parapsychological and psychical research literature of the past few decades is littered with critiques of the paraphysical theories of psychical effects or 'psi', including those maintaining old arguments that some types of psychical effect cannot be reduced to any kind of physical law, those emphasising the untestable nature of theories involving higher dimensions, and those highlighting the failure of signalling models to account for the extraordinary spatial and temporal properties of psychical effects and to provide a plausible physiological justification.[26] Although paraphysicists have accepted the limitations of quantum theories of psi, they remain much more optimistic about the ultimate fruitfulness of this line of enquiry than physicists, many of whom are deeply sceptical of the very interpretations of quantum theory that have been used to create spaces for consciousness and psychical effects.[27]

The difficulty of showing that psychical effects follow from accepted theories in physics has been offered as one reason why today's physicists are ambivalent at best about paraphysics, parapsychology and psychical research.[28] By becoming integrated into the theories of physics, psychical effects would achieve an embeddedness within a scientific culture that is regarded as one of the ways in which scientists are likely to take far greater interest in such effects.[29] The largely negative attitude of today's physicists, however, stems more from the serious misgivings that they share with practitioners of other sciences, and which have plenty of late-nineteenth- and early-twentieth-century precedents: psychical effects still lack convincing scientific proof, they are notoriously difficult to replicate, they have associations with fraudulence, and their existence

---

[26] Carr, 'Worlds Apart', pp. 9–10 and 34; Stokes, *Nature of Mind*, pp. 112–20. See also C. T. K. Chari, 'Some Generalized Theories and Models of Psi: A Critical Evaluation', in Benjamin B. Wolman (ed.), *Handbook of Parapsychology* (New York: Van Nostrand Reinhold Co., 1977), pp. 803–22.

[27] Carr, 'Worlds Apart'; Collins and Pinch, *Frames of Meaning*, pp. 82–9.

[28] Beichler, 'To Be or Not to Be', p. 48.

[29] Trevor Pinch, 'Some Suggestions From Sociology of Science to Advance the Psi Debate', *Behavioural and Brain Sciences*, vol. 10 (1987), pp. 603–5.

seems to threaten many of the assumptions on which the sciences are founded.[30] These misgivings certainly help explain why paraphysics, parapsychology and psychical research still do not enjoy a significant place in scientific textbooks, teaching programmes and the priorities of scientific research funding bodies.[31] Our protagonists would have been painfully familiar with these misgivings, as well as the fact that those who express them most forcefully remain psychologists, stage magicians and science popularisers, who are often at pains to reiterate the old argument that expertise in physics does not constitute expertise in psychics.[32]

Among the most significant recent contributions to the debate on the apparent impasse between physics and psychics are those proposing that the foundations of physics be extended well beyond their relativistic and quantum frameworks. This kind of extension, which might involve the hyperdimensional theories that have pervaded paraphysical thinking for decades, would enable physics to embrace the very questions that most of its current practitioners usually regard as beyond the formal boundaries of the subject, including consciousness, psychical effects, mysticism and religion.[33] Although many physicists are likely to question whether these extensions produce a science that can still legitimately be called physics, those who have long been interested in the physics of consciousness are likely to be more accommodating. Another way of trying to break the impasse between physics and psychics involves speculating on the possible extensions of technology rather than the foundations of physics. Given the current achievements in magnetic resonance imaging and other brain mapping technologies, it is not inconceivable that in the future machines could be built that would correlate visual patterns and thoughts.[34] These patterns could be used to generate electromagnetic waves that might induce corresponding thoughts in other brains (a kind of technological telepathy) or to design brain implants enabling us to instruct computers controlling electrical devices (a kind of technological psychokinesis).

[30] James Alcock, 'Parapsychology: Science of the Anomalous or Search for the Soul', *Behavioural and Brain Sciences*, vol. 10 (1987), pp. 553–65; Carr, 'Worlds Apart', pp. 6–7; Harvey J. Irwin and Caroline A. Watt, *An Introduction to Parapsychology* (Jefferson, NC: McFarland & Co., 5th ed., 2007), chapter 17.

[31] Alcock, 'Parapsychology', p. 554.

[32] This argument is made by Martin Gardner, the American science populariser and notorious critic of paranormal investigations, in *Science Good, Bad, Bogus* (Buffalo, NY: Prometheus Books, 1981), p. 199.

[33] Carr, 'Worlds Apart'; Mary B. Hesse, *Forces and Fields: The Concept of Action at a Distance in the History of Physics* (London: Thomas Nelson and Sons, 1961), p. 302.

[34] Michio Kaku, *Physics of the Impossible* (London: Penguin Books, 2009), chapters 5–6.

The protagonists of this book would have been intrigued by the current state of paraphysics. They would have questioned its tendency to characterise the worldview of the physics of their time as rigidly materialistic and mechanistic and which, unlike the modern physics of relativity and quantum, was completely unable to accommodate mind and spirit. This was an interpretation of physics with which they were all too familiar but which they, along with many other nineteenth- and early-twentieth-century physical scientists inside and outside psychical research, spiritualism and other occultisms, vigorously repudiated.

This book has been about physicists and representatives of other physical sciences who reacted to this perceived materialistic threat differently from most of their scientific peers, even if (most notably in the case of Lodge) they never entirely succeeded. They tried to extend theories and practices of the physical sciences to questions of the genuineness and provenance of phenomena commonly grouped under such headings as psychical or psychic. We have shown that this preoccupied far more physical scientists than we might expect given the strongly psychological character of these phenomena, and which correspondingly pushed them outside the formal boundaries of the physical sciences. We have also seen that what gave them the confidence to extend the physical sciences in these unusual, risky and exciting directions was a feature of their scientific culture often overlooked in twentieth-century representations of the rigid world of Victorian physics. This was the problematic nature of many aspects of Victorian physics – the experimental practices, the theories and concepts, and the wider purposes of physical enquiry – that our protagonists struggled with, and in many cases successfully tackled, for much of their careers. Yet the experience of tackling these difficulties, of making phenomena of the physical world more intelligible and controllable, and less occult and capricious, gave them attitudes towards and resources for other, difficult scientific questions that often proved productive in the case of psychical effects.

Historians have shown how the materialistic threat inspired some of the most creative aspects of nineteenth- and early-twentieth-century physical sciences, most famously Maxwell's demon. The conceptual, theoretical and experimental activities examined in this book are no less worthy of that description. They show nineteenth- and early-twentieth-century physical scientists to be even more confident and creative about their scientific enterprises than we have previously supposed. The claim of current paraphysicists that the extension of physics is the result of creative encounters between the subject and psychics would have been heartily endorsed by our protagonists.[35] For Crookes, Lodge, Varley and others,

---

[35] Carr, 'Worlds Apart', pp. 73–4.

psychical phenomena represented exciting opportunities for extending the domain of established concepts and theories of physics, but they also highlighted the limits of physics and seem to have played some role in nurturing new theories and understandings in the science. William Thomson was probably not completely wrong in suggesting that mesmerism informed Crookes's path to the novel idea of matter in a radiant state; Varley's earlier contribution to the pre-history of the electron – his study of the mechanical effect of electrical discharge – was almost certainly fuelled by a spiritualist preoccupation with the apparent materiality of immaterial agents; and the problem of the mechanisms of telepathy, telekinesis and survival undoubtedly spurred Lodge's major experimental and theoretical contributions to ether physics. But we have also encountered individuals for whom psychical puzzles seem to have reacted more directly on instrumental creativity. The history of psychical research is also the history of attempts by such individuals as Hare, Crookes, Carrington and Edison to build elaborate new machines for measuring, reducing or replacing human agency in the exhibition of psychical effects. These individuals would certainly have recognised the goals of those now imagining technologies that might mechanically reproduce telepathy and psychokinesis.

We should no longer feel embarrassed, titillated or puzzled when contemplating the psychical and occult interests of scientists and engineers from the nineteenth and twentieth centuries. This book has suggested that such interests were of a piece with the scientific and technological enterprises for which our protagonists are justly remembered. The psychical and the occult may well turn out to have been even more significant to these enterprises than we have been able to establish. If that is the case, it will greatly enhance our understanding of the complex sources of scientific creativity in the past and our appreciation of them in the present.

Bibliography

Primary Sources

Archival Collections

John Couch Adams Papers, St John's College Library, Cambridge
Earl of Balfour Papers, National Records of Scotland, Edinburgh
William Fletcher Barrett Papers, Royal Society Archives, London
William Bateson Papers, Cambridge University Library
Edward Bulwer-Lytton Papers, Knebworth House Archive, Hertfordshire
Hereward Carrington Papers, Department of Rare Books and Special Collections, Princeton University Library
College of Psychic Studies Archives, London
William Crookes Papers, Science Museum Group Collection, London
Benjamin Davies Papers, National Library of Wales, Aberystwyth
Ghost Club Archives, British Library, London
George Francis FitzGerald Papers, Royal Dublin Society Library and Archives
Edward Edmund Fournier d'Albe Papers, Private Collection, Christine Fournier d'Albe, Peyrusse-Grande
Kew Observatory Records, National Archives, London
Oliver Lodge Papers, Cadbury Research Library, University of Birmingham
Oliver Lodge Papers, University College London
Oliver Lodge Papers, University of Liverpool Library
Macmillan and Company Ltd Archive British Library, London
Frederic William Henry Myers Papers, Trinity College Library, Cambridge
Niels Bohr Library and Archives, American Institute of Physics, College Park, Maryland
Robert Dale Owen Papers, Indiana State Library, Rare Books and Manuscripts Division, Indianapolis, Indiana
Harry Price Collection, Senate House Library, University of London
Rayleigh Papers, United States Air Force Academy, McDermott Library, Colorado Springs, Colorado
Rayleigh Papers, Private Collection, Terling Place, Terling, Essex
Referees Reports, Royal Society Archives

Epes Sargent Papers, Boston Public Library
Emil Prinz zu Sayn-Wittgenstein-Berleburg Papers, Private Collection, Gerd H. Hövelmann, Marburg
Henry Sidgwick Papers, Trinity College Library, Cambridge
Silvanus Philips Thompson Papers, Archives of Imperial College London
Silvanus Philips Thompson Papers, Institution of Engineering and Technology Archives, London
Society for Psychical Research Archive, Cambridge University Library
Joseph John Thomson Papers, Cambridge University Library
Joseph John Thomson Papers, Trinity College Library, Cambridge
University of Exeter Special Collections Alfred Russel Wallace Papers, British Library, London
Western Manuscripts, Autograph Letters Series, Wellcome Collection, London

Books and Articles

[Anon.], 'About Scientific Spiritualism', *Medium and Daybreak*, vol. 1 (1870), pp. 201–2.

[Anon.], 'Animal "Magnetism"', *Electrician*, vol. 2 (1862), pp. 157–8.

[Anon.], 'Annual Report of the Council for the Year 1932', *JSPR*, vol. 28 (1933–4), pp. 19–29.

[Anon.], 'Anniversary Meeting', *Proceedings of the Royal Society*, vol. 24 (1874–5), pp. 70–102.

[Anon.], 'Anniversary Meeting', *PSPR*, vol. 1 (1882–3), pp. 158–60.

[Anon.], 'Arthur Prince Chattock 1860–1934', *Obituary Notices of Fellows of the Royal Society*, vol. 1 (1932–5), pp. 293–8.

[Anon.], 'Benjamin Brodie on Spiritualism', *Spiritual Magazine*, vol. 1 (1860), pp. 97–103.

[Anon.], 'The British Association', *Spiritualist*, vol. 15 (1879), pp. 109–14.

[Anon.], 'The British Association at Glasgow', *Spiritualist*, vol. 9 (1876), pp. 88–94.

[Anon.], 'Cambridge Branch of the S.P.R.', *JSPR*, vol. 1 (1884–5), pp. 52–3, 180–1.

[Anon.], 'Chemistry', *Popular Science Review*, vol. 1 (1862), pp. 382–9.

[Anon.], 'Constitution and Rules', *PSPR*, vol. 1 (1882–3), pp. 331–6.

[Anon.], 'Correspondence Between Mr. Cromwell F. Varley and Mr. William Crookes', *Spiritual Magazine*, vol. 6 (New Series) (1871), pp. 350–3.

[Anon.], 'Council Meetings', *JSPR*, vol. 1 (1884–5), pp. 259–61.

[Anon.], 'Crookes Brain Waves', *Electrical Engineer*, vol. 23 (1897), pp. 220–1.

[Anon.], 'Discussion on Dr. Stone's Paper', *Journal of the Society of Telegraph Engineers*, vol. 11 (1882), pp. 118–28.

[Anon.], 'Discussion of Professor Lodge's Paper', *JSPR*, vol. 6 (1893–4), pp. 336–45.

[Anon.], 'Edison's Own Secret Spirit Experiments', *Modern Mechanix and Inventions*, October 1933, pp. 34–6.

[Anon.], 'Electric Waves', *Electrical World*, vol. 29 (1897), p. 252.

[Anon.], 'Endowment Fund for Psychical Research', *JSPR*, vol. 11 (1903–4), pp. 44–5.

[Anon.], 'The Ether and Human Survival', *Light*, vol. 45 (1925), p. 198.

[Anon.], 'An Experimental Investigation of Spiritual Phenomena', *Spiritualist*, vol. 1 (1869–71), pp. 180–2.

[Anon.], 'The Force of Gravitation', *Spiritualist*, vol. 1 (1869–71), p. 197.

[Anon.], 'From Matter to Spirit', *London Review*, vol. 7 (1863), pp. 547–8.

[Anon.], 'Further Experiments by Mr. Crookes', *Spiritualist*, vol. 1 (1869–71), p. 177.

[Anon.], 'General Meeting', *JSPR*, vol. 7 (1895–6), pp. 131–8.

[Anon.], 'General Meeting', *JSPR*, vol. 11 (1903–4), pp. 152–7.

[Anon.], 'General Meeting', *JSPR*, vol. 12 (1906–7), pp. 179–84.

[Anon.], 'Gregory's Edition of Reichenbach', *Medical Times*, vol. 21 (1850), pp. 451–2.

[Anon.], 'Gregory's "Letters on Animal Magnetism"', *Mechanics' Magazine*, vol. 54 (1851), pp. 364–70.

[Anon.], *Heaven Opened; Or, Messages for the Bereaved, From Our Little Ones in Glory. Through the Mediumship of F. J. T.* (London: James Burns, 1870).

[Anon.], 'The Heroes of the Atlantic Telegraph Cable', *Illustrated Times*, 25 August 1866, p. 21.

[Anon.], 'Howitt on the Supernatural', *London Quarterly Review*, vol. 21 (1863–4), pp. 27–70.

[Anon.], 'Human Levitation', *Quarterly Journal of Science*, vol. 5 (New Series) (1875), pp. 31–61.

[Anon.], 'Important Interview with Sir William Crookes', *International Psychic Gazette*, vol. 4 (1917), pp. 61–2.

[Anon.], 'Introductory', *Electrician*, vol. 1 (1861), p. 1.

[Anon.], 'La Métapsychique', *Lancet*, vol. 1 (1905), pp. 449–50.

[Anon.], 'Lessons in Humility', *Light*, vol. 17 (1897), p. 78.

[Anon.], 'Lyon v. Home', *Spiritual Magazine*, vol. 3 (New Series) (1868), pp. 241–54.

[Anon.], 'Meeting of Council', *JSPR*, vol. 1 (1884), p. 33.

[Anon.], 'Meetings of Council', *JSPR*, vol. 3 (1887–8), pp. 65–8.

[Anon.], 'Meetings of Council', *JSPR*, vol. 3 (1887–8), pp. 149–50.

[Anon.], 'Meeting of the British Association at Sheffield', *Graphic*, 30 August 1879, p. 1.

[Anon.], 'Memorial to Mr. Myers at Cheltenham College', *JSPR*, vol. 13 (1907–8), pp. 148–52.

[Anon.], 'Mesmeric Hospital Reports', *British Medical Journal*, vol. 2 (1862), pp. 308–9.

[Anon.], 'Miss Florrie Cook', *Two Worlds*, vol. 10 (1897), pp. 173–4, p. 185.

[Anon.], 'Model Engineers. Record Number at Exhibition – The Electric Psychometer', *Electrician*, 9 September 1932, p. 324.

[Anon.], 'Modern Necromancy', *North British Review*, vol. 67 (1861), pp. 110–41.

[Anon.], 'Modern Spiritualism', *Edinburgh Review*, vol. 198 (1903), pp. 304–29.

[Anon.], 'Modern Spiritualism', *Quarterly Review*, vol. 114 (1863), pp. 179–210.

[Anon.], 'Moral Causes', *Electrician*, vol. 2 (1862), pp. 39–40.

[Anon.], 'Mrs. Verrall on Telepathy', *Cambridge Magazine*, vol. 1(1912), p. 111.

[Anon.], 'Mr. Varley and Professor Allen Thomson', *Spiritualist*, vol. 1 (1869–71), p. 194.

[Anon.], 'Mr. William Crookes, F.R.S. on Disembodied Spirits', *Spiritualist*, vol. 1 (1869–71), p. 161.

[Anon.], 'The Mystery of the Tables', *Illustrated London News*, 18 June 1853, pp. 481–2.

[Anon.], 'The National Association of Spiritualists', *Spiritualist*, vol. 6 (1875), pp. 122–6.

[Anon.], 'National Conference of Spiritualists in Liverpool', *Spiritualist*, vol. 3 (1872–3), pp. 291–7.

[Anon.], 'National Jubilee Conference of Progressive Spiritualists', *Medium and Daybreak*, vol. 3 (1872), pp. 341–4.

[Anon.], 'The New Light', *Light*, vol. 26 (1896), pp. 102–3.

[Anon.], 'Notes', *Electrician*, vol. 30 (1892–3), pp. 587–8.

[Anon.], 'Notes by the Way', *Quarterly Transactions of the British College of Psychic Science*, vol. 15 (1937), pp. 342–3.

[Anon.], 'Notes of the Month', *Occult Review*, vol. 28 (1918), pp. 187–203.

[Anon.], 'Obituary', *Chemical News*, vol. 19 (1869), p. 82.

[Anon.], 'Obituary. Desmond G. FitzGerald', *Electrician*, vol. 54 (1904–5), p. 21.

[Anon.], 'Obituary. Sir Oliver Lodge, F.R.S., Sir J. J. Thomson, O.M., F.R.S.', *PSPR*, vol. 46 (1940–1), pp. 209–23.

[Anon.], 'Occultism Reconsidered', *Journal of Science*, vol. 4 (1882), pp. 404–9, 441–6.

[Anon.], 'Odyle, Mesmerism, Electro-Biology', *British and Foreign Medico-Chirurgical Review*, vol. 8 (1851), pp. 378–431.

[Anon.], 'On a New Force, Falsely So Called', *Medical Times and Gazette*, vol. 2 (1871), pp. 99–100.

[Anon.], 'Our Point of View', *Scientific American*, vol. 151 (1934), p. 121.

[Anon.], 'The Possibility of a Future Life', *Quarterly Journal of Science*, vol. 5 (New Series) (1875), pp. 472–86.

[Anon.], 'Preliminary Report of the 'Reichenbach' Committee', *PSPR*, vol. 1 (1882–3), pp. 99–100.

[Anon.], 'The Presentation of the Harrison Testimonial', *Spiritualist*, vol. 8 (1876), pp. 53–7.

[Anon.], 'Professor Barrett on "Thought Reading"', *Medium and Daybreak*, vol. 17 (1886), p. 157.

[Anon.], 'Professor Crookes on Ethereal Bodies', *Spectator*, vol. 78 (1897), pp. 200–1.

[Anon.], 'Professor Faraday and Table-Turning', *Mechanics' Magazine*, vol. 59 (1853), pp. 23–5.

[Anon.], 'Professor Lodge's Theology', *Church Times*, vol. 50 (1908), p. 767.

[Anon.], 'Professor W. R. Grove on Spirits', *Spiritualist*, vol. 1 (1869–71), p. 95.
[Anon.], 'Professor Tyndall and the Spiritualists', *Human Nature*, vol. 2 (1868), pp. 455–6.
[Anon.], 'The Professors and the Conjuror', *Speaker*, vol. 12 (1895), pp. 467–8.
[Anon.], 'Psychic Force', *English Mechanic and World of Science*, vol. 14 (1871), p. 85.
[Anon.], 'Psychic Force', *Saturday Review*, 15 July 1871, p. 83.
[Anon.], '"Psychical Force"', *Medical Times and Gazette*, vol. 1 (1876), pp. 545–6.
[Anon.], 'Psychic Photography', *Borderland*, vol. 3 (1896), pp. 313–21.
[Anon.], 'The Psychological Society on the Fundamental Nature of Matter', *Spiritualist*, vol. 7 (1875), p. 301.
[Anon.], 'The Psychological Society of Great Britain', *Spiritualist*, vol. 15 (1879), p. 235.
[Anon.], 'Radio Psychometry', *Radio Craft*, vol. 5 (1933), p. 264.
[Anon.], 'Reichenbach and the Psychical Research Society', *Journal of Science*, vol. 5 (1883), pp. 313–19.
[Anon.], 'Report of the Council to Members and Associates of the Society for Psychical Research for the Year 1902', *JSPR*, vol. 11 (1903–4), pp. 38–43.
[Anon.], 'Report of the Council for 1908', *JSPR*, vol. 14 (1909–10), pp. 36–40.
[Anon.], 'Report of the General Meeting', *JSPR*, vol. 2 (1885–6), pp. 338–46.
[Anon.], 'Review of Reichenbach's "Researches on Magnetism"', *Athenaeum*, 19 October 1850, pp. 1088–90.
[Anon.], 'Reviews', *Lancet*, vol. 1 (1846), pp. 103–4.
[Anon.], 'Reviews', *Medical Times and Gazette*, vol. 14 (New Series) (1857), pp. 122–3.
[Anon.], 'Science', *Athenaeum*, 28 November 1908, pp. 686–9.
[Anon.], 'Science and Psychical Research', *Nature*, vol. 132 (1933), pp. 945–6.
[Anon.], 'Science and Spirit', *Light*, vol. 33 (1913), p. 246.
[Anon.], 'Scientific Gossip', *Photographic News*, vol. 6 (1862), pp. 3–4.
[Anon.], 'A Scientific Testing of Mr. Home', *Spectator*, vol. 44 (1871), pp. 827–8.
[Anon.], 'Second Thoughts', *Punch*, vol. 157 (1919), p. 333.
[Anon.], 'The Sense of Identity – Materialistic Explanations of Spiritualism', *Spiritual Magazine*, vol. 5 (New Series) (1870), pp. 429–32.
[Anon.], 'Sir Arthur Conan Doyle's Resignation', *JSPR*, vol. 26 (1930), pp. 45–52.
[Anon.], 'Sir Oliver Lodge. Further Tributes', *Times*, 27 August 1940, p. 7.
[Anon.], 'Sir William Crookes on "Invisible Intelligent Beings"', *Light*, vol. 20 (1900), p. 223.
[Anon.], 'Sir William Crookes on Psychical Phenomena', *Light*, vol. 36 (1916), p. 397.
[Anon.], 'Societies and Academies', *Nature*, vol. 30 (1884), pp. 161–4.
[Anon.], 'Society for Psychical Research: Dublin Section', *JSPR*, vol. 14 (1909–10), pp. 63–4.
[Anon.], 'The Society for Psychical Research: Objects of the Society', *PSPR*, vol. 1 (1882–3), pp. 3–6.
[Anon], 'The Society of Telegraph Engineers', *Journal of the Society of Telegraph Engineers*, vol. 1 (1872–3), pp. 19–39.

[Anon.], 'Spirit Rapping', *London Review*, 1 March 1862, pp. 206–7.

[Anon.], 'Spiritualism and Science', *Times*, 26 December 1872, p. 5.

[Anon.], 'Spiritualism in Accord with True Science', *Light*, vol. 5 (1885), p. 464.

[Anon.], 'Spiritualism Viewed by the Light of Modern Science, by W. Crookes, F. R.S.', *Spiritual Magazine*, vol. 5 (New Series) (1870), pp. 375–81.

[Anon.], 'Spiritual Phenomena and Men of Science', *Spiritualist*, vol. 9 (1876), p. 1.

[Anon.], 'Spiritual Science. Spiritualism and the British Association', *Medium and Daybreak*, vol. 22 (1891), pp. 547–9.

[Anon.], 'The Study of Human Nature', *Human Nature*, vol. 1 (1867), pp. 1–5.

[Anon.], 'Swedenborg – No. IV', *English Mechanic*, vol. 2 (1865), pp. 87–8.

[Anon.], 'Sympathetic Vibrations', *Times*, 29 December 1876, p. 3.

[Anon.], 'A Symposium: Why I Became a Spiritualist', *Quarterly Journal of the British College of Psychic Science*, vol. 14 (1935–6), pp. 48–54.

[Anon.], 'Telepathy Again', *Cambridge Magazine*, vol. 7 (1917–18), pp. 97–8.

[Anon.], 'Transactions of the National Association of Spiritualists', *Spiritualist*, vol. 8 (1876), pp. 174–7.

[Anon], 'Transcendental Physics', *Saturday Review*, 11 September 1880, pp. 327–8.

[Anon.], 'The Unseen Universe', *London Quarterly Review*, vol. 44 (1875), pp. 49–83.

[Anon.], 'The Unseen Universe. Mr Crookes FRS, on Materialism', *Light*, vol. 7 (1887), pp. 146–7.

[Anon.], 'W. F. Barrett, F.R.S.E., M.R.I.A., &c', *Light*, vol. 14 (1894), pp. 439–41.

[Anon.], 'What Constitutes a Spiritualist?', *Yorkshire Spiritual Telegraph*, vol. 1 (1856), pp. 127–8.

[Anon.], 'Wider Outlooks of Science', *Light*, vol. 29 (1909), p. 427.

[Anon.], 'Wireless and Mediumship Problems', *Light*, vol. 53 (1933), p. 150.

[Anon.], 'Wireless and the Next World', *Light*, vol. 46 (1926), pp. 544–5.

[Anon.], 'Wireless and Telepathy', *Light*, vol. 47 (1927), p. 78.

Henry Adams, *The Education of Henry Adams* (Boston: Houghton Mifflin Co., 1918).

E[dward] N. [da] C[osta] A[ndrade], 'A Veteran's View of Modern Physics', *Nature*, vol. 114 (1924), pp. 599–601.

E[dward] N. [da] C[osta] A[ndrade], 'Ether and Erdgeist', *Nature*, vol. 116 (1925), pp. 305–6.

John Ashburner, 'Observations upon the Analogies between the Mesmeric and Magnetic Phenomena', *Zoist*, vol. 4 (1846–7), pp. 124–39.

John Ashburner, 'On the Connection Between Mesmerism and Spiritualism, with Considerations on Their Relations to Natural and Revealed Religion and to the Welfare of Mankind', *Supplement to the British Spiritual Telegraph*, vol. 3 (1859), pp. 1–96.

Henry G. Atkinson, 'Animal Magnetism', *Human Nature*, vol. 9 (1877), p. 384.

W. W. Baggally, 'Some Sittings with Carancini', *JSPR*, vol. 14 (1909–10), pp. 193–211.

Arthur J. Balfour, 'Address by the President', *PSPR*, vol. 10 (1894), pp. 2–13.

Arthur J. Balfour, William F. Barrett, William Crookes et al., 'The Memorial, of Which a Copy Is Printed Below, Was Forwarded to Lord Rayleigh on 18th February', *JSPR*, vol. 10 (1901–2), pp. 58–60.

Hipployte Baraduc, *La force vitale: Notre corps vitale fluidique. Sa formule biométrique* (Paris: Georges Carré, 1893).

Thomas P. Barkas, *Outline of Ten Years' Investigations into the Phenomena of Modern Spiritualism* (London: Frederick Pitman, 1862).

William F. Barrett, 'Glaciers and Ice', *Popular Science Review*, vol. 5 (1866), pp. 41–54.

William F. Barrett, 'Note on "Sensitive Flames"', *Philosophical Magazine*, vol. 33 (4th Series) (1867), pp. 216–22.

William F. Barrett, 'On Musical and Sensitive Flames', *Chemical News*, vol. 17 (1868), pp. 220–2.

William F. Barrett, 'Light and Sound: An Examination of Their Reputed Analogy', *Quarterly Journal of Science*, vol. 7 (1870), pp. 1–16.

William F. Barrett, 'A Fragment of Faraday's Electrical Discoveries', in *Science Lectures for the People. Science Lectures Delivered in Manchester. Third and Fourth Series* (Manchester: John Heywood, 1873), pp. 286–303.

[William F. Barrett], 'Spiritualism and Science', *Nonconformist*, vol. 34 (1873), pp. 445–6.

William F. Barrett, 'On the Molecular Changes That Accompany the Magnetisation of Iron, Nickel and Cobalt', *Philosophical Magazine*, vol. 47 (4th Series) (1874), pp. 51–6.

William F. Barrett, 'The Phenomena of Spiritualism', *Nonconformist*, vol. 36 (1875), pp. 1017–20.

William F. Barrett, 'On the Points of Contact Between Magnetism and Light', *Telegraphic Journal*, vol. 4 (1876), pp. 301–2, 319–20.

William F. Barrett, 'On Some Phenomena Associated with Abnormal Conditions of Mind', *Spiritualist*, vol. 9 (1876), pp. 85–8.

William F. Barrett, 'The Demons of Derrygonnelly', *Dublin University Magazine*, vol. 90 (1877), pp. 692–705.

William F. Barrett, 'A Letter from Professor Barrett', *Medium and Daybreak*, vol. 8 (1877), p. 209.

William F. Barrett, 'The Words "Magnetism" and "Electricity" – Their Use and Abuse', *Human Nature*, vol. 9 (1877), pp. 430–1.

William F. Barrett, 'Mr. Edison's Inventions', *Electrician*, vol. 2 (1878–9), pp. 76–7.

William F. Barrett, 'Sensitive Flames as Illustrative of Sympathetic Vibration', in *Science Lectures at South Kensington*, 2 vols. (London: Macmillan and Co., 1879), vol. 2, pp. 183–200.

William F. Barrett, 'Mind-Reading Versus Muscle-Reading', *Nature*, vol. 24 (1881), p. 212.

William F. Barrett, 'Appendix to the Report on Thought-Reading', *PSPR*, vol. 1 (1882–3), pp. 47–64.

William F. Barrett, 'Note on the Alleged Luminosity of the Magnetic Field', *Philosophical Magazine*, vol. 15 (5th Series) (1883), pp. 270–5.

William F. Barrett, 'On a "Magnetic Sense"', *Nature*, vol. 29 (1883–4), pp. 476–7.

William F. Barrett, 'Note on the Existence of a "Magnetic Sense"', *PSPR*, vol. 2 (1884), pp. 56–60.

William F. Barrett, 'Is There a "Magnetic Sense"?', *Dublin University Review*, vol. 1 (1886), pp. 23–34.

William F. Barrett, 'The Society for Psychical Research and Spiritualism', *Light*, vol. 6 (1886), pp. 51–2.

William F. Barrett, 'On Some Physical Phenomena, Commonly Called Spiritualistic, Witnessed by the Author', *PSPR*, vol. 4 (1886–7), pp. 25–42.

William F. Barrett, 'Psychical Research', *Good Words*, vol. 32 (1891), pp. 467–71.

William F. Barrett, 'Sympathetic Vibration', *Good Words*, vol. 32 (1891), pp. 41–6.

William F. Barrett, 'Science and Spiritualism', *Light*, vol. 14 (1894), pp. 539–40, 559–61, 571–2, 583–5, 595–7.

William F. Barrett, 'Dynamic Thought. Part II. The Realm of the Unconscious', *Humanitarian*, vol. 7 (1895), pp. 345–53.

William F. Barrett, 'On the So-Called Divining Rod or *Virgula Divina*', *PSPR*, vol. 13 (1897–8), pp. 2–282.

William F. Barrett, 'On the So-Called Divining Rod', *PSPR*, vol. 15 (1900–1), pp. 130–383.

William F. Barrett, 'Address by the President', *PSPR*, vol. 18 (1903–4), pp. 323–50.

William F. Barrett, *On the Threshold of a New World of Thought* (London: Kegan Paul, Trench, Trübner and Co., 1908).

William F. Barrett, 'Poltergeists: Old and New', *PSPR*, vol. 25 (1911), pp. 377–412.

William F. Barrett, *Psychical Research* (London: Williams and Norgate, 1911).

William F. Barrett, *Swedenborg: The Savant and the Seer* (London: John M. Watkins, 1912).

William F. Barrett, 'Discrete Degrees', *The New Church Magazine*, vol. 33 (1914), pp. 415–25.

William F. Barrett, 'The Spiritual Significance of Nature', *Contemporary Review*, vol. 105 (1914), pp. 791–9.

William F. Barrett, 'The Psychic Factor in Evolution', *Quest*, vol. 9 (1917–18), pp. 177–202.

William F. Barrett, 'Report of Physical Phenomena Taking Place at Belfast with Dr. Crawford's Medium [1916]', *PSPR*, vol. 30 (1918–19), pp. 334–7.

William F. Barrett, *On the Threshold of the Unseen: An Examination of the Phenomena of Spiritualism and of the Evidence for Survival After Death* (London: Kegan Paul, Trench, Trübner and Co., 1920).

William F. Barrett, 'Ectoplasms', *Light*, vol. 41 (1921), p. 347.

William F. Barrett, 'Some Reminiscences of Fifty Years' Psychical Research', *PSPR*, vol. 32 (1923–4), pp. 275–97.

William F. Barrett, 'The Early Years of Psychical Research', *Light*, vol. 44 (1924), p. 395.

William F. Barrett, 'Sur la luminosité du champ magnétique et de certaines personnes qui, d'après le baron Reichenbach, serait perçue par les sensitifs', in *L'état actuel des recherches psychique d'après le travaux du IIme Congrès International tenu à Varsovie en 1923* (Paris: Les Presses Universitaires de France, 1924), 169–73.

William F. Barrett and W. Brown, *Practical Physics: An Introductory Handbook for the Laboratory* (London: Percival and Co., 1892).

William F. Barrett and Frederic W. H. Myers, 'D. D. Home, His Life and Mission', *JSPR*, vol. 4 (1889–90), pp. 101–16.

William F. Barrett, Maxwell H. Close, Frederic W. H. Myers et al., 'First Report of the 'Reichenbach' Committee', *PSPR*, vol. 1 (1882–3), pp. 230–7.

James B. Beal, 'The Emergence of Paraphysics: Research and Applications', in John White (ed.), *Psychic Exploration: A Challenge for Science* (New York: Perigee Books, 1974), pp. 426–46.

Annie Besant and Charles Leadbeater, *Occult Chemistry: Clairvoyant Observations on the Chemical Elements*, ed. by A. P. Sinnett (London: Theosophical Publishing House, 1919).

J. Malcom Bird, 'Some Theoretical Aspects of Psychical Research', *British Journal of Psychical Research*, vol. 1 (1928), pp. 331–9.

Helena Petrovna Blavatsky, *Isis Unveiled: A Master Key to the Mysteries of Ancient and Modern Science and Theology*, 2 vols. (New York: J. W. Bouton, 1877).

[Ernest] Bonnaymé, *La force psychique. L'agent magnétique* (Paris: Librairie du Magnétisme, 1908).

Max Born (ed.), *The Born-Einstein Letters: Correspondence Between Albert Einstein and Max and Hedwig Born from 1916 to 1955, with Commentaries by Max Born*, translated by Irene Born (London: Macmillan, 1971).

William R. Bousfield, 'Telepathy', *Hibbert Journal*, vol. 20 (1921–2), pp. 497–506.

William R. Bousfield, 'Human Survival', *Hibbert Journal*, vol. 22 (1923–4), pp. 501–14.

James Braid, *The Power of the Mind over the Body: An Experimental Enquiry into the Nature and Cause of the Phenomena Attributed by Baron Reichenbach and Others to a 'New Imponderable'* (London: John Churchill, 1846).

Stanley De Brath, 'Relativity', *Light*, vol. 41 (1921), pp. 520–1.

Thomas Brevoir, 'The Religious Heresies of the Working Classes', *Spiritual Magazine*, vol. 6 (1865), pp. 29–32.

[David Brewster], 'Pretensions of Spiritualism – Life of D. D. Home', *North British Review*, vol. 39 (1863), pp. 174–206.

Guy B. Brown, 'Review of Oliver Lodge's Beyond Physics', *Philosophy*, vol. 5 (1930), pp. 624–6.

J. Brown, 'The Psychical Society's Experiments on Reichenbach's Phenomenon', *English Mechanic and World of Science*, vol. 37 (1883), p. 246.

James Burns, 'Professor Tyndall and the Spiritualists', *Human Nature*, vol. 2 (1868), pp. 454–6.

[James Burns], 'Spiritualism and Science', *Medium and Daybreak*, vol. 1 (1870), p. 108.

[James Burns], 'Editorial Comments', *Medium and Daybreak*, vol. 2 (1871), p. 231.

[James Burns], 'Electrical Tests with Miss Cook when Entranced', *Spiritual Magazine*, vol. 9 (New Series) (1874), pp. 161–8, p. 168.

[James Burns], 'A Scientific Séance – The Electrical Test of Mediumship', *Medium and Daybreak*, vol. 6 (1875), pp. 161–3.

Reginald J. Campbell, *A Spiritual Pilgrimage* (London: Williams and Norgate, 1917).

Fritjof Capra, *The Tao of Physics: An Exploration of the Parallels Between Modern Physics and Eastern Mysticism* (London: Wildwood House, 1975).

J. Rand Capron, *Aurorae: Their Characters and Spectra* (London: E. and F. N. Spon, 1879).

Whately Carington, *The Death of Materialism* (London: George Allen and Unwin, 1933).

William B. Carpenter, 'On the Mutual Relations of the Vital and Physical Forces', *Philosophical Transactions of the Royal Society of London*, vol. 140 (1850), pp. 727–57.

William B. Carpenter, 'On the Influence of Suggestion in Modifying and Directing Muscular Movement, Independent of Volition [1852]', *Notices of the Proceedings of the Meetings of the Members of the Royal Institution*, vol. 1 (1851–4), pp. 147–53.

[William B. Carpenter], 'Electro-Biology and Mesmerism', *Quarterly Review*, vol. 93 (1853), pp. 501–57.

[William B. Carpenter], 'Spiritualism and Its Recent Converts', *Quarterly Review*, vol. 131 (1871), pp. 301–53.

William B. Carpenter, *Principles of Mental Physiology* (London: Henry S. King, 1875).

William B. Carpenter, 'Spiritualism', *Spectator*, vol. 49 (1876), pp. 1281–2.

William B. Carpenter, *Mesmerism, Spiritualism, &c. Historically & Scientifically Considered* (London: Longmans, Green, and Co., 1877).

William B. Carpenter, 'Psychological Curiosities of Spiritualism', *Frasers's Magazine*, vol. 16 (1877), pp. 541–64.

William B. Carpenter, 'The Radiometer and Its Lessons', *Nineteenth Century*, vol. 1 (1877), pp. 242–56.

Hereward Carrington, *Eusapia Palladino and Her Phenomena* (London: T. Werner Laurie, 1910).

Hereward Carrington, *Personal Experiences in Spiritualism* (London: T. Werner Laurie, 1913).

Hereward Carrington, 'Sir William Crookes's Psychical Researches', *Electrical Experimenter*, vol. 7 (1919), pp. 407, 440, 442, 444.

Hereward Carrington, 'Are the Dead Trying to Reach Us by Radio?', *Popular Radio*, vol. 1 (1922), pp. 188–93.

Hereward Carrington, 'New York's Laboratory of the Mysterious', *New York Tribune*, 9 April 1922, p. 3.

Hereward Carrington, 'Will We Talk to the Dead by Radio?', *Popular Radio*, vol. 1 (1922), pp. 92–7.

Hereward Carrington, *The Story of Psychic Science* (London: Rider and Co., 1930).

Hereward Carrington, *Laboratory Investigations into Psychic Phenomena* (London: Rider and Co., 1939).

Hereward Carrington, *The American Seances with Eusapia Palladino* (New York: Garrett Publications, 1954).

Arthur P. Chattock, 'Experiments in Thought Transference', *JSPR*, vol. 8 (1897–8), pp. 302–7.

Robert Chambers, *Testimony: Its Posture in the Scientific World* (London: William and Robert Chambers, 1859).

Latimer Clark and Robert Sabine, *Electrical Tables and Formulae* (London: E. & F. N. Spon, 1871).

William Kingdon Clifford, 'The Unseen Universe', *Fortnightly Review*, vol. 17 (1875), pp. 776–93.

Edward Clodd, 'Occultism', *Fortnightly Review*, vol. 107 (1920), pp. 757–68.

Benjamin Coleman, 'Spirit Forms', *Spiritualist*, vol. 4 (1874), p. 177.

C. H. Collings, 'Wireless: Some Facts and Speculations', *Light*, vol. 42 (1922), p. 461.

J. C. Colquhoun, *Isis Revelata: An Inquiry into the Origin, Progress and Present State of Animal Magnetism*, 2 vols. (Edinburgh: Maclachlan and Stewart, 1836).

J. C. Colquhoun, *An History of Magic, Witchcraft and Animal Magnetism*, 2 vols. (London: Longman, Brown, Green and Longmans, 1851).

W. R. C. Coode-Adams, *A Primer of Occult Physics* (London: Theosophical Publishing House, 1927).

F. F. Cook, 'The Relations of Spiritualism to Science', *Light*, vol. 1 (1881), pp. 130–1, 138–9.

Robert Cooper, *Spiritual Manifestations, Including Seven Months with the Brothers Davenport* (London: Heywood and Son, 1867).

Philip R. Coursey, 'Aether the Substratum of the Universe', *Wireless World*, vol. 8 (1920), pp. 37–40.

Jules Courtier, *Documents sur Eusapia Palladino. Rapport sur les séances d'Eusapia Palladino à l'Institut général psychologique en 1905, 1906, 1907, 1908* (Paris: Institut général psychologique, 1908).

Edward W. Cox, *Spiritualism Answered by Science: With the Proofs of a Psychic Force* (London: Longman and Co., 2nd ed., 1872).

Edward Cox, *The Mechanism of Man: An Answer to the Question of What Am I?*, 2 vols. (London: Longmans and Co., 1876–9).

Q. C. A. Craufurd, 'The Unknown Force: An Electrician's View', *Light*, vol. 47 (1927), p. 339.

Q. C. A. Craufurd, 'Vibrations', *Light*, vol. 47 (1927), p. 632.

Q. C. A. Craufurd, 'The Crisis in the SPR', *International Psychic Gazette*, vol. 18 (1930), p. 135.

Q. C. A. Craufurd and Jack Frost, 'Psychic Communication and Wireless: A New Instrument', *Light*, vol. 48 (1928), p. 305.

William Crookes, 'The Breath of Life', *Popular Science Review*, vol. 1 (1862), pp. 91–9.

William Crookes, 'Spiritualism Viewed by the Light of Modern Science', *Quarterly Journal of Science*, vol. 7 (1870), pp. 316–21.

William Crookes, 'Experimental Investigation of a New Force', *Quarterly Journal of Science*, vol. 1 (New Series) (1871), pp. 339–49.

William Crookes, 'Mr. Crookes' Psychic Force', *Echo*, 10 November 1871, p. 2.

William Crookes, *Psychic Force and Modern Spiritualism: A Reply to the 'Quarterly Review' and Other Critics* (London: Longmans, Green and Co., 1871).

William Crookes, 'Some Further Experiments in Psychic Force', *Quarterly Journal of Science*, vol. 1 (New Series) (1871), pp. 471–93.

William Crookes, 'Researches on the Atomic Weight of Thallium', *Philosophical Transactions of the Royal Society of London*, vol. 163 (1873), pp. 277–330.

William Crookes, 'On Attraction and Repulsion Resulting from Radiation', *Philosophical Transactions of the Royal Society of London*, vol. 164 (1874), pp. 501–27.

William Crookes, 'The Last of Katie King. The Photographing of Katie King by the Aid of the Electric Light', *Spiritualist*, vol. 4 (1874), pp. 270–1.

William Crookes, 'Notes of an Enquiry into Phenomena Called Spiritual', *Quarterly Journal of Science*, vol. 3 (New Series) (1874), pp. 77–97.

William Crookes, 'The Outrage at a Spirit-Circle', *Spiritualist*, vol. 4 (1874), p. 71.

William Crookes, *Researches in the Phenomena of Spiritualism* (London: James Burns, 1874).

William Crookes, 'Spirit Forms', *Spiritualist*, vol. 4 (1874), pp. 158–9.

William Crookes, 'On Repulsion Resulting from Radiation. Part II', *Philosophical Transactions of the Royal Society of London*, vol. 165 (1875), pp. 519–47.

William Crookes, 'A Scientific Examination of Mrs. Fay's Mediumship', *Spiritualist*, vol. 6 (1875), pp. 126–8.

William Crookes, 'On Repulsion Resulting from Radiation. Influence of the Residual Gas', *Proceedings of the Royal Society of London*, vol. 25 (1876–7), pp. 136–40.

William Crookes, 'Another Lesson from the Radiometer', *Nineteenth Century*, vol. 1 (1877), pp. 879–87.

William Crookes, 'On Radiant Matter', *Nature*, vol. 20 (1879), pp. 419–23, 436–40.

William Crookes, 'On the Supposed "New Force" of M. J. Thore', *Philosophical Transactions of the Royal Society of London*, vol. 178 (1888), pp. 451–69.

William Crookes, 'Notes of Séances with D. D. Home', *PSPR*, vol. 6 (1889–90), pp. 98–127.

William Crookes, 'Electricity *in Transitu*: From Plenum to Vacuum', *Journal of the Institution of Electrical Engineers*, vol. 20 (1891), pp. 4–49.

William Crookes, 'Some Possibilities of Electricity', *Fortnightly Review*, vol. 51 (New Series) (1892), pp. 173–81.

William Crookes, 'Address by the President', *PSPR*, vol. 12 (1896–7), pp. 338–55.

William Crookes, 'Address by Sir William Crookes', in *Report of the Sixty-Eighth Meeting of the British Association for the Advancement of Science Held at Bristol in September 1898* (London: John Murray, 1899), pp. 3–38.

'C.W.', 'Students' Psychical Society', *The Mermaid: The Journal of the Guild of Undergraduates of the University of Birmingham*, vol. 3 (1906), pp. 68–72.

H. A. Dallas, 'The Etherial Body: Its Nature and Scope', *Light*, vol. 44 (1924), p. 116.

Benjamin Davies, 'Unusual Physical Phenomena', *Liverpool Daily Post*, 10 January 1895, p. 3.

Benjamin Davies, 'Professor Richet's Address', *Liverpool Daily Post*, 11 August 1899, p. 3.

Benjamin Davies, 'Experiments on Levitation', *Light*, vol. 36 (1916), pp. 186–7, 194–5, 202–3.

Charles Maurice Davies, *Mystic London; Or, Phases of Occult Life in the Metropolis* (London: Tinsley Brothers, 1875).

[Sophia De Morgan], *From Matter to Spirit: The Result of Ten Years' Experience in Spirit Manifestations* (London: Longman, Green, Longman, Roberts and Green, 1863).

S[ophia] D[e] M[organ], 'Scientists and Spiritualism', *Light*, vol. 7 (1887), pp. 117–18.

Henry Dircks, 'Science Versus Spiritualism', *Times*, 27 December 1872, p. 10.

Amos E. Dolbear, *Matter, Ether and Motion: The Factors and Relations of Physical Science* (London: Society for the Promotion of Christian Knowledge, 1899).

Arthur Conan Doyle, *The New Revelation* (London: Hodder and Stoughton, 1918).

C. L. Drawbridge (ed.), *The Religion of Scientists: Being Recent Opinions Expressed by Two Hundred Fellows of the Royal Society on the Subject of Religion and Theology* (New York: Macmillan Company, 1932).

Baron Dupotet de Sennevoy, *An Introduction to the Study of Animal Magnetism* (London: Saunders and Otley, 1838).

Mrs C. W. Earle, *Memoirs and Memories* (London: Smith, Elder and Co., 1911).

J. P. Earwaker, 'The New Psychic Force', *Nature*, vol. 4 (1871), pp. 278–9.

Arthur Stanley Eddington, 'The Domain of Physical Science', in Joseph Needham (ed.), *Science, Religion and Reality* (London: Sheldon Press, 1925), pp. 189–218.

Arthur Stanley Eddington, *Science and the Unseen World* (London: George Allen & Unwin, 1929).

'E. G.', 'A Test Séance with Mr. Williams', *Medium and Daybreak*, vol. 3 (1872), p. 318.

Albert Einstein, 'Ether and the Theory of Relativity [1920]', in *Sidelights on Relativity* (New York: E. P. Dutton, 1922), pp. 3–24.

John Elliotson, *Human Physiology* (London: Longman, Orme, Brown, Green and Longman, 1840).

John Elliotson, 'Review of an *Abstract of Researches on Magnetism and Certain Allied Subjects*', *Zoist*, vol. 4 (1846–7), pp. 104–24.

John Elliotson, 'The Departed Spirits', *Zoist*, vol. 11 (1853–4), pp. 191–201.

Michael Faraday, 'Table-Turning', *Times*, 30 June 1853, p. 8.

Michael Faraday, 'Experimental Investigation of Table-Moving', *Athenaeum*, 2 July 1853, pp. 801–3.

Michael Faraday, 'Observations on Mental Education [1854]', in *Experimental Researches in Chemistry and Physics* (London: Richard Taylor and William Francis, 1859), pp. 463–91.

'A Fellow of the Royal Astronomical Society', 'Psychic Force, &c', *English Mechanic and World of Science*, vol. 13 (1871), p. 539.

Arthur Findlay, *On the Edge of the Etheric* (London: Psychic Press, 1931).

G. W. Fisk, 'Criticism of a Radiation Theory as an Explanation of Telepathy or Clairvoyance', *JSPR*, vol. 29 (1935–6), pp. 35–6.

Desmond G. Fitzgerald, 'The "Conservation of Energy" in Relation to Certain Views of the Theosophists', *Spiritualist*, vol. 12 (1878), pp. 249–51.

Desmond G. Fitzgerald, 'Spiritualism and the Society for Psychical Research', *Light*, vol. 6 (1886), pp. 62–3.

George F. FitzGerald, 'Helmholtz Memorial Lecture', in *Memorial Lectures Delivered Before the Chemical Society, 1893–1900* (London: Gurney and Jackson, 1901), pp. 885–912.

E. E. Fournier d'Albe, *The Electron Theory: A Popular Introduction to the New Theory of Electricity and Magnetism* (London: Longmans, Green, and Co., 1906).

E. E. Fournier d'Albe, *New Light on Immortality* (London: Longmans, Green, and Co., 1908).

E. E. Fournier d'Albe, *The Goligher Circle, May to August 1921* (London: John M. Watkins, 1922).

E. E. Fournier d'Albe, *The Life of Sir William Crookes* (London: T. Fisher Unwin, 1923).

T. B. Franklin and V. J. Vickers, 'A Radio Divining Rod', *Modern Wireless*, March 1933, p. 278.

George Fraser, 'The New Psychic Force', *Nature*, vol. 4 (1871), pp. 279–80.

E. E. Free, 'The Ether and the Soul', *Popular Radio*, vol. 8 (1925), p. 280.

E. E. Free, 'Have "Brain Waves" Been Discovered?', *Popular Radio*, vol. 9 (1926), pp. 366–7.

Jack Frost, 'Radio and Psychic Science', *Light*, vol. 46 (1926), p. 221.

Comte Agénor de Gasparin, *Des tables tournantes, du surnaturel en général et des esprits*, 2 vols. (Paris: E. Dentu, 1854).

Hugo Gernsback, 'The Thought Recorder', *Electrical Experimenter*, vol. 7 (1919), pp. 84–5.

Charles Gore, *The New Theology and the Old Religion* (London: John Murray, 1907).

Francis Gotch, 'Some Physiological Aspects of Hypnotism', *Science Progress*, vol. 1 (1897), pp. 511–30.

William Gregory, *Letters to a Candid Inquirer on Animal Magnetism* (London: Taylor, Walton, and Moberly, 1851).

William Robert Grove, *The Correlation of Physical Forces* (London: Longman, Green, Longman, Roberts, & Green, 1862).

William Robert Grove, 'Address', in *Report of the Thirty-Sixth Meeting of the British Association for the Advancement of Science Held at Dundee in September 1867* (London: John Murray, 1868), pp. liii–lxxxii.

J. C. F. Grumbine, *Telepathy: Or, The Science of Thought Transference* (London: L. N. Fowler and Co., 1915).

Fritz Grunewald, *Physicalisch-mediumistische Untersuchungen* (Pfullingen: Johannes Baum, 1920).

Edmund Gurney, 'The Nature of Evidence in Matters Extraordinary', *Fortnightly Review*, vol. 22 (1884), pp. 472–91.

Edmund Gurney, Frederic W. H. Myers and Frank Podmore, *Phantasms of the Living*, 2 vols. (London: Rooms of the Society for Psychical Research, 1886).

Joseph W. Haddock, *Somnolism and Psycheism; Or, the Science of the Soul and the Phenomena of Nervation as Revealed by Mesmerism* (London: James S. Hodson, 2nd ed., 1851).

Viscount Halifax, '"Raymond; Or, Life and Death"', *Church Times*, vol. 72 (1917), pp. 181–2.

Edwin H. Hall, 'Sir Oliver Lodge's British Association Address', *Harvard Theological Review*, vol. 8 (1915), pp. 238–51.

Fio Hara, 'The Advance of Science Towards Occult Teachings', *Theosophical Review*, vol. 37 (1905–6), pp. 548–54.

Emma Hardinge, 'Psychology; Or, the Science of Soul', *Spiritual Magazine*, vol. 1 (New Series) (1866), pp. 385–401.

Emma Hardinge, 'Rules to Be Observed for the Spirit Circle', *Human Nature*, vol. 2 (1868), pp. 48–52.

Emma Hardinge, *Modern American Spiritualism: A Twenty Years' Record of the Communion Between the Earth and the World of Spirits* (New York: Emma Hardinge, 1870).

Emma Hardinge, 'The Scientific Investigation of Spiritualism', *Spiritual Magazine*, vol. 6 (New Series) (1871), pp. 3–17.

Robert Hare, *Experimental Investigation of the Spirit Manifestations* (New York: Partridge and Brittan, 1855).

William H. Harrison, 'The Invisible Photographic Image Considered at Motion', *British Journal of Photography*, vol. 6 (1867), pp. 424–5.

[William H. Harrison], 'Professor Tyndall at a Spirit Circle', *Spiritualist*, vol. 1 (1869–71), pp. 156–7.

William H. Harrison, 'The Work of a Psychological Society', *Spiritualist*, vol. 1 (1869–71), pp. 206–7.

[William H. Harrison], 'Real and Sham Spirit Photographs', *Spiritualist*, vol. 2 (1872), pp. 75–6.

[William H. Harrison], 'Spirit Forms', *Spiritualist*, vol. 3 (1872–3), pp. 451–4.

[William H. Harrison], 'Miss Cook's Mediumship', *Spiritualist*, vol. 4 (1874), pp. 133–4.

William H. Harrison, 'Electrical Tests Popularly Explained', *Spiritualist*, vol. 6 (1875), pp. 135–6.

William H. Harrison, 'New Experiments on Odic Flames from Magnets', *Spiritualist*, vol. 7 (1875), pp. 97–8.

William H. Harrison, 'The Scientific Research Committee of the National Association of Spiritualists', *Spiritualist*, vol. 9 (1876), pp. 193–4.

William H. Harrison, 'Weighing a Medium During the Production of Spiritual Manifestations', *Spiritualist*, vol. 12 (1878), pp. 210–16.

[William H. Harrison], 'Experimental Research in Spiritualism', *Spiritualist*, vol. 19 (1881), p. 162.

William H. Harrison, 'Recent Psychical Researches', *Medium and Daybreak*, vol. 14 (1883), pp. 310–11.

Eduard Haschek, 'Über Leuchterscheinungen des menschlichen Körpers', *Sitzungsberichte der Kaiserlichen Akademie der Wissenschaften. Mathematisch-Naturwissenschaften Klasse*, vol. 123 (1914), pp. 523–32.

Oliver Heaviside, *Electromagnetic Theory Volume II* (London: The 'Electrician' Printing and Publishing Company, 1899).

John F. W. Herschel, *A Preliminary Discourse on the Study of Natural Philosophy* (London: Longman, Rees, Orme, Brown and Green, 1830).

J. W. A. Hickson, 'Sir Oliver Lodge and the Beclouding of Reason', *University Magazine*, vol. 16 (1917), pp. 379–97.

J. Arthur Hill, *Letters from Sir Oliver Lodge: Psychical, Religious, Scientific and Personal* (London: Cassell and Company, 1932).

Richard Hodgson, 'The Value of Evidence for Supernormal Phenomena in the Case of Eusapia Palladino', *JSPR*, vol. 7 (1895–6), pp. 36–55.

Richard Hodgson and S. J. Davey, 'The Possibilities of Mal-Observation and Lapse of Memory from a Practical Point of View', *PSPR*, vol. 4 (1886–7), pp. 381–495.

D. D. Home, *Lights and Shadows of Modern Spiritualism* (London: Virtue and Co., 1877).

J. Page Hopps, 'Professor Tyndall's Excursions into Spiritualism', *Light*, vol. 14 (1894), pp. 67–9.

Edwin J. Houston, 'Cerebral Radiation', *Journal of the Franklin Institute*, vol. 133 (1892), pp. 488–97.

William Howitt, 'A Letter from William Howitt', *Spiritual Magazine*, vol. 2 (1861), pp. 449–56.

Robert Hughes, 'Seeing Things: The Scientists and Spiritualism', *Pearson's Magazine*, vol. 21 (1909), pp. 188–97.

T. N. Hutchinson, 'Radiometers and Radiometers', *Nature*, vol. 13 (1875–6), p. 324.

James Hervey Hyslop, *Psychical Research and the Resurrection* (Boston: Small, Maynard and Company, 1908).

[William Josiah] Irons, 'On the Scientific Conclusions and Theological Inferences of a Work Entitled "The Unseen Universe, or Physical Speculations on a Future State"', *Journal of the Transactions of the Victoria Institute*, vol. 11 (1878), pp. 83–139.

Robert G. Jahn, 'The Persistent Paradox of Psychic Phenomena: An Engineering Perspective', in Charles T. Tart, Harold E. Puthoff and Russell Targ (eds.), *Mind at Large: Institution of Electrical and Electronic Engineers Symposia on the Nature of Extra Sensory Perception* (Charlottesville, VA: Hampton Roads, 2002), pp. 238–316.

E. O. James, 'The Return of Materialism', *Church Times*, vol. 104 (1930), pp. 161–2.

Joseph Jastrow and George Nuttall, 'On the Existence of a Magnetic Sense', *Proceedings of the American Society for Psychical Research*, vol. 1 (1885–9), pp. 116–26.

James Jeans, untitled essay, in *More Points of View: A Second Series of Broadcast Addresses* (London: George Allen & Unwin, 1930), pp. 55–71.

William Stanley Jevons, *The Principles of Science: A Treatise of Logic and Scientific Method*, 2 vols. (London: Macmillan and Co., 1874).

Alice Johnson, 'The Education of the Sitter', *PSPR*, vol. 21 (1907–9), pp. 483–511.

Alice Johnson, 'Note on the Above Paper', *JSPR*, vol. 14 (1909–10), pp. 259–60.

Alice Johnson, 'Mrs Henry Sidgwick's Work in Psychical Research', *PSPR*, vol. 44 (1936), pp. 53–93.

George Lindsay Johnson, *The Great Problem and the Evidence for Its Solution* (London: Hutchinson and Co., 1927).

Pascual Jordan, 'Reflections on Parapsychology, Psychoanalysis and Atomic Physics', *Journal of Parapsychology*, vol. 15 (1951), pp. 278–81.

Waldemar Kaempffert, 'The Duke Experiments in Extra-Sensory Perception', *New York Times*, 10 October 1937, pp. 2 and 21.

Waldemar Kaempffert, 'Searching Out the Mind's Mysteries', *New York Times*, 17 October 1937, pp. 8 and 24.

Waldemar Kaempffert, 'Science in the News', *New York Times*, 25 February 1940, p. 51.

Walter J. Kilner, *The Human Atmosphere; Or the Aura Made Visible by Means of Chemical Screens* (London: Rebman, 1911).

William Kingsland, *The Higher Science* (London: The Theosophical Publication Society, 1889).

William Kingsland, 'Theosophy and Christianity', *Lucifer*, vol. 14 (1894), pp. 335–40.

William Kingsland, 'Natural Law in the Spiritual World', *Theosophical Review*, vol. 26 (1900), pp. 441–50.

William Kingsland, *Scientific Idealism or Matter and Force and Their Relation to Life and Consciousness* (London: London Press Co., 1909).

William Kingsland, *The Physics of the Secret Doctrine* (London: Theosophical Publishing Society, 1910).

William Kingsland, *Rational Mysticism: A Development of Scientific Idealism* (London: George Allen and Unwin, 1924).

William Kingsland, *The Real H. P. Blavatsky. A Study in Theosophy, and the Memoir of a Great Soul* (London: John M. Watkins, 1928).

William Kingsland, *The Art of Life and How to Conquer Old Age* (London: C. W. Daniel Company, 1934).

J[ames] T[homas] K[nowles], 'Brain-Waves – A Theory', *Spectator*, vol. 42 (1869), pp. 135–7.

Samuel P. Langley, 'The History of a Doctrine', *American Journal of Science*, vol. 37 (3rd Series) (1889), pp. 1–23.

Samuel P. Langley, 'The Laws of Nature', in *Annual Report of the Board of Regents of the Smithsonian Institution for the Year Ending June 30, 1901* (Washington, DC: Government Printing Office, 1902), pp. 545–52.

[Dionysius Lardner and Edward Bulwer-Lytton], 'Animal Magnetism', *Monthly Chronicle*, vol. 1 (1838), pp. 289–306; vol. 2 (1838), pp. 11–30.

Joseph Larmor, 'Aether', in *Encyclopaedia Britannica*, 29 vols. (Cambridge University Press, 11th ed., 1910–11), vol. 1, pp. 292–7.

Joseph Larmor, 'Physical Aspects of the Atomic Theory [1908]', in *Mathematical and Physical Papers*, 2 vols. (Cambridge University Press, 1929), vol. 2, pp. 344–71.

Walter Leaf, *Walter Leaf 1852–1897. Some Chapters of Autobiography* (London: John Murray, 1932).

Austin C. Lescarboura, 'Edison's Views on Life and Death', *Scientific American*, 30 October 1920, p. 446, pp. 458–60.

George Henry Lewes, 'The Rappites Exposed', *Leader*, vol. 4 (1853), pp. 261–3.

Justus von Liebig, *Über das Studium der Naturwissenschaften. Eröffnungsrede zu seinen Vorlesungen über Experimental-Chemie im Wintersemester 1852/53* (Munich: Cotta, 1852).

Oliver Lodge, *Elementary Mechanics: Including Hydrostatics and Pneumatics* (London: W. & R. Chambers, 1879).

Oliver Lodge, 'The Ether and Its Functions', *Nature*, vol. 27 (1882–3), pp. 304–6, 328–30.

Oliver Lodge, 'An Account of Some Experiments in Thought-Transference', *PSPR*, vol. 2 (1884), pp. 189–200.

Oliver Lodge, *Modern Views of Electricity* (London: Macmillan and Co., 1889).

Oliver Lodge, 'Account of Sittings with Mrs Piper', *PSPR*, vol. 6 (1889–90), pp. 443–557.

Oliver Lodge, 'Presidential Address to the Liverpool Physical Society', *Proceedings of the Liverpool Physical Society*, vol. 1 (1889–92), pp. 1–8.

Oliver Lodge, 'Some Recent Thought-Transference Experiments', *PSPR*, vol. 7 (1891–92), 374–82.

Oliver Lodge, 'Address', in *Report of the Sixty-First Meeting of the British Association for the Advancement of Science Held in Cardiff in August 1891* (London: John Murray, 1892), pp. 547–57.

Oliver Lodge, 'Thought Transference: An Application of Modern Thought to Ancient Superstitions', *Proceedings of the Literary and Philosophical Society of Liverpool*, vol. 46 (1892), pp. 127–45.

Oliver Lodge, 'Aberration Problems: A Discussion Concerning the Motion of Ether near the Earth, and Concerning the Connection Between Ether and Gross Matter; with Some New Experiments', *Philosophical Transactions of the Royal Society of London*, vol. 184 (1893), pp. 727–804.

Oliver Lodge, 'The Interstellar Ether', *Fortnightly Review*, vol. 53 (New Series) (1893), pp. 856–62.

Oliver Lodge, 'Correspondence', *JSPR*, vol. 6 (1893–4), pp. 270–2.

Oliver Lodge, 'Experience of Unusual Phenomena Occurring in the Presence of an Entranced Person (Eusapia Palladino)', *JSPR*, vol. 6 (1893–4), pp. 306–60.

Oliver Lodge, 'Professor Heinrich Hertz', *JSPR*, vol. 6 (1893–4), pp. 197–9.

Oliver Lodge, 'A Reply to Mr. Barkworth', *JSPR*, vol. 6 (1893–4), pp. 215–17.

Oliver Lodge, 'On the Difficulty of Making Crucial Experiments as to the Source of the Extra or Unusual Intelligence Manifested in Trance-Speech, Automatic Writing, and Other States of Apparent Mental Inactivity', *PSPR*, vol. 10 (1894), pp. 14–24.

Oliver Lodge, 'The Work of Hertz', *Nature*, vol. 50 (1894), pp. 133–9.

Oliver Lodge, 'The Exposure of Eusapia', *Daily Chronicle*, 5 November 1895, p. 3.

Oliver Lodge, 'Unusual Physical Phenomena', *Liverpool Daily Post*, 2 January 1895, p. 7.

Oliver Lodge, 'On the Rays of Lenard and Röntgen', *Electrician*, vol. 36 (1896), pp. 438–40.

Oliver Lodge, 'Address by Professor Lodge', *Light*, vol. 17 (1897), pp. 162–8.

Oliver Lodge, 'Supplement to the Discussion on Mr. Balfour's Paper [1900]', in *Papers Read Before the Synthetic Society*, pp. 334–40.

Oliver Lodge, 'Scope and Tendencies of Physics', in *The 19th Century: A Review of Progress* (London: G. P. Putnam's Sons, 1901), pp. 348–57.

Oliver Lodge, 'Address by the President', *PSPR*, vol. 17 (1901–3), pp. 37–57.

Oliver Lodge, 'In Memory of F. H. W. Myers', *PSPR*, vol. 17 (1901–3), pp. 1–12.

Oliver Lodge, 'The Outstanding Controversy Between Science and Faith', *Hibbert Journal*, vol. 1 (1902–3), pp. 32–61.

Oliver Lodge, 'The Reconciliation Between Science and Faith', *Hibbert Journal*, vol. 1 (1902–3), pp. 209–27.

Oliver Lodge, 'The University in the Modern State', *Nature*, vol. 67 (1902–3), pp. 193–6.

Oliver Lodge, untitled essay dated 20 February 1903, in *Papers Read Before the Synthetic Society*, pp. 385–92.

Oliver Lodge, 'The Survival of Personality', *Quarterly Review*, vol. 198 (1903), pp. 211–29.

Oliver Lodge, 'The Alleged Indifference of Laymen to Religion', *Hibbert Journal*, vol. 2 (1903–4), pp. 235–41.

Oliver Lodge, 'Presidential Address', *PSPR*, vol. 18 (1903–4), pp. 1–21.

Oliver Lodge, untitled paper dated 26 January 1904, in *Papers Read Before the Synthetic Society*, pp. 460–2.

Oliver Lodge, 'Religion, Science and Miracle', *Contemporary Review*, vol. 86 (1904), pp. 798–807.

Oliver Lodge, *University of Birmingham. Lectures to Medical Practitioners on Physics Applied to Medicine* ([University of Birmingham, 1904]).

Oliver Lodge, *Electrons or the Nature and Properties of Negative Electricity* (London: George Bell, 1906).

Oliver Lodge, 'On the Scientific Attitude to Marvels', *Fortnightly Review*, vol. 79 (1906), pp. 460–74.

Oliver Lodge, 'The Immortality of the Soul. Part II. The Permanence of Personality', *Hibbert Journal*, vol. 6 (1907–8), pp. 563–85.

Oliver Lodge, 'The Ether of Space', *Contemporary Review*, vol. 93 (1908), pp. 536–46.

Oliver Lodge, *Man and the Universe: A Study of the Influence of the Advance in Scientific Knowledge upon Our Understanding of Christianity* (London: Methuen and Co., 1908).

Oliver Lodge, 'The Attitude of Science to the Unusual: A Reply to Professor Newcomb', *Nineteenth Century*, vol. 65 (1909), pp. 206–22.

Oliver Lodge, *The Ether of Space* (London: Harper & Brothers, 1909).

Oliver Lodge, *The Survival of Man: A Study in Unrecognised Human Faculty* (London: Methuen and Co., 1909).

Oliver Lodge, 'The Education of an Observer', *JSPR*, vol. 14 (1909–10), pp. 253–8.

Oliver Lodge, 'On the A Priori Argument Against Physical Phenomena', *PSPR*, vol. 25 (1911), pp. 447–54.

Oliver Lodge, *Modern Problems* (London: Methuen and Co., 1912).

Oliver Lodge, '"Uncommon Sense" as a Substitute for Investigation', *Bedrock*, vol. 1 (1912–13), pp. 333–50.

Oliver Lodge, 'Continuity', in *Report of the Eighty-Third Meeting of the British Association for the Advancement of Science. Birmingham: 1913* (London: John Murray, 1914), pp. 3–42.

Oliver Lodge, *The War and After: Short Chapters on Subjects of Serious Practical Import for the Average Citizen in A.D. 1915 Onwards* (London: Methuen and Co., 1915).

Oliver Lodge, *Raymond or Life and Death* (London: Methuen and Co., 1916).

Oliver Lodge, 'The Reality of Psychic Phenomena', *Light*, vol. 37 (1917), pp. 66–7.

Oliver Lodge, 'Ether, Matter and the Soul', *Hibbert Journal*, vol. 17 (1918–19), pp. 252–60.

Oliver Lodge, 'Einstein's Real Achievement', *Fortnightly Review*, vol. 110 (1921), pp. 353–73.

Oliver Lodge, *Ether and Reality: A Series of Discourses on the Many Functions of the Ether of Space* (London: Hodder and Stoughton, 1925).

Oliver Lodge, 'Introduction to the Earl of Dunraven's Record of Experiences with D. D. Home', *PSPR*, vol. 35 (1925), pp. 1–20.

Oliver Lodge, *Evolution and Creation* (London: Hodder and Stoughton, 1926).

Oliver Lodge, *Why I Believe in Personal Immortality* (London: Cassell and Company, 1928).

Oliver Lodge, *Phantom Walls* (London: Hodder and Stoughton, 1929).

Oliver Lodge, *Beyond Physics or the Idealisation of Mechanism* (London: George Allen & Unwin, 1930).

Oliver Lodge, 'The Ether and Relativity', *Nature*, vol. 126 (1930), pp. 804–5.

Oliver Lodge, *Advancing Science: Being Personal Reminiscences of the British Association in the Nineteenth Century* (London: Ernest Benn, 1931).

Oliver Lodge, *Past Years: An Autobiography* (London: Hodder and Stoughton, 1931).

Oliver Lodge, 'The Past and the Future', *PSPR*, vol. 41 (1932–3), pp. 61–74.

Oliver Lodge, *My Philosophy Representing My Views on the Many Functions of the Ether of Space* (London: Ernest Benn, 1933).

Oliver Lodge, *Making of Man* (London: Hodder and Stoughton, 12th ed., 1938).

Oliver Lodge, 'The Life Work of My Friend F. W. H. Myers', *Nature*, vol. 144 (1939), pp. 1027–8.

Arthur Lovell, 'Reichenbach's Researches', *Transactions of the Vril-ya Club*, no. 2 (1904), pp. 1–34.

J. C. Luxmoore, 'The Outrage at a Spirit Circle', *Spiritualist*, vol. 3 (1872–3), p. 491.

William Lynd, 'Thought Transference and Wireless Telegraphy', *Surrey Magazine*, vol. 1 (1900), pp. 24–7.

Ernst Mach, 'The Propensity Toward the Marvellous', *Open Court*, vol. 14 (1900), pp. 539–50.

W. H. Mallock, 'The Gospel of Mr. F. W. H. Myers', *Nineteenth Century and After*, vol. 53 (1903), pp. 628–44.

W. H. Mallock, 'Sir Oliver Lodge on Life and Matter', *Fortnightly Review*, vol. 80 (1906), pp. 33–47.

James Marchant (ed.), *Alfred Russel Wallace: Letters and Reminiscences* (London: Cassell and Co., 1916).

Henry Margenau, 'Physics and Psychic Research', *Newsletter of the Parapsychology Foundation*, 1956, pp. 14–15.

John Nevil Maskelyne, *Modern Spiritualism: A Short Account of Its Rise and Progress, with Some Exposures of So-Called Spirit Media* (London: Frederick Warne and Co., 1875).

Charles Carleton Massey, 'Spiritualism and Men of Science', *Spiritualist*, vol. 9 (1876), pp. 21–2.

James Clerk Maxwell, 'On Action at a Distance', *Nature*, vol. 7 (1872–3), pp. 323–5, 341–3.

James Clerk Maxwell, 'Idiotic Imps [1853]', in Lewis Campbell and William Garnett, *The Life of James Clerk Maxwell* (London: Macmillan and Co., 2nd ed., 1884), pp. 341–3.

James Clerk Maxwell, 'Molecules', *Nature*, vol. 8 (1873), pp. 437–41.

James Clerk Maxwell, 'Ether', in W. D. Niven (ed.), *The Scientific Papers of James Clerk Maxwell* (Cambridge University Press, 1890), vol. 2, pp. 763–75.

Herbert Mayo, *On the Truths Contained in Popular Superstitions with an Account of Mesmerism* (London: William Blackwood, 1851).

Joseph McCabe, *The Religion of Sir Oliver Lodge* (London: Watts and Co., 1914).

Robert A. McConnell, 'Physical or Non-Physical', *Journal of Parapsychology*, vol. 11 (1947), pp. 111–17.

John G. McKendrick, 'Human Electricity', *Fortnightly Review*, vol. 51 (1892), pp. 634–41.

Charles Mercier, *Spiritualism and Sir Oliver Lodge* (London: Watts and Co., 1919).

Franz A. Mesmer, *Mémoire sur la découverte du magnétisme animal* (Geneva: P. F. Didot la jeune, 1779).

Franz A. Mesmer, *Mémoire de F. A. Mesmer, docteur en médecine, sur ses découvertes* (Paris: Fuchs, 1799).

Jeffrey Mishlove, *The Roots of Consciousness: Psychic Liberation Through History, Science and Experience* (New York: Random House, 1975).

Conwy Lloyd Morgan, 'Supernormal Psychology', *Nature*, vol. 35 (1886–7), pp. 290–2.

Tudor A. Morgan, 'The Ether and Spiritual Science', *Occult Review*, vol. 50 (1929), pp. 83–9.

[William Stainton Moses], 'Notes by the Way', *Light*, vol. 8 (1888), pp. 361–2.

Hugo Münsterberg, 'Psychology and Mysticism', *Atlantic Monthly*, vol. 83 (1899), pp. 67–85.

Hugo Münsterberg, *American Problems: From the Point of View of a Psychologist* (New York: Moffat, Yard and Company, 1910).

Frederic W. H. Myers, 'On Alleged Movements of Objects, Without Contact, Occurring Not in the Presence of a Paid Medium', *PSPR*, vol. 7 (1891–2), pp. 383–94.

Frederic W. H. Myers, 'The Subliminal Consciousness', *PSPR*, vol. 7 (1891–2), pp. 298–355; vol. 8 (1892), pp. 436–535.

Frederic W. H. Myers, *Human Personality and its Survival of Bodily Death*, 2 vols. (London: Longmans, Green, and Co., 1903).

Simon Newcomb, 'Address of the President', *Proceedings of the American Society for Psychical Research*, vol. 1 (1889–9), pp. 63–86.

Simon Newcomb, 'Modern Occultism', *Nineteenth Century*, vol. 65 (1909), pp. 126–39.

Robert Dale Owen, *Footfalls on the Boundary of Another World* (London: Trübner and Sons, 1860).

'P.', 'On the Odic Principle of Reichenbach', *London Journal of Arts, Sciences and Manufactures*, vol. 38 (1851), pp. 124–32, 193–9.

Papers Read Before the Synthetic Society, 1896–1908 ([London]: Spottiswoode and Co., 1909).

Bernard Partridge, 'Mr. Punch's Personalities. XXX – Sir Oliver Lodge, F.R.S.', *Punch*, 24 November 1926, p. 585.

C. W. Pearce, 'Mr. Crookes's Experiments', *Spiritualist*, vol. 1 (1869–71), p. 190.

Karl Pearson (ed.), *The Life, Letters and Labours of Francis Galton*, 3 vols. (Cambridge University Press, 1914–30).

'Perplexed', 'Is Sir William Crookes a Spiritualist?', *Light*, vol. 25 (1905), p. 35.

William H. Pickering, 'A Research on the Reality of Reichenbach's Flames', *Proceedings of the American Society for Psychical Research*, vol. 1 (1885–9), p. 127.

J. G. Piddington, 'Presidential Address', *PSPR*, vol. 34 (1924), pp. 131–52.

Frank Podmore, *Modern Spiritualism: A History and Criticism*, 2 vols. (London: Methuen and Co., 1902).

Post-Prandial Proceedings of the Cavendish Society (Cambridge: Bowes and Bowes, 6th ed., 1926).

J. H. Powell, *Spiritualism: Its Facts and Phases* (London: F. Pitman, 1864).

J. H. Poynting, 'Appendix. Biographical Sketch', in Oliver Lodge, *Mind and Matter: An Address Delivered in the Town Hall, Birmingham* (Birmingham: Birmingham and Midland Institute, 1904), pp. 29–35.

J. H. Poynting and J. J. Thomson, *A Textbook of Physics. Heat* (London: Charles Griffin, 1904).

S. Tolver Preston, 'On the Importance of Experiments in Relation to the Mechanical Theory of Gravitation', *Philosophical Magazine*, vol. 11 (5th Series) (1881), pp. 391–3.

Thomas Preston, *The Theory of Light* (London: Macmillan and Co., 4th ed., 1912).

Andrija Puharich (ed.), *The Iceland Papers: Select Papers on Theoretical and Experimental Research on the Physics of Consciousness* (Amherst, WI: Essentia Research Associates, 1979).

William Ramsay, 'Mr. Barkworth and "The Unthinkable"', *JSPR*, vol. 6 (1893–4), pp. 254–5.

Rapport des commissaires de la Société Royale de Médecine, nommés par le roi, pour faire l'examen du magnétisme animal (Paris: Chez Moutard, 1794).

Lord Rayleigh [John William Strutt], 'On the Invisibility of Small Objects in Bad Light', *Proceedings of the Cambridge Philosophical Society*, vol. 4 (1880–3), p. 324.

Lord Rayleigh [John William Strutt], 'Further Observations upon Liquid Jets, in Continuation of Those Recorded in the Royal Society's 'Proceedings' for March and May, 1879', *Proceedings of the Royal Society of London*, vol. 34 (1882–3), pp. 130–45.

Lord Rayleigh [John William Strutt], *The Theory of Sound*, 2 vols. (London: Macmillan and Co., 2nd ed., 1894).

Lord Rayleigh [John William Strutt], 'Presidential Address', *PSPR*, vol. 19 (1919–20), pp. 276–90.

Lord Rayleigh [Robert John Strutt], *John William Strutt: Third Baron Rayleigh* (London: Edward Arnold, 1924).

[Lord Rayleigh] [Robert John Strutt], 'MM. Osty's Investigations of Rudi Schneider', *Nature*, vol. 133 (1934), pp. 747–9.

Lord Rayleigh [Robert John Strutt], 'Presidential Address: The Problem of Physical Phenomena in Connection with Psychical Research', *PSPR*, vol. 45 (1939), pp. 1–18.

Lord Rayleigh [Robert John Strutt], 'The Question of Lights Supposed to Have Been Observed Near the Poles of a Magnet', *PSPR*, vol. 45 (1939), pp. 19–24.

Lord Rayleigh [Robert John Strutt], *The Life of J. J. Thomson* (Cambridge University Press, 1942).

G. H. Reddalls, 'Spirit Apparatus', *English Mechanic and World of Science*, vol. 21 (1875), p. 283.

Herbert S. Redgrove, 'Mathematics and Psychical Research', *Psychic Research Quarterly*, vol. 1 (1920–1), pp. 220–34.

Jules Regnault, 'Odic Phenomena and the New Radiations', *Annals of Psychical Science*, vol. 1 (1905), pp. 145–63.

Karl von Reichenbach, *Abstract of 'Researches on Magnetism and on Certain Allied Subjects', Including a New Imponderable*, translated and abridged from the German by William Gregory (London: Taylor and Walton, 1846).

Karl von Reichenbach, *Physicalisch-physiologische Untersuchungen über Die Dynamide der Magnetismus, der Electricität, der Wärme, des Lichtes, der Krystallisation, des Chemismus in ihren Beziehungen zur Lebenskraft*, 2 vols. (Braunschweig: Friedrich Vieweg and Son, 1849–50).

Karl von Reichenbach, *Physico-Physiological Researches on the Dynamics of Magnetism, Electricity, Heat, Light, Crystallisation, and Chemism in Their Relations to the Vital Force*, translated and edited by John Ashburner (London: Hippolyte Baillière, 1850).

Karl von Reichenbach, *Researches on Magnetism, Electricity, Heat, Light, Crystallisation, and Chemical Attraction, in Their Relations to the Vital Force*, translated and edited by William Gregory (London: Taylor, Walton and Moberly, 1850).

Karl von Reichenbach, 'Zur Intensität der Lichterscheinungen', *Annalen der Physik und Chemie*, vol. 112 (1861), pp. 459–68.

Karl von Reichenbach, *Odische Begebenheiten zu Berlin in dem Jahren 1861 und 1862* (Berlin: G. H. Schroeder, 1862).

Karl von Reichenbach, *Die odische Lohe und einige Bewegungserscheinungen als neuendeckte Formen des odischen Prinzips in der Natur* (Vienna: Wilhelm Braumüller, 1867).

Report on Spiritualism of the Committee of the London Dialectical Society (London: Longmans, Green, Reader and Dyer, 1871).

Joseph Banks Rhine, *Extra-Sensory Perception* (Boston: Boston Society for Psychical Research, 1934).

Benjamin Ward Richardson, 'A Theory of a Nervous Atmosphere', *Medical Times and Gazette*, vol. 1 (1871), pp. 507–9.

Charles Richet, *Traité de métapsychique* (Paris: Librairie Félix Alcan, 1922).

Philip J. Risdon, 'Psychic Phenomena and Wireless', *Popular Wireless Weekly*, vol. 1 (1922), pp. 237–8.

Philip J. Risdon, 'There Is No Ether?', *Popular Wireless Weekly*, vol. 1 (1922), p. 145.

'R. M. N.', 'Psychography', *Journal of Science*, vol. 7 (1885), pp. 143–9.

Edward Coit Rogers, *Philosophy of the Mysterious Agents, Human and Mundane* (Boston: John P. Jewett and Company, 1853).

Royal College of Science for Ireland: Its Origin and Development (Dublin University Press, 1923).

J. H. Rush, 'Some Considerations as to a Physical Basis of ESP', *Journal of Parapsychology*, vol. 7 (1943), pp. 44–9.

John O. N. Rutter, *Human Electricity: The Means of Its Development* (London: John W. Parker, 1854).

Helen de G. Salter and Isabel Newton, 'Obituary – Alice Johnson', *PSPR*, vol. 46 (1940–1), pp. 16–22.

George Sandby, 'The Mesmerisation and Movement of Tables', *Zoist*, vol. 11 (1853–4), pp. 175–85.

Epes Sargent, *Planchette, Or, the Despair of Science* (Boston: Roberts Brothers, 1869).

Epes Sargent, *Proof Palpable of Immortality* (Boston: Colby and Rich, 1876).

Epes Sargent, *The Scientific Basis of Spiritualism* (Boston: Colby and Rich, 1881).

Ferdinand C. S. Schiller, 'Psychology and Psychical Research', *PSPR*, vol. 14 (1898–9), pp. 348–65.

Ferdinand C. S. Schiller, 'The Future of the S.P.R.', *JSPR*, vol. 10 (1901–2), pp. 74–7.

Helmut Schmidt, 'Psychokinesis', in John White (ed.), *Psychic Exploration: A Challenge for Science* (New York: Perigee Books, 1974), pp. 179–93.

Arthur Schuster, *Biographical Fragments* (London: Macmillan and Co., 1932).

George F. C. Searle, 'The Modern Conception of the Universe', in *Pan-Anglican Papers. Being Problems for Consideration at the Pan-Anglican Congress, 1908. Religion and Science* (London: Society for the Promotion of Christian Knowledge, 1908), pp. 1–8.

J. W. Sharpe, 'Photographing the Unseen', *Light*, vol. 21 (1901), pp. 429–30.

A[rthur] S[idgwick] and E[leanor] M[ildred] S[idgwick], *Henry Sidgwick: A Memoir* (London: Macmillan and Co., 1906).

Eleanor Mildred Sidgwick, 'Mr. Eglinton', *JSPR*, vol. 2 (1886), pp. 282–334.

[Eleanor Mildred] Sidgwick, 'Results of a Personal Investigation into the Physical Phenomena of Spiritualism', *PSPR*, vol. 4 (1886–7), pp. 45–74.

[Eleanor Mildred Sidgwick], 'Correspondence', *JSPR*, vol. 6 (1893–4), p. 214.

E[leanor] M[ildred] S[idgwick], Review of Psicologia e "Spiritismo", *JSPR*, vol. 21 (1908–9), pp. 516–25.

[Eleanor Mildred] Sidgwick, 'Presidential Address', *PSPR*, vol. 22 (1908), pp. 1–18.

[Eleanor Mildred] Sidgwick, 'Phantasms of the Living', *PSPR*, vol. 33 (1923), pp. 23–429.

[Eleanor Mildred] Sidgwick, 'Experimental Telepathy: The Need of Further Experiments', in *L'état actuel des recherches psychique d'après le travaux du IIme Congrès International tenu à Varsovie en 1923* (Paris: Les Presses Universitaires de France, 1924), 174–80.

Eleanor Sidgwick, 'The Society for Psychical Research. A Short Account of Its History and Work on the Occasion of the Society's Jubilee, 1932', *PSPR*, vol. 41 (1932–3), pp. 1–26.

Henry Sidgwick, 'Remarks', *PSPR*, vol. 4 (1886–7), pp. 103–6.

Henry Sidgwick, 'The Canons of Evidence in Psychical Research', *PSPR*, vol. 6 (1889–90), pp. 1–6.

Henry Sidgwick, 'Eusapia Palladino', *JSPR*, vol. 7 (1895–7), pp. 230–1.

'Sigma' [John T. Sprague], 'Modern Spiritualism', *English Mechanic and World of Science*, vol. 19 (1874), p. 122.

'Sigma' [John T. Sprague], 'Perceptive Powers of Men and Other Animals', *English Mechanic and World of Science*, vol. 20 (1874), p. 223.

W. Whately Smith, *A Theory of the Mechanism of Survival* (London: Kegan Paul, Trübner, Trench and Co., 1920).

John Spiller, 'Mr. Crookes's "Psychic Force"', *Echo*, 6 November 1871, p. 1.

John T. Sprague, 'Luminosity of Magnets', *English Mechanic and World of Science*, vol. 37 (1883), p. 233.

J. B. Stallo, *The Concepts and Theories of Modern Physics* (New York: D. Appleton and Co., 1884).

W. T. Stead, 'The Response to the Appeal. From Prelates, Pundits and Persons of Distinction', *Borderland*, vol. 1 (1893), pp. 10–23.

Charles P. Steinmetz, 'There Are No Ether Waves', *Popular Radio*, vol. 1 (1922), pp. 161–6.

Frederick Stephens, 'Science and the Unseen World', *Light*, vol. 49 (1929), pp. 338–9, 356–7.

Balfour Stewart, *The Recent Developments of Cosmical Physics: A Lecture* (London: Macmillan and Co., 1870).

Balfour Stewart, 'Mr. Crookes on the "Psychic Force"', *Nature*, vol. 4 (1871), p. 237.

Balfour Stewart, *Lessons in Elementary Physics* (London: Macmillan and Co., 1872).

Balfour Stewart, 'Note on Thought-Reading', *PSPR*, vol. 1 (1882–3), pp. 35–42.

Balfour Stewart, 'President's Address', *PSPR*, vol. 3 (1885), pp. 64–8.

Balfour Stewart, 'Address', *PSPR*, vol. 4 (1886–7), pp. 262–7.

Balfour Stewart, 'Note on the Above Paper', *PSPR*, vol. 4 (1886–7), pp. 42–4.

Balfour Stewart and J. Norman Lockyer, 'The Sun as a Type of the Material Universe', *Macmillan's Magazine*, vol. 18 (1868), pp. 246–57, 319–27.

[Balfour Stewart and Peter Guthrie Tait], *The Unseen Universe; Or Physical Speculations on a Future State* (London: Macmillan and Co., 1875).

[Balfour Stewart and Peter Guthrie Tait], *The Unseen Universe; Or Physical Speculations on a Future State* (London: Macmillan and Co., 3rd ed., 1875).

Balfour Stewart and Peter Guthrie Tait, *Paradoxical Philosophy: A Sequel to the Unseen Universe* (London: Macmillan and Co., 1878).

George G. Stokes, *Natural Theology: The Gifford Lectures Delivered Before the University of Edinburgh in 1893* (London: Adam and Charles Black, 1893).

William H. Stone, 'The Persecution of Dr. Stone – Medical Electricity', *English Mechanic and World of Science*, vol. 18 (1873), p. 312.

William H. Stone, 'Reichenbach's Experiments', *English Mechanic and World of Science*, vol. 20 (1874), p. 120.

William H. Stone, 'The Radiometer', *Popular Science Review*, vol. 17 (1878), pp. 164–81.

William H. Stone, 'Hysteria and Hystero-Epilepsy', *Saint Thomas's Hospital Reports*, vol. 11 (1880), pp. 85–102.

William H. Stone, 'Some Effects of Brain Disturbance on the Handwriting', *Saint Thomas's Hospital Reports*, vol. 12 (1882), pp. 67–75.

William H. Stone, 'The Physiological Bearing of Electricity on Health', *Journal of the Society of Telegraph Engineers*, vol. 13 (1884), pp. 415–32.

William H. Stone, 'Abstracts of the Lumleian Lectures on the Electrical Condition of the Human Body: Man as a Conductor and Electrolyte. Lecture III', *Lancet*, vol. 1 (1886), pp. 863–5.

William H. Stone and W. J. Kilner, 'On Measurement in the Medical Application of Electricity', *Journal of the Society of Telegraph Engineers*, vol. 11 (1882), pp. 107–28.

George Johnstone Stoney, 'On Crookes's Radiometer', *Philosophical Magazine*, vol. 1 (5th Series) (1876), pp. 177–81.

Frederick J. M. Stratton, 'Review', *JSPR*, vol. 14 (1909), pp. 78–9.

Frederick J. M. Stratton, 'Psychical Research – A Lifelong Interest', *PSPR*, vol. 50 (1954), pp. 135–52.

Frederick J. M. Stratton and P. Phillips, 'Some Experiments with the Sthenometer', *JSPR*, vol. 12 (1905–6), pp. 335–9.

René Sudre, 'Psychical Research and the New Physics', *Psychic Research*, vol. 24 (1930), pp. 117–21.

[J. W. N. Sullivan], 'Ether and Reality', *Times Literary Supplement*, 14 May 1925, p. 325.

A. H. Tabrum, *Religious Beliefs of Scientists* (London: North London Christian Evidence League, 2nd ed., 1913).

Peter G. Tait, 'Address', *Report of the Forty-First Meeting of the British Association for the Advancement of Science; Held at Edinburgh in August 1871* (London: John Murray, 1872), pp. 1–8.

E. S. Talbot, 'Sir Oliver Lodge on "The Re-Interpretation of Christian Doctrine"', *Hibbert Journal*, vol. 2 (1903–4), pp. 649–61.

Hans Thirring, 'The Position of Science in Relation to Psychical Research', *British Journal of Psychical Research*, vol. 1 (1927), pp. 165–81.

Henry Edward Thompson, 'Grasping a Spirit', *Medium and Daybreak*, vol. 4 (1873), pp. 598–9.

Jane Smeal Thompson and Helen G. Thompson, *Silvanus Philips Thompson: His Life and Letters* (London: T. Fisher Unwin, 1920).

Silvanus P. Thompson, 'A Physiological Effect of an Alternating Magnetic Field', *Philosophical Transactions of the Royal Society of London*, vol. 82B (1909–10), pp. 396–8.

Silvanus P. Thompson, *The Life of William Thomson, Baron Kelvin of Largs*, 2 vols. (London: Macmillan and Co., 1910).

Silvanus P. Thompson, *A Not Impossible Religion* (London: Bodley Head, 1917).

Allen Thomson, 'Address to the Biology Section', in *Report of the Forty-First Meeting of the British Association for the Advancement of Science; Held at Edinburgh in August 1871* (London: John Murray, 1872), pp. 114–22.

J. J. Thomson, 'On the Passage of Electricity Through Hot Gases', *Philosophical Magazine*, vol. 29 (5th Series) (1890), pp. 441–9.

J. J. Thomson, 'On the Effect of Electrification and Chemical Action on a Steam-Jet, and of Water Vapour on the Discharge of Electricity Through Gases', *Philosophical Magazine*, vol. 36 (5th Series) (1893), pp. 313–27.

J. J. Thomson, 'Address', in *Report of the Seventy-Ninth Meeting of the British Association for the Advancement of Science. Winnipeg: 1909* (London: John Murray, 1910), pp. 3–29.

J. J. Thomson, *Recollections and Reflections* (London: G. Bell, 1936).

William Thomson, 'The Six Gateways of Knowledge [1883]', in William Thomson, *Popular Lectures and Addresses*, 3 vol. (London: Macmillan and Co., 1889), vol. 1, pp. 253–99.

William M. Thornton, *The Scientific Background of the Christian Creeds* (Newcastle-on-Tyne: Andrew Reid and Company, 1930).

Rudolf Tischner, *Telepathy and Clairvoyance*, translated by W. D. Hutchinson (London: Kegan Paul, Trench, Trübner and Co., 1925).

Chauncy Hare Townshend, *Facts in Mesmerism, with Reasons for a Dispassionate Inquiry into It* (London: Longman, Orme, Brown, Green and Longmans, 1840).

'Treadle', 'Radiometers', *English Mechanic and World of Science*, vol. 22 (1876), p. 558.

Alexander P. Trotter, 'Illumination and Light', *Journal of the Institution of Electrical Engineers*, vol. 64 (1926), pp. 367–71.

John Trowbridge, 'Telepathy', *Nation*, vol. 76 (1903), pp. 308–9.

Ivor Tuckett, *The Evidence for the Supernatural: A Critical Study Made with 'Uncommon Sense'* (London: Kegan Paul, Trench, Trübner & Co., 1911).

Ivor Tuckett, 'Psychical Investigators and "The Will to Believe"', *Bedrock*, vol. 1 (1912–13), pp. 180–204.

George W. De Tunzelmann, *A Treatise on Electricity and the Problem of the Universe* (London: Charles Griffin and Company, 1910).

[John Tyndall], 'Science and the Spirits', *Reader*, vol. 4 (1864), pp. 725–6.

John Tyndall, 'Miracles and Special Providences', *Fortnightly Review*, vol. 7 (1867), pp. 645–60.

John Tyndall, *Sound. A Course of Eight Lectures Delivered at the Royal Institution of Great Britain* (London: Longmans, Green, and Co., 1867).

John Tyndall, 'Scientific Materialism [1868]', in *Fragments of Science*, vol. 2, pp. 75–90.

John Tyndall, 'On Prayer', *Contemporary Review*, vol. 20 (1872), pp. 763–6.

John Tyndall, 'The Belfast Address [1874]', in *Fragments of Science*, vol. 2, pp. 135–201.

John Tyndall, *Fragments of Science: A Series of Detached Essays, Addresses and Reviews* (London: Longmans, Green and Co., 7th ed., 1889).

George N. M. Tyrrell, *Science and Psychical Phenomena* (London: Methuen and Co., 1939).

University College, Liverpool. Calendar for the Session 1886–1887 (Liverpool: Adam Holden, 1886).

F. L. Usher and F. P. Burt, 'Thought Transference (Some Experiments in Long Distance Thought-Transference)', *Annals of Psychical Science*, vol. 8 (1909), pp. 561–600.

P. H. Van der Weyde, 'On Mr. Crookes's Further Experiments on Psychic Force', *Journal of the Franklin Institute*, vol. 92 (1871), pp. 423–6.

P. H. Van der Weyde, 'On the Psychic Force', *Scientific American*, 23 September 1871, p. 197.

Cromwell F. Varley, 'Spiritualism', *Human Nature*, vol. 3 (1869), pp. 367–71.

Cromwell F. Varley, 'Electricity, Magnetism and the Human Body', *Spiritualist*, vol. 1 (1869–71), pp. 137–8.

Cromwell F. Varley, 'Some Experiments on the Discharge of Electricity Through Rarefied Media and the Atmosphere', *Proceedings of the Royal Society of London*, vol. 19 (1870–1), pp. 236–42.

Cromwell F. Varley, 'Evidence of Mr. Varley', in *Report on Spiritualism*, pp. 157–72.

Cromwell F. Varley, 'Psychic Force', *English Mechanic and World of Science*, vol. 14 (1872), pp. 454–5.

Cromwell F. Varley, 'Electrical Experiments with Miss Cook when Entranced', *Spiritualist*, vol. 4 (1874), pp. 134–5.

Cromwell F. Varley, 'The Reality of Spiritual Phenomena', *Spiritualist*, vol. 9 (1876), pp. 265–6.

Leonid L. Vasiliev, *Experiments in Distant Influence: Discoveries by Russia's Foremost Parapsychologist* (London: Wildwood House, 1976).

Cesar de Vesme, 'Human Brain Waves', *Light*, vol. 54 (1934), p. 115.

Alfred Russel Wallace, *The Scientific Aspect of the Supernatural* (London: F. Farrah, 1866).

Alfred Russel Wallace, *Miracles and Modern Spiritualism* (London: James Burns, 1875).

Alfred Russel Wallace, *My Life: A Record of Events and Opinions*, 2 vols. (London: Chapman and Hall, 1905).

René Warcollier, *La télépathie recherches expérimentales* (Paris: Librairie Félix Alcan, 1921).

Walter White, *The Journals of Walter White* (London: Chapman and Hall, 1898).

W. Matthieu Williams, 'Science and Spiritualism [1871]', in *Science in Short Chapters* (London: Chatto and Windus, 1882), pp. 237–51.

Beckles Willson, *Occultism and Common Sense* (London: T. Werner Laurie, 1908).

Robert W. Wood, 'The *n*-Rays', *Nature*, vol. 70 (1904), pp. 530–1.

Elwood Worcester, 'The Intrepid Pioneer', *Journal of the American Society for Psychical Research*, vol. 14 (1920), pp. 501–10.

Wilhelm Wundt, 'Spiritualism as a Scientific Question', *Popular Science Monthly*, vol. 15 (1879), pp. 577–93.

Emile Yung, 'On the Errors of Our Sensations: A Contribution to the Study of Illusion and Hallucination', *Philosophical Magazine*, vol. 15 (5th Series) (1883), pp. 259–70.

Gustavus G. Zerffi, *Spiritualism and Animal Magnetism: A Treatise on Dreams, Second Sight, Somnambulism, Magnetic Sleep, Spiritual Manifestations, Hallucinations and Spectral Visions* (London: Robert Hardwicke, 1871).

Johann Karl Friedrch Zöllner, *Transcendental Physics* (translated by Charles Carleton Massey) (London: W. H. Harrison, 1880).

Secondary Sources

James Alcock, 'Parapsychology: Science of the Anomalous or Search for the Soul', *Behavioural and Brain Sciences*, vol. 10 (1987), pp. 553–65.

Carlos Alvarado, 'On the Centenary of Frederic W. H. Myers's *Human Personality and its Survival of Bodily Death*', *Journal of Parapsychology*, vol. 68 (2003), pp. 3–43.

Rollo Appleyard, *Pioneers of Electrical Communication* (London: Macmillan and Co., 1930).

Rollo Appleyard, *The History of the Institution of Electrical Engineers 1871–1931* (London: The Institution of Electrical Engineers, 1931).

Stathis Arapostathis and Graeme Gooday, *Patently Contestable: Electrical Technologies and Inventor Identities on Trial in Britain* (Cambridge, MA: MIT Press, 2013).

Egil Asprem, 'Pondering Imponderables: Occultism in the Mirror of Late Classical Physics', *Aries*, vol. 11 (2011), pp. 129–65.

Egil Asprem, *The Problem of Disenchantment: Scientific Naturalism and Esoteric Discourse 1900–1939* (Leiden: Brill, 2014).

Egil Asprem, 'Science and the Occult', in Christopher Partridge (ed.), *The Occult World* (Abingdon: Routledge, 2015), pp. 710–19.

Allan Barham, 'Dr. W. J. Crawford, His Work and Legacy in Psychokinesis', *JSPR*, vol. 55 (1988), pp. 113–38.

Logie Barrow, *Independent Spirits: Spiritualism and English Plebeians, 1850–1910* (London: Routledge and Kegan Paul, 1986).

Ruth Barton, 'John Tyndall, Pantheist: A Rereading of the Belfast Address', *Osiris*, vol. 3 (1987), pp. 111–34.

Ruth Barton, 'Just Before *Nature*: The Purposes of Science and the Purposes of Popularization in English Popular Science Journals of the 1860s', *Annals of Science*, vol. 55 (1998), pp. 1–33.

Barbara J. Becker (ed.), *Selected Correspondence of William Huggins*, 2 vols. (London: Routledge, 2014).

Harold Begbie, *Master Workers* (London: Methuen and Co., 1905).

James E. Beichler, 'To Be or Not to Be! A "Paraphysics" for the New Millennium', *Journal of Scientific Exploration*, vol. 15 (2001), pp. 33–56.

Bernadette Bensaude-Vincent and Christine Blondel (eds.), *Des savants face à l'occulte 1870–1940* (Paris: Éditions la Découverte, 2002).

Ernest Benz, *The Theology of Electricity* (Eugene, OR: Pickwick Publications, 2009).

Arthur S. Berger, *Lives and Letters in American Parapsychology: A Biographical History, 1850–1987* (Jefferson, NC: McFarland and Company, 1988).

Theodore Besterman, *A Bibliography of Sir Oliver Lodge F.R.S.* (Oxford University Press, 1935).

Fabio Bevilacqua, 'Helmholtz's *Über die Erhaltung der Kraft*: The Emergence of a Theoretical Physicist', in David Cahan, *Hermann von Helmholtz*, pp. 293–333.

Mark Blacklock, *The Emergence of the Fourth Dimension: Higher Spatial Thinking in the Fin de Siècle* (Oxford University Press, 2018).

Christine Blondel, 'Eusapia Palladino: la méthode experimentale et la "diva des savants"', in Bernadette Bensaude-Vincent and Christine Blondel, *Des savants face à l'occulte*, pp. 143–71.

Peter J. Bowler, *Reconciling Science and Religion: The Debate in Early-Twentieth-Century Britain* (Chicago University Press, 2001).

Peter J. Bowler, *Science for All: The Popularisation of Science in Early Twentieth-Century Britain* (Chicago University Press, 2009).

Robert M. Brain, 'Materialising the Medium: Ectoplasm and the Quest for Supra-Normal Biology in *Fin-de-Siècle* Science and Art', in Anthony Enns and Shelley Trower (eds.), *Vibratory Modernism* (Basingstoke: Palgrave Macmillan, 2013), pp. 115–44.

Ann Braude, *Radical Spirits: Spiritualism and Women's Rights in Nineteenth-Century America* (Boston: Beacon Press, 1989).

W. G. Bringmann and R. D. Tweney (eds.), *Wundt Studies* (Toronto: C. J. Hogrefe, 1980).

William H. Brock, *Justus von Liebig: The Chemical Gatekeeper* (Cambridge University Press, 1997).

William H. Brock, *William Crookes (1832–1919) and the Commercialization of Science* (Farnham: Ashgate, 2008).

William H. Brock and Geoffrey Cantor (eds.), *The Correspondence of John Tyndall Volume 5* (University of Pittsburgh Press, 2018).

William H. Brock and Roy M. Macleod (eds.), *Natural Knowledge in Social Context: Journals of Thomas Archer Hirst (1830–1892)* (London: Mansell, 1980).

John Hedley Brooke, *Science and Religion: Some Historical Perspectives* (Cambridge University Press, 1991).

M. Brady Brower, *Unruly Spirits: The Science of Psychic Phenomena in Modern France* (Urbana, IL: University of Illinois Press, 2010).

Alan Willard Brown, *The Metaphysical Society: Victorian Minds in Crisis* (New York: Columbia University Press, 1947).

Jed Z. Buchwald, *From Maxwell to Microphysics: Aspects of Electromagnetic Theory in the Last Quarter of the Nineteenth Century* (Chicago University Press, 1985).

Jed Z. Buchwald, *The Creation of Scientific Effects: Heinrich Hertz and Electric Waves* (Chicago University Press, 1994).

Jed Z. Buchwald and Robert Fox (eds.), *The Oxford Handbook of the History of Physics* (Oxford University Press, 2013).

Jed Z. Buchwald and Sungook Hong, 'Physics', in David Cahan (ed.), *From Natural Philosophy to the Sciences: Writing the History of Nineteenth-Century Science* (Chicago University Press, 2003), pp. 163–95.

Frederick Burkhardt, James A. Secord, Janet Browne et al. (eds.), *The Correspondence of Charles Darwin, vol. 20 1872* (Cambridge University Press, 2013).

Georgina Byrne, *Modern Spiritualism and the Church of England, 1850–1939* (Woodbridge: Boydell Press, 2010).

David Cahan, 'An Institutional Revolution in German Physics, 1865–1914', *Historical Studies in the Physical Sciences*, vol. 15 (1985), pp. 1–65.

David Cahan (ed.), *Hermann von Helmholtz and the Foundations of Nineteenth Century Science* (Berkeley, CA: University of California Press, 1993).

Bruce F. Campbell, *Ancient Wisdom Revived: A History of the Theosophical Movement* (Berkeley, CA: University of California Press, 1980).

Kenneth Caneva, *Julius Robert Mayer and the Conservation of Energy* (Princeton University Press, 1993).

Geoffrey Cantor, 'The Theological Significance of Ethers', in Geoffrey Cantor and M. J. S. Hodge, *Conceptions of Ether*, pp. 135–55.

Geoffrey Cantor, *Michael Faraday: Sandemanian and Scientist* (London: Macmillan, 1991).

Geoffrey Cantor, *Quakers, Jews and Science: Religious Responses to Modernity and the Sciences in Britain, 1650–1900* (Oxford University Press, 2005).

Geoffrey Cantor and M. J. S. Hodge (eds.), *Conceptions of Ether: Studies in the History of Ether Theories 1740–1900* (Cambridge University Press, 1981).

Geoffrey Cantor and Sally Shuttleworth (eds.), *Science Serialized: Representations of Science in Nineteenth-Century Periodicals* (Cambridge, MA: The MIT Press, 2004).

Bernard Carr, 'Worlds Apart? Can Psychical Research Bridge the Gulf Between Matter and Mind?', *PSPR*, vol. 59 (2008), pp. 1–96.

Bret E. Carroll, *Spiritualism in Antebellum America* (Bloomington, IN: Indiana University Press, 1997).

Bret E. Carroll, '"A Higher Power to Feel": Spiritualism, Grief, and Victorian Manhood', *Men and Masculinities*, vol. 3 (2000), pp. 3–29.

John J. Cerullo, *The Secularization of the Soul: Psychical Research in Modern Britain* (Philadelphia, PA: Institute for the Study of Human Issues, 1982).

James Chadwick, 'Frederick John Marrian Stratton', *Biographical Memoirs of Fellows of the Royal Society*, vol. 7 (1961), pp. 280–93.

C. T. K. Chari, 'Some Generalized Theories and Models of Psi: A Critical Evaluation', in Benjamin B. Wolman (ed.), *Handbook of Parapsychology* (New York: Van Nostrand Reinhold Co., 1977), pp. 803–22.

Georges Charpak and Henri Bloch, *Debunked! ESP, Telekinesis, Other Pseudoscience* (Baltimore, MD: Johns Hopkins University Press, 2004).

Clement Chéroux, 'Photographs of Fluids: An Alphabet of Invisible Rays', in Chéroux et al., *Perfect Medium*, pp. 114–25.

Clement Chéroux, Andreas Fischer, Pierre Apraxine, Denis Canguilhem and Sophie Schmit, *The Perfect Medium: Photography and the Occult* (New Haven, CT: Yale University Press, 2004).

Bruce Clarke, *Energy Forms: Allegory and Science in the Era of Classical Thermodynamics* (Ann Arbor, MI: University of Michigan Press, 2001).

Edwin Clarke and L. S. Jacyna, *Nineteenth-Century Origins of Neuroscientific Concepts* (Berkeley, CA: University of California Press, 1987).

Imogen Clarke, 'Ether at the Crossroads of Classical and Modern Physics', in Jaume Navarro, *Ether and Modernity*, pp. 14–29.

Nani N. Clow, 'The Laboratory of Victorian Culture: Experimental Physics, Industry and Pedagogy in the Liverpool Laboratory of Oliver Lodge, 1881–1900' (unpublished doctoral dissertation, Harvard University, 1999, UMI 9949744).

H. M. Collins, *Changing Order: Replication and Induction in Scientific Practice* (London: Sage, 1985).

H. M. Collins and T. J. Pinch, *Frames of Meaning: The Social Construction of Extraordinary Science* (London: Routledge and Kegan Paul, 1982).

Deborah J. Coon, 'Testing the Limits of Sense and Science: American Experimental Psychologists Combat Spiritualism, 1880–1920', *American Psychologist*, vol. 47 (1992), pp. 143–51.

Roger Cooter, *The Cultural Meaning of Popular Science: Phrenology and the Organisation of Consent in Nineteenth-Century Britain* (Cambridge University Press, 1984).

Roger Cooter, 'The Conservatism of "Pseudoscience"', in Patrick Grim (ed.), *Philosophy of Science and the Occult* (Albany, NY: State University of New York Press, 1990), pp. 156–69.

Adam Crabtree, *From Mesmer to Freud: Magnetic Sleep and the Roots of Psychological Healing* (New Haven, CT: Yale University Press, 1993).

Andrew Cunningham and Nicholas Jardine (eds.), *Romanticism and the Sciences* (Cambridge University Press, 1990).

W. C. D. Dampier, 'William Robert Bousfield 1854–1943', *Obituary Notices of Fellows of the Royal Society*, vol. 4 (1942–4), pp. 570–6.

Kurt Danziger, 'Mid-Nineteenth Century British Psycho-Physiology: A Neglected Chapter in the History of Psychology', in William R. Woodward and Mitchell G. Ash (eds.), *The Problematic Science: Psychology in Nineteenth Century Thought* (New York: Praeger, 1982), pp. 119–46.

Kurt Danziger, *Constructing the Subject: Historical Origins of Psychological Research* (Cambridge University Press, 1990).

Olivier Darrigol, *Electrodynamics from Ampère to Einstein* (Oxford University Press, 2000).

Lorraine Daston, 'British Responses to Psycho-Physiology, 1860–1900', *Isis*, vol. 69 (1978), pp. 192–208.

Lorraine Daston and Peter Galison, *Objectivity* (New York: Zone Books, 2007).

Owen Davies, *The Haunted: A Social History of Ghosts* (Basingstoke: Palgrave Macmillan, 2007).

Gowan Dawson and Bernard Lightman (eds.), *Victorian Scientific Naturalism: Community, Identity, Continuity* (Chicago University Press, 2014).

Robert K. DeKosky, 'William Crookes and the Fourth State of Matter', *Isis*, vol. 67 (1976), pp. 36–60.

Robert K. DeKosky, 'William Crookes and the Quest for the Absolute Vacuum', *Annals of Science*, vol. 40 (1983), pp. 1–18.

Shannon Delorme, 'Physiology or Psychic Powers? William Benjamin Carpenter and the Debate over Spiritualism in Victorian Britain', *Studies in History and Philosophy of Biological and Biomedical Sciences*, vol. 48 (2014), pp. 57–66.

Adrian Desmond and James Moore, *Darwin* (London: Michael Joseph, 1991).

Ursula DeYoung, *A Vision of Modern Science: John Tyndall and the Role of the Scientist in Victorian Culture* (Basingstoke: Palgrave Macmillan, 2011).

Matthias Dörries, 'Balances, Spectroscopes and the Reflexive Nature of Experiment', *Studies in History and Philosophy of Science*, vol. 25 (1994), pp. 1–36.

Susan J. Douglas, *Listening In: Radio and the American Imagination* (Minneapolis, MN: University of Minnesota Press, 2004).

Simon During, *Modern Enchantments: The Cultural Power of Secular Magic* (Cambridge, MA: Harvard University Press, 2002).

Gordon Epperson, *The Mind of Edmund Gurney* (Madison, NJ: Fairleigh Dickinson University Press, 1997).

Isobel Falconer, 'Corpuscles, Electrons and Cathode Rays: J. J. Thomson and the Discovery of the Electron', *British Journal for the History of Science*, vol. 20 (1987), pp. 241–76.

Franz Ferzak, *Karl Freiherr von Reichenbach* (Munich: Franz Ferzak World and Space Publications, 1987).

Martin Fichman, *An Elusive Victorian: The Evolution of Alfred Russel Wallace* (Chicago University Press, 2004).

Patrick Fuentès, 'Camille Flammarion et les force naturelles inconnues', in Bernadette Bensaude-Vincent and Christine Blondel, *Des savants face à l'occulte*, pp. 105–21.

Jill Galvan, *The Sympathetic Medium: Feminine Channelling, the Occult, and Communication Technologies, 1859–1919* (Ithaca, NY: Cornell University Press, 2010).

Martin Gardner, *Science Good, Bad, Bogus* (Buffalo, NY: Prometheus Books, 1981).

Alan Gauld, *The Founders of Psychical Research* (London: Routledge, 1968).

Alan Gauld, *A History of Hypnotism* (Cambridge University Press, 1992).

Hannah Gay, 'Invisible Resource: William Crookes and His Circle of Support', *British Journal for the History of Science*, vol. 29 (1996), pp. 311–36.

Hannah Gay, 'Science, Scientific Careers and Social Exchange: The Diaries of Herbert McLeod, 1885–1900', *History of Science*, vol. 46 (2008), pp. 457–96.

Thomas F. Gieryn, *Cultural Boundaries of Science: Credibility on the Line* (Chicago University Press, 1999).

Joscelyn Godwin, *The Theosophical Enlightenment* (Albany, NY: State University of New York Press, 1994).

Stanley Goldberg, 'In Defense of the Ether: The British Response to the Special Theory of Relativity, 1905–1911', *Historical Studies in the Physical Sciences*, vol. 2 (1970), pp. 89–125.

Graeme J. N. Gooday, 'Precision Measurement and the Genesis of Physics Teaching Laboratories in Victorian Britain', *British Journal for the History of Science*, vol. 25 (1990), pp. 25–51.

Graeme J. N. Gooday, 'Teaching Telegraphy and Electrotechnics in the Physics Laboratory: William Ayrton and the Creation of an Academic Space for Electrical Engineering in Britain 1873–1884', *History of Technology*, vol. 13 (1991), pp. 73–111.

Graeme J. N. Gooday, 'Instrumentation and Interpretation: Managing and Representing the Working Environments of Victorian Experimental Science', in Bernard Lightman (ed.), *Victorian Science in Context* (Chicago University Press, 1997), pp. 409–37.

Graeme J. N. Gooday, *The Morals of Measurement: Accuracy, Irony and Trust in Late Victorian Electrical Practice* (Cambridge University Press, 2004).

Graeme J. N. Gooday, 'Sunspots, Weather and the *Unseen Universe*: Balfour Stewart's Anti-Materialist Representations of "Energy" in British Periodicals', in Geoffrey Cantor and Sally Shuttleworth, *Science Serialized*, pp. 111–47.

Graeme J. N. Gooday, 'Periodical Physics in Britain: Institutional and Industrial Contexts, 1870–1900', in Gowan Dawson, Bernard Lightman, Sally Shuttleworth and Jonathan Topham (eds.), *Constructing Scientific Communities: Science Periodicals in Nineteenth Century Britain*, forthcoming.

David Gooding, 'A Convergence of Opinion on the Divergence of Lines: Faraday and Thomson's Discussion of Diamagnetism', *Notes and Records of the Royal Society of London*, vol. 36 (1982), pp. 243–59.

David Gooding, 'In Nature's School: Faraday as an Experimentalist', in David Gooding and Frank A. J. L. James, *Faraday Rediscovered*, pp. 105–35.

David Gooding and Frank A. J. L. James (eds.), *Faraday Rediscovered: Essays on the Life and Work of Michael Faraday, 1791–1867* (London: Macmillan, 1985).

Nicholas Goodrick-Clarke, *The Western Esoteric Traditions: A Historical Introduction* (New York: Oxford University Press, 2008).

Michael Gordin, *A Well-Ordered Thing: Dmitrii Mendeleev and the Shadow of the Periodic Table* (New York: Basic Books, 2004).

Allen W. Grove, 'Röntgen's Ghosts: Photography, X-Rays and the Victorian Imagination', *Literature and Medicine*, vol. 16 (1997), pp. 141–73.

Anna Guagnini, 'Ivory Towers? The Commercial Activity of British Professors of Engineering and Physics, 1880–1914', *History and Technology*, vol. 33 (2017), pp. 70–108.

Cathy Gutierrez, *Plato's Ghost: Spiritualism in the American Renaissance* (New York: Oxford University Press, 2009).

Vance D. Hall, 'The Contribution of the Physiologist William Benjamin Carpenter (1813–1885) to the Development of the Principles of the Correlation of Forces and the Conservation of Energy', *Medical History*, vol. 23 (1979), pp. 129–55.

Trevor Hamilton, *Immortal Longings: F. W. H. Myers and the Search for Life After Death* (Exeter: Imprint Academic, 2009).

Wouter Hanegraaff, *Esotericism and the Academy: Rejected Knowledge in Western Culture* (Cambridge University Press, 2012).

P. M. Harman (ed.), *The Scientific Letters and Papers of James Clerk Maxwell*, 3 vols. (Cambridge University Press, 1990–2002).

P. M. Harman, *The Natural Philosophy of James Clerk Maxwell* (Cambridge University Press, 1998).

H. M. Harrison, *Voyager in Space and Time: The Life of John Couch Adams* (Lewes: The Book Guild, 1994).

Peter Harrison, *The Territories of Science and Religion* (Chicago University Press, 2015).

David Harvey, *Photography and Spirit* (London: Reaktion Books, 2007).

Jenny Hazelgrove, *Spiritualism and British Society Between the Wars* (Manchester University Press, 2000).

Daniel R. Headrick, *The Invisible Weapon: Telecommunications and International Politics, 1851–1945* (New York: Oxford University Press, 1991).

John L. Heilbron, 'Fin-de-Siècle Physics', in Carl Gustaf Bernhard, Elisabeth Crawford and Per Sörbom (eds.), *Science, Technology and Society in the Time of Alfred Nobel* (Oxford: Pergamon Press, 1982), pp. 51–73.

John L. Heilbron, 'Weighing Imponderables and Other Quantitative Science Around 1800', *Historical Studies in the Physical Sciences*, vol. 24 (1993), pp. 1–33, 35–277, 279–337.

P. M. Heimann, 'The *Unseen Universe*: Physics and the Philosophy of Nature in Victorian Britain', *British Journal for the History of Science*, vol. 6 (1972), pp. 73–9.

Linda D. Henderson, *The Fourth Dimension and Non-Euclidean Geometry in Modern Art* (Princeton University Press, 1983).

Linda D. Henderson, 'Vibratory Modernism: Boccioni, Kupka, and the Ether of Space', in Bruce Clarke and Linda D. Henderson (eds.), *From Energy to Information: Representation in Science and Technology, Art and Literature* (Stanford, CA: Stanford University Press, 2002), pp. 126–49.

Klaus Hentschel, *Mapping the Spectrum: Techniques of Representation in Research and Teaching* (Oxford University Press, 2002).

David J. Hess, *Science in the New Age: The Paranormal, Its Defenders and Debunkers and American Culture* (Madison, WI: University of Wisconsin Press, 1993).

Mary B. Hesse, *Forces and Fields: The Concept of Action at a Distance in the History of Physics* (London: Thomas Nelson and Sons, 1961).

Arne Hessenbruch (ed.), *The Readers' Guide to the History of Science* (London: Fitzroy Dearborn, 2000).

Erwin N. Hiebert, 'The State of Physics at the Turn of the Century', in Mario Bunge and William R. Shea (eds.), *Rutherford and Physics at the Turn of the Century* (New York: Dawson and Science History Publications, 1979), pp. 3–22.

[Julie] Dunglas Home, *D. D. Home: His Life and Mission*, ed. by Arthur Conan Doyle (London: Kegan Paul, Trench, Trübner and Co., 1921).

Sungook Hong, 'Controversy over Voltaic Contact Phenomena, 1862–1900', *Archive for History of Exact Sciences*, vol. 47 (1994), pp. 233–89.

Sungook Hong, 'Efficiency and Authority in the "Open Versus Closed Transformer Controversy"', *Annals of Science*, vol. 52 (1995), pp. 49–76.

Jill Howard, '"Physics and Fashion": John Tyndall and His Audiences in Mid-Victorian Britain', *Studies in History and Philosophy of Science*, vol. 35 (2004), pp. 729–58.

John N. Howard, 'Eleanor Mildred Sidgwick and the Rayleighs', *Applied Optics*, vol. 3 (1964), pp. 1120–2.

Jeff Hughes, 'Occultism and the Atom: The Curious Story of Isotopes', *Physics World*, September 2003, pp. 31–5.

Bruce J. Hunt, 'Experimenting on the Ether: Oliver J. Lodge and the Great Whirling Machine', *Historical Studies in the Physical Sciences*, vol. 16 (1986), pp. 111–34.

Bruce J. Hunt, *The Maxwellians* (Ithaca, NY: Cornell University Press, 1991).

Bruce J. Hunt, 'The Ohm Is Where the Art Is: British Telegraph Engineers and the Development of Electrical Standards', *Osiris*, vol. 9 (1994), pp. 48–63.

Bruce J. Hunt, 'Scientists, Engineers and Wildman Whitehouse: Measurement and Credibility in Early Cable Telegraphy', *British Journal for the History of Science*, vol. 29 (1996), pp. 155–69.

Brian Inglis, *Science and Parascience: A History of the Paranormal 1914–1939* (London: Hodder and Stoughton, 1984).

Harvey J. Irwin and Caroline A. Watt, *An Introduction to Parapsychology* (Jefferson, NC: McFarland & Co., 5th ed., 2007).

Eric Ives, Diane Drummond and Leonard Schwarz, *The First Civic University: Birmingham 1880–1980. An Introductory History* (University of Birmingham, 2000).

Myles W. Jackson, *Spectrum of Belief: Joseph von Fraunhofer and the Craft of Precision Optics* (Cambridge, MA: MIT Press, 2000).

Frank A. J. L. James, 'The Study of Spark Spectra, 1835–1859', *Ambix*, vol. 30 (1983), pp.137–62.

Frank A. J. L. James, '"The Optical Mode of Investigation": Light and Matter in Faraday's Natural Philosophy', in David Gooding and Frank A. J. L. James, *Faraday Rediscovered*, pp. 137–61.

Frank A. J. L. James, 'The Practical Problems of "New" Experimental Science: Spectro-Chemistry and the Search for Hitherto Unknown Chemical Elements in Britain 1860–1869', *British Journal for the History of Science*, vol. 21 (1988), pp. 181–94.

Frank A. J. L. James (ed.), *The Correspondence of Michael Faraday Volume 5 November 1855–October 1860* (London: Institution of Engineering and Technology, 2008).

Frank A. J. L. James, *Michael Faraday: A Very Short Introduction* (Oxford University Press, 2010).

Frank A. J. L. James (ed.), *The Correspondence of Michael Faraday Volume 6 November 1860–August 1867* (London: Institution of Engineering and Technology, 2012).

Richard R. John, *Network Nation: Inventing American Telecommunications* (Cambridge, MA: Harvard University Press, 2010).

George M. Johnson, *Mourning and Mysticism in First World War Literature and Beyond* (Basingstoke: Palgrave, 2015).

K. Paul Johnson, *The Masters Revealed: Madame Blavatsky and the Myth of the Great White Lodge* (Albany, NY: State University of New York Press, 1994).

Martyn Jolly, *Faces of the Living Dead: The Belief in Spirit Photography* (London: British Library, 2006).

W. P. Jolly, *Sir Oliver Lodge* (London: Constable, 1974).

Christa Jungnickel and Russell McCormmach, *Intellectual Mastery of Nature: Theoretical Physics from Ohm to Einstein*, 2 vols. (Chicago University Press, 1986).

David Kaiser, *How the Hippies Saved Physics: Science, Counterculture, and the Quantum Revival* (New York: W. W. Norton, 2011).

Michio Kaku, *Physics of the Impossible* (London: Penguin Books, 2009).

Robert H. Kargon, *Science in Victorian Manchester: Enterprise and Expertise* (Manchester University Press, 1977).

Dong-Won Kim, *Leadership and Creativity: A History of the Cavendish Laboratory, 1871–1919* (Dordrecht: Kluwer Academic Publishers, 2002).

Martin J. Klein, 'Mechanical Explanation at the End of the Nineteenth Century', *Centaurus*, vol. 17 (1972), pp. 58–82.

Rene Kollar, *Searching for Raymond: Anglicanism, Spiritualism and Bereavement Between the Two World Wars* (Lanham, MD: Lexington Books, 2000).

Malcom J. Kottler, 'Alfred Russel Wallace, the Origin of Man and Spiritualism', *Isis*, vol. 65 (1974), pp. 145–92.

Sophie Lachapelle, *Investigating the Supernatural: From Spiritism and Occultism to Psychical Research and Metapsychics in France, 1853–1931* (Baltimore, MD: Johns Hopkins University Press, 2011).

Sophie Lachapelle, *Conjuring Science: A History of Scientific Entertainment and Stage Magic in Modern France* (London: Palgrave Macmillan, 2015).

Peter Lamont, *The First Psychic: The Peculiar Mystery of a Notorious Victorian Wizard* (London: Little, Brown, 2005).

Peter Lamont, *Extraordinary Beliefs: A Historical Approach to a Psychological Problem* (Cambridge University Press, 2013).

John L. Lewis (ed.), *Promoting Physics and Supporting Physicists: The Physical Society and the Institute of Physics 1874–2002* (Bristol: Institute of Physics Publishing, 2003).

Bernard Lightman, 'Victorian Sciences and Religions: Discordant Harmonies', *Osiris*, vol. 16 (2001), pp. 343–66.

Bernard Lightman, 'Scientists as Materialists in the Periodical Press: Tyndall's Belfast Address', in Geoffrey Cantor and Sally Shuttleworth, *Science Serialized*, pp. 199–237.

Bernard Lightman, *Victorian Popularizers of Science: Designing Nature for New Audiences* (Chicago University Press, 2007).

Bernard Lightman, 'Lodge and the New Physics, 1919–1933', in Graeme Gooday and James Mussell (eds.), *Oliver Lodge: Continuity and Communication, 1875–1940* (University of Pittsburgh Press, in press).

William C. Lubenow, 'Intimacy, Imagination and the Inner Dialectics of Knowledge Communities: The Synthetic Society, 1896–1908', in Martin J. Daunton (ed.), *The Organisation of Knowledge in Victorian Britain* (Oxford University Press, 2005), pp. 357–70.

Roger Luckhurst, *The Invention of Telepathy 1870–1901* (Oxford University Press, 2002).

W. J. Mander, *British Idealism: A History* (Oxford University Press, 2011).

Seymour H. Mauskopf, 'Marginal Science', in R. G. Olby, G. N. Cantor, J. R. R. Christie and M. J. S. Hodge (eds.), *Companion to the History of Modern Science* (London: Routledge, 1990), pp. 869–85.

Seymour H. Mauskopf and Michael R. McVaugh, *The Elusive Science: Origins of Experimental Psychical Research* (Baltimore, MD: Johns Hopkins University Press, 1980).

Shane McCorristine, *Spectres of the Self: Thinking About Ghosts and Ghost-Seeing in England, 1750–1920* (Cambridge University Press, 2010).

R. G. Medhurst (ed.), *Crookes and the Spirit World: A Collection of Writings by or Concerning the Work of Sir William Crookes, O.M., F.R.S. in the Field of Psychical Research* (London: Souvenir Press, 1972).

R. G. Medhurst and K. M. Goldney, 'William Crookes and the Physical Phenomena of Mediumship', *PSPR*, vol. 54 (1964), pp. 25–157.

John Warne Monroe, *Laboratories of Faith: Mesmerism, Spiritism and Occultism in Modern France* (Ithaca, NY: Cornell University Press, 2008).

James R. Moore, 'Wallace in Wonderland', in Charles H. Smith and George Beccaloni, *Natural Selection and Beyond*, pp. 353–67.

Mark S. Morrisson, *Modern Alchemy: Occultism and the Emergence of Atomic Theory* (New York: Oxford University Press, 2007).

Iwan Rhys Morus, *Frankenstein's Children: Electricity, Exhibition and Experiment in Early-Nineteenth-Century London* (Princeton University Press, 1998).

Iwan Rhys Morus, 'The Measure of Man: Technologizing the Victorian Body', *History of Science*, vol. 37 (1999), pp. 249–82.

Iwan Rhys Morus, *When Physics Became King* (Chicago University Press, 2005).

Iwan Rhys Morus, 'Worlds of Wonder: Sensation and the Victorian Scientific Performance', *Isis*, vol. 101 (2010), pp. 806–16.

Iwan Rhys Morus, *Shocking Bodies: Life, Death and Electricity in Victorian England* (Stroud: History Press, 2011).

Iwan Rhys Morus, 'Physics and Medicine', in Jed Z. Buchwald and Robert Fox, *Oxford Handbook of the History of Physics*, pp. 679–97.

Iwan Rhys Morus, *William Robert Grove: Victorian Gentleman of Science* (Cardiff: University of Wales Press, 2017).

Ornella Moscucci, *The Science of Woman: Gynaecology and Gender in England, 1800–1929* (Cambridge University Press, 1990).

Albert E. Moyer, *American Physics in Transition: A History of Conceptual Change in the Late Nineteenth Century* (Los Angeles: Tomash Publishers, 1983).

Albert E. Moyer, *A Scientist's Voice in American Culture: Simon Newcomb and the Rhetoric of Scientific Method* (Berkeley, CA: University of California Press, 1992).

Falk Müller, *Gasentladungsforschung im 19. Jahrhundert* (Berlin: Verlag für Geschichte der Naturwissenschaften und der Technik, 2004).

Simone Müller-Pohl, *Wiring the World: The Social and Cultural Creation of Global Telegraph Networks* (New York: Columbia University Press, 2015).

Greg Myers, 'Nineteenth Century Popularizations of Thermodynamics and the Rhetoric of Social Prophecy', *Victorian Studies*, vol. 29 (1985), pp. 35–66.

Paul J. Nahin, *Oliver Heaviside: The Life, Work, and Times of an Electrical Genius of the Victorian Age* (Baltimore, MD: Johns Hopkins University Press, 1988).

Michael Nahm, 'The Sorcerer of Coblenzl and His Legacy: The Life of Baron Karl Ludwig von Reichenbach, His Work and Its Aftermath', *Journal of Scientific Exploration*, vol. 26 (2012), pp. 381–407.

Simone Natale, 'A Cosmology of Invisible Fluids: Wireless, X-Rays and Psychical Research around 1900', *Canadian Journal of Communication*, vol. 36 (2011), pp. 263–75.

Simone Natale, *Supernatural Entertainments: Victorian Spiritualism and the Rise of Modern Media Culture* (University Park, PA: Pennsylvania State University Press, 2016).

Jaume Navarro (ed.), *Ether and Modernity: The Recalcitrance of an Epistemic Object in the Twentieth Century* (Oxford University Press, 2018).

Jaume Navarro, 'Ether and Wireless: An Old Medium into New Media', in Jaume Navarro, *Ether and Modernity*, pp. 130–54.

Geoffrey K. Nelson, *Spiritualism and Society* (London: Routledge, 1968).

Richard Noakes, 'Telegraphy Is an Occult Art: Cromwell Fleetwood Varley and the Diffusion of Electricity to the Other World', *British Journal for the History of Science*, vol. 32 (1999), pp. 421–59.

Richard Noakes, '"Instruments to Lay Hold of Spirits": Technologising the Bodies of Victorian Spiritualism', in Iwan Rhys Morus (ed.), *Bodies/Machines* (Oxford: Berg, 2002), pp. 125–63.

Richard Noakes, 'Spiritualism, Science and the Supernatural in Mid-Victorian Britain', in Nicola Bown, Carolyn Burdett and Pamela Thurschwell (eds.), *The Victorian Supernatural* (Cambridge University Press, 2004), pp. 23–43.

Richard Noakes, '"The Bridge Which Is Between Physical and Psychical Research": William Fletcher Barrett, Sensitive Flames and Spiritualism', *History of Science*, vol. 42 (2004), pp. 419–64.

Richard Noakes, 'Ethers, Religion and Politics in Late-Victorian Physics: Beyond the Wynne Thesis', *History of Science*, vol. 43 (2005), pp. 415–55.

Richard Noakes, 'Cromwell Varley FRS, Electrical Discharge and Spiritualism', *Notes and Records of the Royal Society*, vol. 61 (2007), pp. 5–21.

Richard Noakes, 'The "World of the Infinitely Little": Connecting Physical and Psychical Realities Circa 1900', *Studies in History and Philosophy of Science*, vol. 39 (2008), pp. 323–34.

Richard Noakes, 'Industrial Research at the Eastern Telegraph Company, 1872–1929', *British Journal for the History of Science*, vol. 47 (2014), pp. 119–46.

Richard Noakes, 'Thoughts and Spirits by Wireless: Imagining and Building Psychic Telegraphs in America and Britain, Circa 1900–1930', *History of Technology*, vol. 32 (2016), pp. 137–58.

Richard Noakes, 'Making Space for the Soul: Oliver Lodge, Maxwellian Psychics and the Etherial Body', in Jaume Navarro, *Ether and Modernity*, pp. 88–106.

Mary Jo Nye, 'N-Rays: An Episode in the History and Psychology of Science', *Historical Studies in the Physical Sciences*, vol. 11 (1980), pp. 125–56.

Kathryn Olesko and Frederic L. Holmes, 'Experiment, Quantification and Discovery: Helmholtz's Early Physiological Researches', in David Cahan, *Hermann von Helmholtz*, pp. 50–108.

Donald Opitz, '"Not Merely Wifely Devotion": Collaborating in the Construction of Science at Terling Place', in Annette Lykknes, Donald L. Opitz and Brigitte van Tiggelen (eds.), *For Better or for Worse? Collaborative Couples in the Sciences* (Heidelberg: Birkhäuser, 2012), pp. 33–56.

Janet Oppenheim, *The Other World: Spiritualism and Psychical Research in Britain, 1850–1914* (Cambridge University Press, 1985).

Laura Otis, *Networking: Communicating with Bodies and Machines in the Nineteenth Century* (Ann Arbor, MI: University of Michigan Press, 2011).

Alex Owen, *The Darkened Room: Women, Power and Spiritualism in Late Victorian England* (London: Virago, 1989).

Alex Owen, *The Place of Enchantment: British Occultism and the Culture of the Modern* (Chicago University Press, 2004).

Robert Park, *Voodoo Science: The Road from Foolishness to Fraud* (Oxford University Press, 2000).

Trevor Pinch, 'Some Suggestions From Sociology of Science to Advance the Psi Debate', *Behavioural and Brain Sciences*, vol. 10 (1987), pp. 603–5.

Reginé Plas, *Naissance d'une science humaine, la psychologie: Les psychologues et de 'le merveilleux psychique'* (Presses Universitaires de Rennes, 2000).

Mary Poovey, *Uneven Developments: The Ideological Work of Gender in Mid-Victorian England* (Chicago University Press, 1988).

Harry Price, *Fifty Years of Psychical Research: A Critical Survey* (London: Longmans, Green and Co., 1939).

Courtenay Grean Raia, 'From Ether Theory to Ether Theology: Oliver Lodge and the Physics of Immortality', *Journal of the History of the Behavioural Sciences*, vol. 43 (2007), pp. 19–43.

Chitra Ramalingam, 'Natural History in the Dark: Seriality and the Electric Discharge in Victorian Physics', *History of Science*, vol. 48 (2010), pp. 371–98.

W. J. Reader (with Rachel Lawrence, Sheila Nemet and Geoffrey Tweedale), *The Institution of Electrical Engineers, 1871–1971* (London: Institution of Electrical Engineers, 1987).

Graham Richards, 'Edward Cox, the Psychological Society of Great Britain (1875–1879) and the Meanings of an Institutional Failure', in G. C. Bunn, A. D. Lovie and G. D. Richards (eds.), *Psychology in Britain: Historical Essays and Personal Reflections* (Leicester: British Psychological Society, 2001), pp. 33–53.

Robert J. Richards, *The Romantic Conception of Life: Science and Philosophy in the Age of Goethe* (Chicago University Press, 2002).

Jessica Riskin, *Science in the Age of Sensibility: The Sentimental Empiricists of the French Enlightenment* (Chicago University Press, 2002).

R. G. Roberts, 'The Training of an Industrial Physicist: Oliver Lodge and Benjamin Davies, 1882–1940' (unpublished doctoral thesis, University of Manchester, 1984).

David Robertson, 'The Radical Who Shaped the Future', *IEE Review*, May 2002, pp. 31–6.

John D. Root, 'Science, Religion and Psychical Research: The Monistic Thought of Sir Oliver Lodge', *Harvard Theological Review*, vol. 71 (1978), pp. 245–63.

Pascal Rousseau, *Cosa mentale: Art et télépathie au XXe siècle* (Paris: Gallimard, 2015).

Peter Rowlands, *Oliver Lodge and the Liverpool Physical Society* (Liverpool University Press, 1990).

Martin J. S. Rudwick, *The Great Devonian Controversy: The Shaping of Scientific Knowledge Among Gentlemanly Specialists* (Chicago University Press, 1985).

Cynthia Russett, *Sexual Science: The Victorian Construction of Womanhood* (Cambridge, MA: Harvard University Press, 1989).

Jennifer Ruth, '"Gross Humbug" or "The Language of Truth": The Case of the *Zoist*', *Victorian Periodicals Review*, vol. 32 (1999), pp. 299–323.

Diethard Sawicki, *Leben mit den Toten: Geisterglauben und die Enstehung des Spiritismus in Deutschland 1770–1900* (Paderborn: Ferdinand Schöningh, 2002).

Simon Schaffer, 'Late Victorian Metrology and Its Instrumentation: A Manufactory of Ohms', in Robert Bud and Susan E. Cozzens (eds.), *Invisible Connections: Instruments, Institutions, and Science* (Bellingham, WA: SPIE Optical Engineering Press, 1992), pp. 24–55.

Simon Schaffer, 'Rayleigh and the Establishment of Electrical Standards', *European Journal of Physics*, vol. 15 (1994), pp. 277–85.

Simon Schaffer, 'Accurate Measurement Is an English Science', in M. Norton Wise (ed.), *The Values of Precision* (Princeton University Press, 1995), pp. 135–72.

Simon Schaffer, 'Where Experiments End: Tabletop Trials in Victorian Astronomy', in Jed Z. Buchwald (ed.), *Scientific Practice: Theories and Stories of Doing Physics* (Chicago University Press, 1995), pp. 257–99.

Simon Schaffer, 'Physics Laboratories and the Victorian Country House', in Crosbie Smith and Jon Agar (eds.), *Making Space for Science: Territorial Themes in the Shaping of Knowledge* (Basingstoke: Macmillan, 1998), pp. 149–80.

Andreas Schmidt, 'The Most Disrespectful Camera in the World', in Clement Chéroux et al., *Perfect Medium*, pp. 72–91.

Bart Schultz, *Henry Sidgwick: Eye of the Universe. An Intellectual Biography* (Cambridge University Press, 2004).

Jeffrey Sconce, *Haunted Media: Electronic Presence from Telegraphy to Television* (Durham, NC: Duke University Press, 2007).

J. Barton Scott, 'Miracle Publics: Theosophy, Christianity, and the Coulomb Affair', *History of Religions*, vol. 49 (2009), pp. 172–96.

William Seabrook, *Doctor Wood: Modern Wizard of the Laboratory* (New York: Harcourt Books, 1941).

James A. Secord, *Controversy in Victorian Geology: The Cambrian-Silurian Dispute* (Princeton University Press, 1986).

Steven Shapin and Simon Schaffer, *Leviathan and the Air Pump: Hobbes, Boyle and the Experimental Life* (Princeton University Press, 1985).

Susan Sheets-Pyenson, 'Popular Science Periodicals in Paris and London: The Emergence of a Low Scientific Culture, 1820–1875', *Annals of Science*, vol. 42 (1985), pp. 549–72.

Terry Shinn, 'The French Science Faculty System, 1808–1914: Institutional Change and Research Potential in Mathematics and Physical Sciences', *Historical Studies in the Physical Sciences*, vol. 10 (1979), pp. 271–332.

S. E. D. Shortt, 'Physicians and Psychics: The Anglo-American Medical Response to Spiritualism', *Journal of the History of Medicine and Allied Sciences*, vol. 39 (1984), pp. 339–55.

Elaine Showalter, *The Female Malady: Women, Madness and English Culture, 1830–1980* (London: Virago, 1985).

H. Otto Sibum, 'Reworking the Mechanical Value of Heat: Instruments of Precision and Gestures of Accuracy in Early Victorian England', *Studies in History and Philosophy of Science*, vol. 26 (1995), pp. 73–106.

Ethel Sidgwick, *Mrs. Henry Sidgwick: A Memoir by Her Niece* (London: Sidgwick and Jackson, 1938).

Josep Simon, 'Physics Textbooks and Textbook Physics in the Nineteenth and Twentieth Century', in Jed Z. Buchwald and Robert Fox, *Oxford Handbook of the History of Physics*, pp. 651–78.

Charles H. Smith, 'Wallace, Spiritualism, and Beyond: "Change", or "No Change"', in Charles H. Smith and George Beccaloni, *Natural Selection and Beyond*, pp. 391–423.

Charles H. Smith and George Beccaloni (eds.), *Natural Selection and Beyond: The Intellectual Legacy of Alfred Russel Wallace* (Oxford University Press, 2008).

Crosbie Smith, *The Science of Energy: A Cultural History of Energy Physics in Victorian Britain* (London: Athlone, 1998).

Crosbie Smith and M. Norton Wise, *Energy and Empire: A Biographical Study of Lord Kelvin* (Cambridge University Press, 1989).

Gordon Smith, *Beyond Reasonable Doubt: The Case for Supernatural Phenomena in the Modern World* (London: Coronet, 2018).

Roger Smith, 'The Human Significance of Biology: Carpenter, Darwin and the *vera causa*', in U. C. Knoepflmacher and G. B. Tennyson (eds.), *Nature and the Victorian Imagination* (Berkeley, CA: University of California Press, 1977), pp. 216–30.

Roger Smith, *Free Will and the Human Sciences, 1870–1910* (London: Pickering and Chatto, 2013).

Andreas Sommer, 'Psychical Research and the Origins of American Psychology: Hugo Münsterberg, William James and Eusapia Palladino', *History of the Human Sciences*, vol. 25 (2012), pp. 23–44.

Andreas Sommer, 'Crossing the Boundaries of Mind and Body: Psychical Research and the Origins of Modern Psychology' (unpublished doctoral thesis, University College London, 2013).

Andreas Sommer, 'Normalizing the Supernormal: The Formation of the "Gesellschaft fur Psychologische Forschung" ("Society for Psychological Research"), c. 1886–1890', *Journal of the History of Behavioural Sciences*, vol. 49 (2013), pp. 18–44.

Richard Staley, *Einstein's Generation: The Origins of the Relativity Revolution* (Chicago University Press, 2008).

Matthew Stanley, *Practical Mystic: Religion, Science and A. S. Eddington* (Chicago University Press, 2007)

Matthew Stanley, *Huxley's Church and Maxwell's Demon: From Theistic Science to Naturalistic Science* (Chicago University Press, 2015).

Klaus B. Staubermann, 'Tying the Knot: Skill, Judgement and Authority in the 1870s Leipzig Spiritistic Experiments', *British Journal for the History of Science*, vol. 34 (2001), pp. 67–79.

Victor Stenger, *Physics and Psychics: The Search for a World Beyond the Senses* (Buffalo, NY: Prometheus Books, 1990).

Ian B. Stewart, 'E. E. Fournier d'Albe's Fin de siècle: Science, Nationalism and Philosophy in Britain and Ireland', *Cultural and Social History*, vol. 14 (2017), pp. 599–620.

Douglas M. Stokes, *The Nature of Mind: Parapsychology and the Role of Consciousness in the Physical World* (Jefferson, NC: McFarland and Co., 1997), pp. 101–30.

Jeremy Stolow, 'The Spiritual Nervous System: Reflections on a Magnetic Cord Designed for Spirit Communication', in Jeremy Stolow (ed.), *Deus in Machina: Religion, Technology and the Things in Between* (New York: Fordham University Press, 2013), pp. 83–113.

P. Strange, 'Two Electrical Periodicals: *The Electrician* and *The Electrical Review*, 1880–1890', *IEE Proceedings*, vol. 132 (1985), pp. 574–81.

Guy R. Strutt, 'Robert John Strutt, Fourth Baron Rayleigh', *Applied Optics*, vol. 3 (1964), pp. 1105–12.

Lynn G. De Swarte, *Thorson's Principles of Spiritualism* (London: Thorson's, 1999).

Loyd Swenson, *The Ethereal Aether: A History of the Michelson-Morley-Miller Aether-Drift Experiments, 1880–1930* (Austin, TX: University of Texas Press, 1972).

Eugene Taylor, *William James on Consciousness Beyond the Margin* (Princeton University Press, 1996).

Joanna Timms, 'Ghost-Hunters and Psychical Research in Interwar England', *History Workshop Journal*, vol. 74 (2012), pp. 88–104.

Daniel P. Thurs and Ronald L. Numbers, 'Science, Pseudo-Science and Science Falsely So-Called', in Massimo Pigliucci and Maatern Boudry (eds.), *Philosophy of Pseudoscience: Reconsidering the Demarcation Problem* (Chicago University Press, 2013), pp. 121–44.

Morris W. Travers, *The Life of Sir William Ramsay* (London: Edward Arnold, 1956).

Marlene Tromp, *Altered States: Sex, Nation, Drugs, and Self-Transformation in Victorian Spiritualism* (Albany, NY: State University of New York Press, 2006).

Jennifer Tucker, *Nature Exposed: Photography as Eyewitness in Victorian Science* (Baltimore, MD: Johns Hopkins University Press, 2005).

Frank M. Turner, *Between Science and Religion: The Reaction to Scientific Naturalism in Late Victorian England* (New Haven, CT: Yale University Press, 1974).

Frank M. Turner, *Contesting Cultural Authority: Essays in Victorian Intellectual Life* (Cambridge University Press, 1993).

Takahiro Ueyama, 'Capital, Profession and Medical Technology: The Electrotherapeutic Institutes and the Royal College of Physicians, 1888–1922', *Medical History*, vol. 41 (1997), pp. 150–81.

K. G. Valente, '"Who Will Explain the Explanation?": The Ambivalent Reception of Higher Dimensional Space in the British Spiritualist Press, 1875–1900', *Victorian Periodicals Review*, vol. 41 (2008), pp. 124–49.

Elizabeth R. Valentine, 'Spooks and Spoofs: Relations Between Psychical Research and Academic Psychology in the Interwar Period', *History of the Human Sciences*, vol. 25 (2012), pp. 67–90.

Shiv Visvanathan, 'Alternative Science', *Theory, Culture and Society*, vol. 23 (2006), pp. 164–9.

Andrew Warwick, *Masters of Theory: Cambridge and the Rise of Mathematical Physics* (Chicago University Press, 2003).

Roland Wenzlhuemer, *Connecting the Nineteenth Century World: The Telegraph and Globalization* (Cambridge University Press, 2013).

Michael Whitworth, 'Transformations of Knowledge in Oliver Lodge's *Ether and Reality*', in Jaume Navarro, *Ether and Modernity*, pp. 30–44.

Barry H. Wiley, *The Indescribable Phenomenon: The Life and Mysteries of Annie Eva Fay* (Seattle, WA: Hermetic Press, 2005).

John Peregrine Williams, 'The Making of Victorian Psychical Research: An Intellectual Elite's Approach to the Spirit World' (unpublished doctoral thesis, University of Cambridge, 1984).

David B. Wilson, 'The Thought of Late-Victorian Physicists: Oliver Lodge's Ethereal Body', *Victorian Studies*, vol. 15 (1971), pp. 29–48.

David B. Wilson, *Kelvin and Stokes: A Comparative Study in Victorian Physics* (Bristol: Adam Hilger, 1987).

David B. Wilson, 'On the Importance of Eliminating *Science* and *Religion* from the History of Science and Religion: The Cases of Oliver Lodge, J. H. Jeans and A. S. Eddington', in Jitse M. van der Meer (ed.), *Facets of Faith and Science Volume 1* (Lanham, MA: University Press of America, 1996), pp. 27–47.

David B. Wilson, 'Arbiters of Science: George Gabriel Stokes and Joshua King', in Kevin C. Knox and Richard Noakes (eds.), *From Newton to Hawking: A History of Cambridge University's Lucasian Professors of Mathematics* (Cambridge University Press, 2003), pp. 295–342.

Alison Winter, *Mesmerized: Powers of Mind in Victorian Britain* (Chicago University Press, 1998).

Jay Winter, *Sites of Memory, Sites of Mourning: The Great War in European Cultural History* (Cambridge University Press, 1995).

M. Norton Wise, 'German Concepts of Force, Energy and the Electromagnetic Ether: 1845–1880', in Geoffrey Cantor and M. J. S. Hodge, *Conceptions of Ether*, pp. 269–307.

Heather Wolffram, *The Stepchildren of Science: Psychical Research and Parapsychology in Germany, c. 1870–1939* (Amsterdam: Rodopi, 2009).

Brian Wynne, 'Physics and Psychics: Science, Symbolic Action and Social Control in Late Victorian England', in Barry Barnes and Steven Shapin (eds.), *Natural Order: Historical Studies of Scientific Culture* (Beverly Hills, CA: Sage, 1979), pp. 167–87.

Index

Printed in the United States
By Bookmasters